Antimicrobials in Food Science and Technology

The demands of producing high-quality, pathogen-free food rely increasingly on natural sources of antimicrobials to inhibit food spoilage organisms, foodborne pathogens, and toxins. The recent developments and innovations of new antimicrobials from natural sources for a wide range of applications require that knowledge of traditional sources for food antimicrobials is combined with the latest technologies in identification, characterization, and applications. This book explores novel, natural sources of antimicrobials as well as the latest developments in using well-known antimicrobials in food, covering antimicrobials derived from microbial sources, animal-derived products, plants, and value-added products. This book includes the development and use of natural antimicrobials for processed and fresh food products. New and emerging technologies concerning antimicrobials are also discussed.

This book considers recent developments and innovations in food technology in combating infectious diseases and explores advances in antimicrobial constituents and their applications in the fight against microbes. In addition, it also provides a variety of photographs, diagrams, and tables to help illustrate the material. The novel strategies to combat antimicrobial resistance are also described, emphasizing collaborative measures of control. Advanced topics in the volume include food processing, food security, preservation, nutritional analysis, quality control, and maintenance as well as good manufacturing practices in the food industries.

Students, research scientists, academicians, and policy makers can benefit from *Antimicrobials in Food Science and Technology* as a resource that addresses microbial biotechnology, food microbiology, fermentation technology, ethnopharmacology, toxicology, microbial/medicinal plant products, and all disciplines related to antimicrobial research.

Features of the book:

- Covers all food antimicrobials, natural and synthetic, with up-to-date research on each type
- Recent references on every conceivable food antimicrobial
- Describes recent laws and regulatory guidelines in the selection of appropriate additives for specific food products
- Includes innovations in natural antimicrobial value-added products
- Offers current and future applications of emergent antimicrobial technologies and the use of multifactorial food preservation with antimicrobials
- Details methods to improve antimicrobial properties to have a longer service life in combating infection

Current Trends in Antimicrobial Research

Series Editor: Arti Gupta

Antimicrobials in Pharmaceutical and Medicinal Research

Edited by Arti Gupta and Ram Prasad

Antimicrobials in Food Science and Technology

Edited by Arti Gupta and Ram Prasad

For more info: https://www.routledge.com/Current-Trends-in-Antimicrobial-Research/book-series/CTAR

Antimicrobials in Food Science and Technology

Edited by

Arti Gupta and Ram Prasad

CRC Press
Taylor & Francis Group
Boca Raton London New York

CRC Press is an imprint of the
Taylor & Francis Group, an **informa** business

Cover image credit: © Shutterstock

First edition published 2024
by CRC Press
6000 Broken Sound Parkway NW, Suite 300, Boca Raton, FL 33487-2742

and by CRC Press
4 Park Square, Milton Park, Abingdon, Oxon, OX14 4RN

CRC Press is an imprint of Taylor & Francis Group, LLC

Library of Congress Cataloging-in-Publication Data

Names: Gupta, Arti (Professor of biotechnology) editor. | Prasad, Ram (Professor of microbiology) editor.
Title: Antimicrobials in food science and technology/edited by Arti Gupta and Ram Prasad.
Other titles: Current trends in antimicrobial research.
Description: First edition. | Boca Raton, FL: CRC Press, 2024. | Series: Current trends in antimicrobial research | Includes bibliographical references and index.
Identifiers: LCCN 2023026304 (print) | LCCN 2023026305 (ebook) | ISBN 9781032215563 (hardback) | ISBN 9781032215570 (paperback) | ISBN 9781003268949 (ebook)
Subjects: MESH: Anti-Infective Agents | Food Additives | Food Safety—methods | Biological Products | Food Microbiology
Classification: LCC RM267 (print) | LCC RM267 (ebook) | NLM QV 250 | DDC 615.7/92—dc23/eng/20230727
LC record available at https://lccn.loc.gov/2023026304
LC ebook record available at https://lccn.loc.gov/2023026305

ISBN: 9781032215563 (hbk)
ISBN: 9781032215570 (pbk)
ISBN: 9781003268949 (ebk)

DOI: 10.1201/9781003268949

Typeset in Times
by Deanta Global Publishing Services, Chennai, India

Contents

Acknowledgements

First and foremost, we would like to thank GOD for his never-ending grace, mercy, and provision during what ended up being one of the toughest times of our life.

We are indebted to the many people who helped to bring this book to light. I wish to thank series editors Gustavo Molina and Vijai Kumar Gupta; Marc Gutierrez and Bonita Glanville-Morris, senior production editor; and editorial assistants Jyotsna Jangra and Renu Upadhyaya, Taylor & Francis/CRC Press, for their generous assistance, constant support, and patience in initializing the series volume. Editors are very thankful to Vice Chancellor Faizabad University, Ayodhya, and Mahatma Gandhi Central University, Motihari, for their kind support and constant encouragement. Special thanks are due to my well-wishers, family members, and friends.

Dr. Arti Gupta gives special thanks to her exquisite husband Mr. Saket Agrawal and her son Reyansh who encourage her to start the work, preserve, and put everything together. Dr. Ram Prasad gives his special regards to his family for their constant invaluable support, and motivation to publish it. We both also give special thanks to all faculty colleagues and senior faculties of our team.

Arti Gupta
Shri Avadh Raj Singh Smarak Degree College,
Gonda, UP, India

and

Ram Prasad
Mahatma Gandhi Central University,
Motihari, Binhar, India

Preface

In recent years, several investigations have been showing searching the antimicrobial activity of natural products. Natural antimicrobials are secondary metabolites that can be found in plants, animals, and microorganisms. In food applications, these natural antimicrobial compounds could be influenced by food components, processing, and storage. On the other hand, increasing levels of consumer concerns about chemical antimicrobials and increasing resistance of pathogenic microbes have turned the attention of scientific communities towards studies on the potential antimicrobial activities of natural products. Natural antimicrobials are generally documented as safe, and they appear to be the most promising solution for microbial resistance and could best meet consumers' demands for healthier foods. In this context natural antimicrobial compounds are gaining great interest from research and industry, due to the potential to provide quality and safety benefits, with a reduced impact on human health. In addition, utilization of natural active agents promotes the accepted criteria of food sustainability. This book explores the recent findings about natural antimicrobial compounds from plants, animals, and microorganisms which could be used to control spoilage and pathogenic microorganisms in food products.

The first chapter by Kingkaew and Tanasupawat reviews the application of antimicrobial peptides from lactic acid bacteria in food preservation and human health. Chapter 2 by Al Fattah Amara highlights the fermented food microbes and their preservative byproducts. In Chapter 3, Manju Meena et al. describe applications of antimicrobial agents in food science and technology. Extensive studies on antimicrobials in pharmaceutical and medical research are discussed by Chauhan et al. in Chapter 4, and Chapter 5 highlights food preservation by natural antimicrobials: current trends and future prospective. In Chapter 6, Sharma et al. emphasize on efficacy of antimicrobial substances in food safety and quality. In Chapter 7, Sheikh et al. describe plants as antimicrobial agents in food technology, and in Chapter 8, Mishra et al. highlight the antimicrobial resistance and its consequential role in food security. Finally, in Chapter 9, Bhuvaneswari et al. highlight approaches on the use of natural antimicrobial agents and their applications as biopreservatives.

Arti Gupta and Ram Prasad

Editor biographies

Arti Gupta, Ph.D., obtained her doctorate from Mahatma Jyotiba Phule Rohilkhand University, Bareilly, India, in 2010 in Animal Science. Dr. Gupta's current research interests include animal biotechnology, molecular plant biotechnology, molecular animal biotechnology, bioprocess technology, and microbiology. Since July 2014, Dr. Gupta has been employed at Shri Avadh Raj Singh Smarak Degree College, Gonda, Uttar Pradesh, India, and has been engaged in editorial work with Springer Nature. Dr. Gupta has 10 years of teaching and research experience in Plant and Animal Biotechnology and Microbial Biotechnology; in 2023–24 she served as external examiner for central evaluation at MJP Rohilkhand Univeristy, Bareilly.

Dr. Gupta has been awarded University Topper (Gold Medal), M.Sc. (Biotech.) by Ch.C.S. University, Meerut; Young Scientist Award (Gold Medal) by the Zoological Society of India, Lucknow; Best Poster Presenter by *Asian Journal of Experimental Science*, Jaipur; Best Poster Presenter by International Consortium of Contemporary Biologists (ICCB) and Madhawai-Shyam Educational Trust, Ranchi; Fellowship Award by International Consortium of Contemporary Biologists (FICCB) and Madhawi Shayam Educational Trust (FMSET); and Dr. V.P. Agarwal (Gold Medal) by D.A.V. (P.G.) College, Muzaffarnagar. She has published 1 monograph, 7 book chapters, around 50 National and International research papers, and 9 edited books; she has presented 28 abstracts in National and International Symposia/Seminars/Conferences/Workshops. Dr. Gupta has lifetime membership of the Indian Science Congress Association, Biotech Research Society of India, Zoological Society of India, and International Consortium of Contemporary Biologists. She has served as a member of several editorial boards. Previously, Dr. Gupta was employed as Teaching Personnel, G.B. Pantnagar University, Pantnagar, Uttarakhand, India, and Visiting Academic Researcher, Biorefining and Advanced Material Research Center (BAMRC), Scotland's Rural College (SRUC), University of Edinburgh, Scotland, United Kingdom.

Ram Prasad, Ph.D., is associated with the Department of Botany, Mahatma Gandhi Central University, Motihari, Bihar, India. His research interests include applied and environmental microbiology, plant-microbe interactions, sustainable agriculture, and nanobiotechnology. Dr. Prasad has more than 275 publications (total citations 13840 with an h-index 59, i10-index 189) to his credit, including research papers, review articles, and book chapters, as well as seven patents issued or pending; and he has edited or authored several books. Dr. Prasad has 14 years of teaching experience and has been awarded the Young Scientist Award and Prof. J.S. Datta Munshi Gold Medal by the International Society for Ecological Communications; Fellowship of Biotechnology Research Society of India; Fellow of Agricultural Technology Development Society, India; Fellow of the Society for Applied Biotechnology; the American Cancer Society UICC International Fellowship for Beginning Investigators, USA; Outstanding Scientist Award in the field of Microbiology; BRICPL Science Investigator Award and Research Excellence Award. He has served as a member of several editorial boards. Previously, Dr. Prasad served as Assistant Professor at Amity University Uttar Pradesh, India; Visiting Assistant Professor, Whiting School of Engineering, Department of Mechanical Engineering at Johns Hopkins University, Baltimore, United States; and Research Associate Professor at the School of Environmental Science and Engineering, Sun Yat-sen University, Guangzhou, China.

Contributors

Krishnendu Acharya
Molecular and Applied Mycology and Plant Pathology Laboratory, Centre of Advanced Study, Department of Botany, University of Calcutta, 35, Ballygunge Circular Road, Kolkata 700019, West Bengal, India.

A. A. Amara
Protein Research Department, Genetic Engineering, and Biotechnology Research Institute, City of Scientific Research and Technological Applications, Universities and Research Center District, New Borg El-Arab, P.O. Box: 21934 Alex, Egypt

Akshitha R Amin
Dept of Botany, St Aloysius College, (Autonomous), Mangaluru

Bhagyalakshmi
Dept of Botany, St Aloysius College, (Autonomous), Mangaluru

P. Bhuvaneswari
Assistant Professor, PG and Research Department of Microbiology, Cauvery College for Women (Autonomous), Trichy.

M. Chamundeeswari
Department of Biotechnology, St. Joseph's College of Engineering, OMR, Chennai-600119, India

Diwakar Chauhan
Department of Chemistry, School of Basic and Applied Sciences, Galgotias University, Greater Noida 203201 (UP) India

Monika Chauhan
Department of Forensic Science, School of Basic and Applied Sciences, Galgotias University, Greater Noida 203201 (UP) India

Bhavna Choudhary
Division of Life Sciences, School of Basic and Applied Sciences, Galgotias University, Greater Noida 203201 (UP) India

B.S. Dhanya
Department of Biotechnology, Udaya School of Engineering, Vellamodi, Nagercoil- 629204, India

Lovlish Gupta
School of Forensic Sciences, National Forensic Sciences University, Institutional Area, Sector 3, Rohini, Delhi 110085

Somanjana Khatua
Department of Botany, Faculty of Science, University of Allahabad, Prayagraj 211002, Uttar Pradesh, India

Engkarat Kingkaew
Department of Biochemistry and Microbiology, Faculty of Pharmaceutical Sciences, Chulalongkorn University, Bangkok 10330, Thailand

Ajay Kumar
Department of Life Science, School of Basic and Applied Sciences, Galgotias University, Greater Noida 203201 (UP) India

Sanchali Kundu
Department of Botany, Krishnagar Government College, Krishnagar 741101, West Bengal, India.

Shaily Mahur
Division of Life Sciences, School of Basic and Applied Sciences, Galgotias University, Greater Noida 203201 (UP) India

V. Manju Meena
Masters of Food Science, Chapman University, Orange, California-92866

Jyothi Miranda
Dept of Botany, St Aloysius College, (Autonomous), Mangaluru

P.F. Mishel
Department of Botany, Bharathidasan University, Trichy.

Smita Mishra
Lab of Microbial technology and Plant Pathology, Department of Botany, Dr. Harisingh Gour Central University, Sagar (M.P.) – 470003, India.

K.R. Preethy
Department of Biotechnology, St. Joseph's College of Engineering, OMR, Chennai-600119, India

Apeksha Rathi
Lab of Microbial Technology and Plant Pathology, Department of Botany, Dr. Harisingh Gour Central University, Sagar (M.P.) – 470003, India.

Sweta Sain
Division of Life Sciences, School of Basic and Applied Sciences, Galgotias University, Greater Noida 203201 (UP) India

Pragati Saini
Division of Life Sciences, School of Basic and Applied Sciences, Galgotias University, Greater Noida 203201 (UP) India

Sandhya Sharma
Division of Life Sciences, School of Basic and Applied Sciences, Galgotias University, Greater Noida 203201 (UP) India

Sana Sheikh
Dept of Botany, St Aloysius College, (Autonomous) , Mangaluru

P.F. Steffi
Assistant Professor, PG and Research Department of Microbiology, Cauvery College for Women (Autonomous), Trichy.

Somboon Tanasupawat
Department of Biochemistry and Microbiology, Faculty of Pharmaceutical Sciences, Chulalongkorn University, Bangkok 10330, Thailand

B. Thirumalaiyammal
Department of Microbiology, Cauvery College for Women (Autonomous), Annamalai Nagar, Trichy, Tamilnadu.

Deepak Vyas
Lab of Microbial Technology and Plant Pathology, Department of Botany, Dr. Harisingh Gour Central University, Sagar (M.P.) – 470003, India.

1 Application of antimicrobial peptides from lactic acid bacteria in food preservation and human health

Engkarat Kingkaew and Somboon Tanasupawat

1.1 INTRODUCTION

Lactic acid bacteria (LAB) in genera *Bifidobacterium*, *Lactobacillus*, *Lactococcus*, *Leuconostoc*, *Enterococcus*, *Pediococcus*, *Streptococcus*, and *Weissella* (Linares et al. 2017, Lavermicocca, Reguant, and Bautista-Gallego 2021) produce attractive ribosomal peptides known as "Bacteriocin" which have antimicrobial properties (Qin et al. 2019). They play important roles in food and clinical application (Linares et al. 2017, Ayivi et al. 2020). Bacteriocins have been recognized as a potential replacement for chemicals and antibiotics owing to their low toxicity, lack of adverse effects, and proteinaceous structure (Ng et al. 2020).

From the perspective of food quality and safety, the presence of pathogens, the diversity of supply sources, and the food patterns of customers who choose fresher and more organic food items without preservatives or salt are global challenges at present. To ensure the quality and safety of food products, improve consumer health, and limit the presence of antibiotic residues and the spread of resistance genes, it is essential to investigate alternatives to chemical additives, salt, and antibiotics (Soltani et al. 2021, Karlo et al. 2023). The application of lactic acid bacteria in food preservation is described as biopreservation, a natural process of using controlled undesirable microorganisms as an alternative for shelf-life extension and food preservation. LAB bacteriocins function as preservative agents due to the inhibitory effects they have on foodborne pathogens and food-spoilage microorganisms (Perez, Zendo, and Sonomoto 2014). Thus, biopreservation is one of the numerous advantages acquired from lactic acid bacteria in food safety/spoilage.

Infectious diseases are linked to various human diseases for clinical application and other health benefits. Antibiotics have been influential in treating infectious diseases; nevertheless, the widespread misuse of antibiotics has led to the emergence of a substantial worldwide antibiotic resistance issue, necessitating the development of alternative antimicrobial medicines (Ventola 2015). Therefore, bacteriocin can be used as an alternative treatment option for infectious diseases. Because numerous bacteriocins may limit a variety of harmful bacteria, including some antibiotic-resistant isolates, bacteriocins might aid in preventing pathogenic infections (Altveş, Yildiz, and Vural 2019). In addition, bacteriocins have also exhibited antiproliferative activities on various cancer cells and regulating effects on immunity and inflammation, indicating that they also possess anti-inflammatory and anticancer properties. Thus, bacteriocins have also a promising impact for application in human health (Huang et al. 2021).

This chapter presents the bacteriocin classification, LAB producers, application of bacteriocins in food preservation, food packaging containing antimicrobial activity, and clinical and health benefits of LAB bacteriocin. Though there are several publications on bacteriocins and human health in the past few years, systematic reviews in this area of research are still insufficient. Hence, it

DOI: 10.1201/9781003268949-1

is necessary to summarize the bacteriocins produced from diverse bacteria and their therapeutic impacts on health. Moreover, this chapter also describes the impact of crucial factors on the antimicrobial activity of bacteriocins and provides a conclusion and future perspective.

1.2 BACTERIOCIN AND ITS CLASSIFICATION

Bacteriocins are ribosomal peptides synthesized by various groups of bacteria, mainly lactic acid bacteria (LAB). These LAB bacteriocins are interesting due to their proven history of safe usage and the generally regarded as safe (GRAS) and qualified presumption of safety (QPS) status that most LAB possess. Bacteriocins contain bacteriostatic or bactericidal activity (Prudencio, dos Santos, and Vanetti 2015). Bacteriocins work by impeding the cell wall and the membrane of target organisms, either by suppressing cell wall synthesis or causing pore formation, consequently resulting in growth inhibition and/or mortality. Even though several new classifications seem great and applicable, they will take some time to become widely accepted. Therefore, in this chapter, bacteriocins are classified based on biosynthesis and biological activity, as previously reported by Cotter, Hill, and Ross (2005) in Table 1.1

1.2.1 Class I Bacteriocin (Lantibiotics)

The lantibiotics (lanthionine-containing antibiotics) are small peptides (19–38 amino acids in length; MW <5 kDa) that possess unusual post-translational modified residues (i.e., lanthionine or β-methyllanthionine residues). These unique residues generate covalent bridges between amino acids, which result in internal molecular rings and give lantibiotics their characteristic structural functions (Jack and Jung 2000, Pag and Sahl 2002, Xie and Van Der Donk 2004). Nisin is the most well-known Class I bacteriocin synthesized by several *Lactococcus lactis* strains. Briefly, nisin attaches to the lipid II molecules on the bacterial cell wall and forms the nisin-lipid II complex; consequently, it can translocate the cell membrane and generate pore formation (Delves-Broughton and Weber 2011). Remarkably, nisin has been extensively applied in food and clinical applications (Cotter, Hill, and Ross 2005).

1.2.2 Class II Bacteriocins (Non-Lanthionine-Containing Bacteriocins)

1.2.2.1 Class IIa

Class II bacteriocins are larger than bacteriocin class I. These bacteriocins have MW <10 kDa, are heat-stable, and are not subject to extensive post-translational modification (Perez, Zendo, and

TABLE 1.1
Bacteriocins classification scheme (Cotter, Hill, and Ross 2005).

Class	Remarks	Examples
I		
Lanthionine-containing bacteriocin (lantibiotics)	Includes both single- and two-peptide lantibiotics	**Single peptide:** nisin, mersacidin, and lacticin 481 **Two-peptide:** lacticin 3147 and cytolysin
II		
Non-lanthionine-containing bacteriocins	**Heterogeneous class of small peptides:** • **Subclass a:** pediocin-like • **Subclass b:** two-peptide • **Subclass c:** cyclic • **Subclass d:** non-pediocin single linear peptides	**IIa:** pediocin PA1, leucocin A **IIb:** lactacin F **IIc:** enterocin AS48, reuterin 6 **IId:** lactococcin A, divergicin A

Sonomoto 2014). Their major inhibition roles are the triggering membrane permeabilization and the subsequent leakage of vital compounds from target bacteria (Nes et al. 1996, Van Belkum and Stiles 2000, Diep and Nes 2002). The class IIa or pediocin-like bacteriocins show a potentially specific activity against *Listeria monocytogenes* (Montville and Chen 1998, Héchard and Sahl 2002). For instance, pediocin PA-1 is a well-known class IIa bacteriocin and is mainly produced from *Pediococcus* strains. The remarkable characteristics of pediocin-like bacteriocins are disulphide bridges and common YGNG(V/L) motif (pediocin box) (Johnsen, Fimland, and Nissen-Meyer **2005**). The C-terminal domains are less conserved and are considered to define the nonlisterial antimicrobial spectrum. Pediocin PA-1 carries out its inhibitory effects by permeabilizing the cytoplasmic membrane of the receptor bacteria, leading to ion and other molecule leakages. Previous research has shown that pediocin PA-1 interacts with the extracellular loop of the mannose-phosphotransferase system (MPTs) which is located in the membrane (Sun, Song, and Zheng 2015). The conserved N-terminal region of pediocin PA-1 mediates the initial binding with MPTs by electrostatic interaction. After that, the hydrophobic C-terminal region penetrates the target cell membrane, playing a vital role in receptor membrane leakage.

Furthermore, based on differences in the C-terminus, Class-IIa bacteriocins may be categorized into three groups (Johnsen, Fimland, and Nissen-Meyer 2005):

Subgroup I: enterocin A, pediocin PA-1/AcH, and sakacin P
Subgroup II: mesentericin Y105 and leucocin A
Subgroup III: carnobacteriocin B2 and curvacin A

1.2.2.2 Class IIb

Class IIb or two-peptide bacteriocins are composed of two distinct and unique peptide molecules that require an equal peptide ratio to exhibit appropriate antimicrobial activity; however, when each peptide is considered separately (Garneau, Martin, and Vederas 2002). For instance, lactacin F and lactococcin G are produced by *Lc. johnsonii* and *Lc. lactis* strains, respectively. These compounds are two-peptide bacteriocins with a narrow spectrum of inhibition against *Clostridium, Enterococcus faecalis*, and lactobacilli (Woraprayote et al. 2016).

1.2.2.3 Class IIc

Bacteriocin class IIc is characterized by covalently bonded N- and C-termini, leading to a cyclic structure (Kawai, Kemperman, et al. 2004, Maqueda et al. 2004). Compared to linear peptides, their circular structure gives higher structural stability, thermal stress tolerance, and better stability against proteolytic degradation (Perez, Zendo, and Sonomoto 2014). In addition, based on the mature peptide's calculated isoelectric point (pI) (Woraprayote et al. 2016). Therefore, class IIc can be subdivided into two subclasses (Kawai, Kemperman, et al. 2004, Kawai, Ishii, et al. 2004, Maqueda et al. 2004);

- **Subclass c(i) (high pI value; ≈ 10):** enterocin AS48 and carnocyclin A
- **Subclass c(ii) (low pI value; ≈ 7):** reutericin 6 and gassericin A

1.2.2.4 Class IId

Class IId bacteriocins contain no substantial sequence similarity with other class II bacteriocins; in other words, they are "one-peptide linear, nonpediocin-like" (Cotter, Hill, and Ross 2005). The subgroup of bacteriocins contains a typical signal peptide secreted by the producer cell's general secretory (*sec*-) pathway (Nes et al. 1996). *sec*-dependent bacteriocins produced and were translocated with a typical N-terminal signal sequence of a *sec*-dependent pathway (Cintas et al. 2000) and leaderless bacteriocins that are atypically produced without N-terminal leader sequences and are active instantaneously following translation (Klaenhammer 1993). For instance, lacticin Q is synthesized from *Lc. lactis* QU 5 (Fujita et al. 2007), and *Weissella hellenica* QU 13 synthesized weissellicins Y and M (Woraprayote et al. 2016).

1.3 ANTIMICROBIAL PEPTIDES (AMP) DERIVED FROM LAB

Bifidobacterium strains have contributed to their use as probiotics (World Health 2002), and some are also known to synthesize AMP (Martinez et al. 2013). For example, *Bf. lactis* Bb-12 and *Bf. longum* Bb-46 produced bifilact Bb-12 and bifilong Bb-46, respectively. These bacteriocins exhibited inhibitory effects on *Bacillus cereus, E. coli, Salmonella typhimurium*, and *Staphylococcus aureus* (Saleh and El-Sayed 2004). Besides, bifidocin B was synthesized by *Bf. bifidum* NCFB 1454. This compound suppressed the growth of *Listeria* spp., *Bacillus* spp., *Clostridium* spp., and some LAB, *Leuconostoc* spp., *Enterococcus* spp., and *Pediococcus* spp. (Yildirim and Johnson 1998). Bifidin was produced from *Bf. bifidum* NCDC 1452 and bifilong from *Bf. longum* could inhibit Gram-negative and Gram-positive bacteria (Anand, Srinivasan, and Rao 1984). *Carnobacterium* isolated from environments, meat, seafood, dairy, and food products (Leisner et al. 2007) was the source of bacteriocin (Davidson and Zivanovic 2003). *Cb. piscicola* LV17A could produce carnobacteriocin A to inhibit several LAB as well as *Listeria* strains (Worobo et al. 1994). Furthermore, piscicolin 61 obtained from *Cb. piscicola* LV61 effectively limited the development of *Cb. divergens, Cb. piscicola, Cb. mobile*, and *Listeria monocytogenes* (Schillinger, Stiles, and Holzapfel 1993).

Enterococcus strains were found in numerous sources such as environmental animals and humans, plants, and foods (Giraffa 2014). The regulatory status of commensal potential *Enterococcus* candidates could be applied in food, feed, and probiotic use (Hanchi et al. 2018). Enterocins from enterococci were primarily used in food for the prevention of food pathogenic contamination (i.e., suppressing and reducing the growth of *Bacillus, Listeria*, and *Salmonella* strains) and also used in clinical application (i.e., limiting the growth of pathogen-associated diseases and also multidrug-resistant (MDR)) (Dillon 2014, Yusuf 2018, Almeida-Santos et al. 2021). Enterocin L50A, L50B, and Q from *En. faecium* L50 could limit the development of *Pc. acidilactici, En. faecium, Lc. lactis* subsp. *cremoris*, and *Lb. sakei* (Cintas et al. 2000). Moreover, enterocin EJ97 from *En. faecalis* EJ97 could lower the viable cell of *B. macroides* and *B. maroccanus* (Sánchez-Hidalgo et al. 2003). Enterocin B from *En. faecium* T136 suppressed and prevented the growth of several pathogens such as *L. monocytogenes, L. innocua, Cl. sporogenes, Cl. tyrobutyricum, S. aureus, S. carnosus*, and *Propionibacterium* sp. (Casaus et al. 1997). Enterocin 1071 from *En. faecalis* BFE 1071 could inhibit *L. innocua, Micrococcus* spp., *Peptostreptococcus aerogenes, Pro. Freudenreichii* subsp. *shermanii*, and *St. agalactiae* (Balla et al. 2000). *En. faecalis* K-4 could produce enterocin SE-K4, to suppress the development of *B. subtilis, Cl. beijerinckii*, and *L. monocytogenes* (Eguchi et al. 2001). Enterocin E50-52 from *En. faecium* NRRLB-30746 showed inhibitory effects on *Campylobacter jejuni, Yersinia* spp., *Salmonella* spp., *E. coli* O157:H7, *Shigella dysenteriae, Morganella morganii, Staphylococcus* spp., and *Listeria* spp. (Svetoch et al. 2008).

Lactobacillus strains isolated from dairy products, fermented foods, the environment, and humans and animals (Parente et al. 2010) were associated with promoting health benefits to the host (Bosch et al. 2012). They have outstanding probiotic characteristics that have been applied in the development of food and medicinal fields (Parente et al. 2010). Several strains produced bacteriocins (da Silva Sabo et al. 2014). For example, curvacin DF126 from *Lb. curvatus* DF126 potentially inhibited *L. monocytogenes*. *Lb. plantarum* 423 produced plantaricin 423, which could control the development of *L. monocytogenes* (Dicks, Mellett, and Hoffman 2004). Moreover, lactocin 705 obtained from *Lb. casei* CRL 705 could limit the generation of several foodborne pathogens (Cuozzo et al. 2000).

Lactococcus strains derived from food and those derived from human and animal sources serve essential functions in food preservation and medicinal applications, respectively (Yerlikaya 2019). *Lc. lactis* strain is generally recognized as safe (GRAS) by the Food and Drug Administration (FDA) (Song et al. 2017). The most well-known AMP from *Lactococcus* spp. is called nisin. For example, nisin Z could inhibit *L. monocytogenes, Cl. perfringens, B. cereus*, and *S. aureus* (Noonpakdee et al. 2003). Nisin Q synthesized from *Lc. lactis* 61-14 could reduce the growth of *L. monocytogenes, Cl. perfringens, B. cereus*, and *S. aureus* (Zendo et al. 2003). Furthermore, nisins provide attractive health benefits such as anticancer and immunomodulation activities shown in the next section (Wu, Hu, and Cao 2007, Preet et al. 2015, Mouritzen et al. 2019, Jia et al. 2019).

Leuconostoc strains were the starter cultures in the production of dairy products and food fermentation (Feiner 2006). *Ln. mesenteroides* E131 from fermented sausage could produce bacteriocin, which showed an effective inhibitory effect on *L. monocytogenes* (a pathogen causing food spoilage) (Drosinos et al. 2005). Leucocin A and leucocin C from *Ln. carnosum* 4010 could inhibit the development of *L. monocytogenes* during the storage period of vacuum-packed meat products (Budde et al. 2003). Mesenterocin 5 from *Ln. mesenteroides* UL5 controlled the growth of *L. monocytogenes* (Daba et al. 1991), while mesentericin Y105 from *Ln. mesenteroides* Y105 could limit the development of *L. monocytogenes* (Héchard et al. 1992).

Pediococcus strains can be isolated from foods or drink sources and Pediocins are bacteriocins produced by several pediococci (Raccach 2014). This bacteriocin contained anti-listeria activity and could apply as a starter in fermented meat (*Nham*) (Kingcha et al. 2012). In addition, film packaging containing pediocin PA-1/AcH might dramatically decrease *L. monocytogenes* growth on the surface of the meat (Woraprayote et al. 2013). Moreover, pediocin provided several beneficial human health benefits (in the next section), such as preventing *Listeria* infection and antiproliferation of cancer cells.

In the past, *St. thermophilus* was widely used as a starter culture in yogurt and cheese production (Uriot et al. 2017). However, some *Streptococcus* strains that produced bacteriocin played beneficial roles in preventing food pathogen contamination and human pathogen infection. For example, bovicin 255 synthesized from *St. gallolyticus* LRC0255 could suppress the growth of *Butyrivibrio fibrisolvens* and *Lb. ruminis* (Whitford et al. 2001). In addition, ubericin A produced from *St. uberis* could limit the generation of *L. monocytogenes*, *L. grayi*, *L. seeligeri*, *L. ivanovii*, *En. faecalis*, and *En. hirae* (Heng et al. 2007).

Several strains of *Weissella* were isolated from animal skin, milk, and faeces; from saliva, breast milk, faeces, and vagina of humans; and from plants, vegetables, and fermented foods (Fusco et al. 2015). Some *W. cibaria* strains are being studied as probiotics because of their great potential for treating periodontitis (Kang et al. 2011). Furthermore, *Weissella* strains could produce bacteriocin such as weissellicins Y and M from *W. hellenica* QU 13. This bacteriocin showed inhibitory effects on *Kocuria rhizophila*, *Weissella* spp., *Bacillus* spp., *Pediococcus* spp., and *Streptococcus* spp. (Masuda et al. 2012). Bacteriocin-producing strains and their antimicrobial activity are summarized in Table 1.2.

1.4 APPLICATIONS OF BACTERIOCINS IN FOOD PRESERVATION

1.4.1 Anti-Foodborne Pathogens

Food and beverage spoilage is always a critical issue in the food industry. Some microbes contaminate food and drinks, resulting in their deterioration. It might ruin the taste of foods and beverages and induce diverse foodborne diseases (Villalobos-Delgado et al. 2019, Bhalla et al. 2019; Reddy et al. 2022). Foodborne illnesses are divided into two types: food intoxication and food infection (Erhabor, Erhabor, and McGaw 2019, Drolia and Bhunia 2019). *S. aureus* produced enterotoxin, which can cause human intestinal inflammation (Castro, Silva, and Teixeira 2018). In addition, diarrhoea is caused by *Salmonella* infection (Kuang et al. 2018).

Chemical preservative agents have been widely used, but they can trigger several adverse effects on human health. To avoid and prevent health risks, there is a substantial growing demand for chemical-free and natural food preservation solutions (Zanetti et al. 2018). Consequently, numerous studies have lately been conducted on biopreservatives that can inhibit and prevent spoilage bacteria (Wayah and Philip 2018, Johnson et al. 2018, Linares-Morales et al. 2018, Ben Said et al. 2019). Bacteriocin (i.e., nisin, pediocin, enterocin, and leucocin) is the most notable biopreservative applied in food industries to prevent food spoilage (Johnson et al. 2018). The bacteriocin advantages

TABLE 1.2
Bacteriocins from LAB and their antimicrobial activity.

Bacteriocins	Producing strain	Antimicrobial against	References
Bifilact Bb-46	*Bf. longum* Bb-46	*S. aureus*, *Sa. typhimurium*, *B. cereus* and *E. coli*	Saleh and El-Sayed (2004)
Bifidin I	*Bf. infantis* BCRC 14602	*Sa. enteritidis*, *Sa. typhimurium*, *E. coli*, *S. aureus*, *Sh. dysenteriae*, *Cl. butyricum*, *B. subtilis*, and *B. cereus*	Cheikhyoussef et al. (2009),Cheikhyoussef et al. (2010)
Bisin	*Bf. longum* DJO10A	*St. thermophilus*, *Cl. perfringens*, *S. epidermidis*, *B. subtilis*, *Serratia marcescens*, and *E. coli* DH5α	Lee, Li, and O'Sullivan (2011)
Carnocyclin A	*Cb. maltaromaticum* UAL307	*Cb. maltaromaticum*, *Cb. divergens*, *Brochothrix campestris*, *Bro. thermosphacta*, *L. monocytogenes*, *Ln. mesenteroides*, and *S. aureus*	Martin-Visscher et al. (2008)
Bacteriocin T8	*En. faecium* T8	*En. faecalis*, *Lb. sakei*, *Propionibacterium* sp.	De Kwaadsteniet et al. (2006)
Enterocin AS-48	*En. faecalis* A-48-32	*L. monocytogenes* and *Salmonella* spp.	Ananou et al. (2010)
Enterocin C	*En. faecalis* C901	*Enterococcus* spp., *Lb. sakei*, and *Lactococcus* spp.	Maldonado-Barragán et al. (2009)
Sakacin C2	*Lb. sakei* C2	Anti-*Listeria* and anti-*Enterobacteriaceae*	Gao, Li, and Liu (2014)
Sakacins P and X	*Lb. curvatus* MBSa2	*Listeria* spp.	Barbosa et al. (2015)
Sakacins X, T and P	*Lb. curvatus* 54M16	*Staphylococci* and *Enterobacteriaceae*	Casaburi et al. (2016)
Fermencin SA715	*Lb. fermentum* GA715	*Pseudomonas aeruginosa*, *M. luteus*, *E. coli*, *S. aureus*, and *Corynebacterium* spp.	Wayah and Philip (2018)
Nisin A	Several strains of *Lc. lactis* subsp. *lactis*	A wide range of clinical and food-borne pathogens	Field et al. (2010)
Nisin F	*Lc. lactis* F10	*Lb. curvatus*, *Lb. plantarum*, *Lb. reuteri*, and *Staphylococcus* spp.	De Kwaadsteniet, Ten Doeschate, and Dicks (2008)
Lactocyclicin Q	*Lactococcus* sp. QU12	*M. luteus*, *L. monocytogenes*, *L. innocua*, *S. aureus*, and *E. coli*	Sawa et al. (2009)
Garvicin KS	*Lc. garvieae* KS1546	*L. monocytogenes*, *S. aureus*, and *B. cereus*	Chi and Holo (2018)
Lactococcin 972	*Lc. lactis* IPLA972	*Lactococcus* spp.	Madera et al. (2009)
Pediocin PA-1/AcH	*Pc. pentosaceus* BCC 3772	*L. monocytogenes*	Kingcha et al. (2012)
Nisin H	*St. hyointestinalis* DPC6484	*B. cereus*, *En. faecalis*, *St. agalactiae*, *St. bovis*, *St. gallolyticus*, *L. innocua*, *L. monocytogenes*, *S. aureus*, *St. hyointestinalis*, and *St. uberis*	O'Connor et al. (2015)
Nisin P	*St. agalactiae* DPC7040	*En. faecium*, *S. aureus*, *S. capitis*, *S. simulans*, and *S. epidermidis*	Garcia-Gutierrez et al. (2020)
Salivaricin A5	*St. salivarius* DPC 6490	MRSA, *Cutibacterium acnes*, and *En. Faecium*	Ellis, Ross, and Hill (2019)
Bac7293A and Bac7293B	*W. hellenica* BCC 7293	*L. monocytogenes*, *S. aureus*, *Sa. typhimurium*, *E. coli*, *P. aeruginosa*, and *Aeromonas hydrophila*	Woraprayote et al. (2018)

are the following: (1) extend food shelf life, (2) offer additional safety under high-temperature processing and other critical points, (3) reduce the foodborne pathogens risks, and (4) decrease the economic losses (Thomas, Clarkson, and Delves-Broughton 2000, Gálvez et al. 2007). Moreover, bacteriocins may be supplied to foodstuffs without modifying their organoleptic qualities since they are odourless, tasteless, and colourless (Perez, Zendo, and Sonomoto 2014). Table 1.3 displays a list of bacteriocins applied in food preservation.

1.4.1.1 Nisin

Nisin is the bacteriocin that has been permitted for use in food factories by the US Food and Drug Administration (FDA) and the World Health Organization (WHO) (Delves-Broughton 2005). Nisin

TABLE 1.3
Bacteriocins from LAB and application in food.

Bacteriocin	Producer strains	Application in food	Antimicrobial against	References
Bacteriocin 7293A and 7293B	*W. hellenica* BCC 7239	Meat and Meat products	*A. hydrophila*, *E. coli*, *Sa. typhimurium*, and *P. aeruginosa*	Woraprayote et al. (2015)
Carnocyclin A, carnobacteriocin BM1, Piscicolin 126	*Cb. maltaromaticum* UAL307	Meat and Milk products	*E. coli*, *P. aeruginosa*, and *Sa. typhimurium*	Martin-Visscher et al. (2011)
Enterocin L50A and B	*En. faecium* L50	Non-alcoholic and alcoholic beer	*Lb. brevis* and *Pc. damnosus*	Hata, Sato, and Morimitsu (2016)
Enterocin MR-10A	*En. faecalis* N1-33	Custard cream	*B. cereus*	Hata, Sato, and Morimitsu (2016)
Enterocin A and B	*En. faecium* WHE 81	Munster cheese	*L. monocytogenes*	Khan, Flint, and Yu (2010)
Lactococcin BZ	*Lc.* subsp. *lactis* BZ	Skim and full-fat milk	*L. monocytogenes*	Yildirim et al. (2016)
Leucocin K7	*Ln. mesenteroides* K7	Milk	*L. monocytogenes*	Shi et al. (2016)
Leucocin A	*Ln. gelidum* UAL187	Milk products, fresh meat, and sausage Meat	*L. monocytogenes* *Cb. divergens*	Balay, Gänzle, and McMullen (2018)
Leucocin B	*Ln. carnosum* Ta11a	Meat and meat products	*L. monocytogenes*	Hwang et al. (2018)
Nisin	*Lc.* subsp. *lactis*	Meat and sausages Milk and milk products Bakery, culinary, dairy products, and beverages	*Cl. botulinum* and *L. monocytogenes* *Cl. botulinum*, *Cl. perfringens*, and *B. cereus* *B. cereus*, *L. monocytogenes*, and *Cl. botulinum*	da Costa et al. (2019)Ibarra-Sánchez et al. (2020)Ibarra-Sánchez et al. (2020)
Pediocin 34	*Pc. pentosaceus* 34	Meat and milk products	*L. monocytogenes*	(Kaur, Singh, and Malik 2013)
Pentocin 31-1	*Lb. pentosus* 31-1	Tray-packed chilled pork	*L. monocytogenes* and *P. fluorescens*	Zhang et al. (2010)
Pentocin 31-1	*Lb. pentosus* 31-1	Tray-packed chilled pork	*L. monocytogenes* and *P. fluorescens*	Zhang et al. (2010)
Sakacin Q	*Lb. curvatus* ACU-1	Cooked meat	*Listeria* spp.	Rivas et al. (2014)
Sukacin	*Lb. sakei*	Meat products	*L. monocytogenes*	Saraoui et al. (2018)

is mainly produced by *Lactococcus* spp. and *Streptococcus* spp. Nisin(s) desirably expresses the limitation effects against foodborne bacterial groups (i.e., *B. cereus*, *Cl. botulinum*, *L. monocytogenes*, and *S. aureus*). Several studies determined the synergistic benefits of the combination of nisin and antibiotics, which could also enhance inhibition activity (Zhao et al. 2016). For instance, nisin suppressed the growth of *L. monocytogenes* and *S. aureus* in cheese products to extend their storage life (Silva, Silva, and Ribeiro 2018). The study of Gharsallaoui et al. (2016) described that nisin supplementation in milk and milk products could prevent and avoid spoilage by inhibiting the growth of *B. cereus*, *Cl. botulinum*, and *Cl. perfringens*. Moreover, nisin Z showed great antipathogenic activity to *L. monocytogenes* strain 13 and 699 (MIC: 1 and 0.2 μg/ml, respectively) (Fu et al. 2018), methicillin-resistant *S. aureus* (MRSA) (MIC: 4.17 μg/ml) (Raygada and Levine 2009).

1.4.1.2 Enterocin

Practically, there are two methods for applying enterocin as a food preservative, including the use of enterocin-producing strains and the use of partial or pure enterocin. For example, enterocin-producing strains are used as artisanal cheese starters (Terzić-Vidojević et al. 2015). Recently, ready-to-eat foods without food processing are the primary cause of foodborne pathogen contamination, which has led to the development of human health risks (Luna-Guevara et al. 2019). To overcome this problem, Khan, Flint, and Yu (2010) found that *En. faecalis* A-48-32 produced enterocin AS-48, which could be against several foodborne pathogens such as *B. cereus*, *B. macroides*, *Paenibacillus* spp., and *S. aureus* in fresh vegetables. Besides, 10 μg/ml of enterocin AS-48 controlled and suppressed *B. cereus*, *B. macroides*, *Pa. polymyxa*, *Pa. amylolyticus*, and *S. aureus* (Grande Burgos et al. 2014, Le Lay et al. 2016). Besides, the use of enterocin AS-18 also was used in canned fruits/foods, soybean sprouts products, and chilled vegetables (Grande Burgos et al. 2014). Dairy products are always easily contaminated because they are well-known as rich source of nutrients. Thus, several studies applied enterocin in dairy products (Gupta et al. 2016, da Costa et al. 2019, Sharma, Rashid, and Kaur 2020). For instance, *L. monocytogenes* in cottage cheese was inhibited by 4.57 μg/ml of enterocin A and enterocin 416K1, which were produced by *En. faecium* CTC 492 and *En. casseliflavus* IM 416K1, respectively (Vimont et al. 2017). The combination of enterocin A and enterocin B could be used as biopreservative agents in Munster cheese to control *L. monocytogenes* contamination (Aspri et al. 2017). In addition, enterocin CCM 4231 could prevent *S. aureus* and *L. monocytogenes* contamination by preserving Saint-Paulin cheese skimmed milk, yogurt, and soy milk (Santos et al. 2018, Choyam et al. 2019). Furthermore, enterocin is widely used in fisheries, meats, and beverages (Singh 2018). Enterocins A and B are derived from *En. faecium* CTC492 that exhibited antipathogenic effects in dry fermented sausages and cooked ham, such as *L. innocua*, L. monocytogenes, and *Lb. sakei* (Castellano et al. 2017). Enterocin 1071 A and B, derived from *En. faecalis* BFE 1071, were found to have inhibitory effects on *L. innocua*, *S. epidermis*, and *Proteus vulgaris* in the fish spread processing (Arbulu et al. 2016). The study of García et al. (2004) stated that enterocin EJ97 from *En. faecalis* EJ97 was utilized to suppress *B. macroides* and *B. maroccanus* in the manufacture of vegetable puree (Zucchini). Furthermore, enterocin L50A and B from *En. faecium* L50 has a crucial role in preventing the contamination of non-alcoholic and alcoholic beer by *Lb. brevis* and *Pc. damnosus* (Hata, Sato, and Morimitsu 2016).

1.4.1.3 Pediocin

Pediocin prevented and limited the microorganisms that might cause food spoilage, such as *Cl. perfringens* and *L. monocytogenes* (Nieto-Lozano et al. 2010). Currently, pediocin could be used as biopreservatives in food industries in the form of pediocin-producing starter or partial/purified pediocin (Silva, Silva, and Ribeiro 2018). The research of Nieto-Lozano et al. (2002) reported that *Pc. acidilactici* MCH14 synthesized pediocin PA-1, which was applied in salad dressing, fresh beef, vacuum-packed beef, dried sausage, fermented meat products, cottage cheese, chicken, and fish fillets to suppress the growth of *Lb. bifermentans*, *Ln. mesenteroides*, *L. monocytogenes*, and *Cl. perfringens* (Nieto-Lozano et al. 2002, Zhang et al. 2009). Furthermore, pediocin 34, derived from

Pc. pentosaceus 34, was shown to have an inhibitory effect on *L. monocytogenes* in milk and meat (Kaur, Singh, and Malik 2013).

1.4.1.4 Leucocin

Leucocin A, synthesized by *Ln. gelidum* UAL187, showed antimicrobial effects on *L. monocytogenes* in milk products, fresh meat, and sausage (Sawa et al. 2010). The study of Balay et al. (2017) showed that several strains of *L. monocytogenes* were inhibited by leucocin (MIC > 2,200 μM). In addition, leucocin K7 from *Ln. mensenteroides* K7 was reported to have an inhibitory effect on *L. monocytogenes* with a 28 mg/ml (MIC) in milk (Fu et al. 2018). Furthermore, *Ln. pseudomesenteroides* KM432Bz produced leucocin B-KM432Bz. This compound preserved and prevented Spanish-style dry fermented sausage spoilage and meat products from the contamination of *L. innocua* CIP 80.11. and *L. monocytogenes* CIP 82.110 (Hwang et al. 2018, Tirloni et al. 2019).

1.4.1.5 Other bacteriocins

Moreover, *Lb. sakei* synthesized sukacin was applied to limit the generation of *L. monocytogenes* in meat products (Saraoui et al. 2018). Additionally, *W. hellenica* BCC 7239 produced bacteriocins 7293A and 7293B. These compounds showed inhibitory effects on *P. aeruginosa*, *A. hydrophila*, *Sa. typhimurium*, and *E. coli* in meat and meat products (da Costa et al. 2019). Lactocin AL705 and Lactocin 705 are produced from *Lb. curvatus* CRL705. These two bacteriocins contained anti-*Listeria* and anti-*Br. thermosphacta* activity in vacuum-packed fresh beef, respectively (Castellano and Vignolo 2006).

1.4.2 Food Packaging Containing Antimicrobial Activity

Direct bacteriocin supplementation into food products may result in some loss of activity due to the dilution effect, decreasing bacteriocin efficacy. The application of bacteriocin-containing packaging technologies might be more practical due to the gradual bacteriocin distribution from the packaging to the product surface, consequently assisting in the maintenance of effective concentrations (Quintavalla and Vicini 2002; Chausali et al. 2022). Furthermore, this technique required lower bacteriocin concentration than direct addition to the food products. For instance, Neetoo, Ye, and Chen (2008) study showed that the application of nisin-coated synthetic films on vacuum-packed cold-smoked salmon resulted in a significant decrease in the survival rate of *L. monocytogenes* when done at a specific storage temperature. This technique was advantageous to other techniques of bacteriocin application because bacteriocins could be degraded or lose their activity when they directly interacted with food materials (Pilevar et al. 2020). Besides, the enterocin 416K1 was entrapped in polyvinyl alcohol (PVOH)-based coatings applied to poly (ethylene terephthalate) (PET) films, and their efficiency in controlling the generation of *L. monocytogenes* ATCC 19117 (Degli Esposti et al. 2018). Furthermore, bacteriocin 32Y produced from *Lb. curvatus* 32Y was applied to polythene films (PE films) (Mauriello et al. 2004). This antimicrobial packaging showed limitation activities on the development of *L. monocytogenes* during the storage of meat products. The other food packaging containing antimicrobial activity is shown in Table 1.4.

1.5 CLINICAL AND HEALTH BENEFIT APPLICATIONS

1.5.1 Antibacterial Activity

Antibiotic administration has several adverse effects and causes of gut microbiota imbalance, resulting in various risks in humans. As a result of antibiotic abuse, the problem of multiple drug resistance bacteria has developed significance (Joerger 2003). Bacteriocins are an alternative treatment for human health because they are non-pathogenic and non-toxic (Acuña et al. 2012). In addition,

TABLE 1.4
Food packaging containing antimicrobial activity.

Antibacterial packaging material	Antimicrobial against	References
Bac7293-PLA/SP	*L.* monocytogenes, *S. aureus*, *P. aeruginosa*, *A. hydrophila*, *E. coli*, and *Sa. typhimurium*	Woraprayote et al. (2018)
Enterocin 416K1-PVOH	*L. monocytogenes*	Degli Esposti et al. (2018)
Lactocin 705 and lactocin AL705 mixture/synthetic polymer-based and wheat gluten-based films	*L. innocua* while *Lb. plantarum* was slightly inhibited	Massani et al. (2014)
Nisin and EDTA-BC	*E. coli* and *S. aureus*	Dos Santos et al. (2018)
Nisin-LLPDE	*L. innocua* and *M. luteus*	Perna (2016)
Nisin Z and lauroyl arginate ethyl (LAE)-bilayer thermoplastic starch/ PBAT	*Sa. typhymurium*, and *V. parahaemolyticus*	Pattanayaiying et al. (2019)
Nisin-PET/LLPDE	Total aerobic bacteria and *S. aureus*	Ayers (2018)
Nisin-PE and/or HPMC	*Br. thermosphacta* *S. aureus*	Malhotra, Keshwani, and Kharkwal (2015)
Nisin-CAF-lauric alginate (LAE)	Shiga toxin-producing *E. coli* (STEC), *Salmonella* spp., *L. monocytogenes*, and *S. aureus*.	Hassan, A. H., and Cutter, C. N. (2020)
Pediocin ALTA® 2351-cellulose acetate film	*L. innocua* and *Salmonella* spp.	Santiago-Silva et al. (2009)
Sakacin A-PE-coated paper	*L. innocua*	Barbiroli et al. (2017)

Abbreviation: BC, bacterial cellulose; CAF, composite antimicrobial film, made from biopolymer-poured PE film; HPMC, hydroxypropyl methylcellulose; LLPDE, linear low-density polyethylene; PBAT, polybutylene adipate terephthalate; PE, polyethylene; PET, polyethylene terephthalate; PLA/SP; poly(lactic acid) and sawdust particle and PVOH, polyvinyl alcohol

bacteriocins can express the activity in the nanomolar unit, high-specific activity, and particular mechanisms of action (van Heel, Montalban-Lopez, and Kuipers 2011). Notably, bacteriocin may reduce cross-resistance development risks (Soltani et al. 2021). Bacteriocins have been shown to suppress critical pathogens such as Shiga toxin-producing *E. coli* (STEC), enterotoxigenic *E. coli* (ETEC), and methicillin-resistant *S. aureus* (MRSA), VRE, *Agrobacterium*, and *Brenneria* spp. (Grinter, Milner, and Walker 2012, Cotter, Ross, and Hill 2013). Because of their abundance and diversity, bacteriocins can be adapted for infection prevention, antibiotic resistance, and other therapies (Vidya 2020).

1.5.1.1 Nisin

Nisin is also used in clinical applications (Zhao et al. 2016). Nowadays, the population of antibiotic resistance pathogens is significantly increasing, such as *S. aureus* and MRSA, a cause of skin infection (Raygada and Levine 2009). In the treatment of skin diseases, nisin-containing wound dressings have been shown to be effective in reducing *S. aureus* growth and promoting wound healing (Agnes Mary and Giri Dev 2015). In addition, *Cl. difficile* triggers colon inflammation, and it can rapidly adapt and resist several antibiotics. To solve this problem, nisin could alleviate colon inflammation by the *Cl. difficile* elimination (Ghose and Kelly 2015). Besides, dental caries is a global health problem. Nisin was shown to inhibit activity against oral pathogens that caused dental caries, such as *St. sanguinis*, *St. sobrinus*, and *St. mutans* (Tong et al. 2010). Moreover, candidemia (bloodstream and mucosal infected with *Candida albicans*) could be treated with nisin Z (Le Lay et al. 2008). Additionally, nisin helped treat gastric ulcers and pneumonia by suppressing the growth of *Helicobacter pylori* and *P. aeruginosa*, respectively (Song et al. 2018). Nisin is safe to use in humans at lower dosages than 83.25 mg/kg and less than 66.7 mg/kg in the mouse model. However, at concentrations of 300–400 mg/mL, nisin may have a human contraceptive effect (Kitagawa, Otani, and Inai 2019).

1.5.1.2 Lacticin

Lacticin has a broad spectrum (Mirkovic et al. 2016). Lacticin 3147 from *Lc. lactis* subsp. *lactis* DPC3174 was potentially applied to treat prosthetic joints, surgical sites, and skin infections (Morgan et al. 2005). According to Ellis, Ross, and Hill (2019), lacticin 3147 showed a tremendous inhibitory effect on MRSA ST291. Moreover, *Cutibacterium acnes* (opportunistic pathogen) can infect during the surgery. However, the growth of *Cu. acnes* was inhibited by 2.50 µg/ml of lacticin (Ellis, Ross, and Hill 2019). For dental caries prevention, 1280–5120 AU/ml of lacticin 3147 can partially inhibit and suppress the growth of *St. mutans* (oral pathogen and plaque formation) (O'Connor et al. 2006). Furthermore, according to Titze and Krömker (2020), *H. pylori* plays a primary role in gastric cancer, gastric ulcers, and gastritis development. Lacticins BH5 and A164, synthesized by *Lc. lactis* subsp. *lactis* BH5 and *Lc. lactis* subsp. *lactis* A164, were the most effective bacteriocins among lacticins that might be developed to treat gastric ulcers caused by *H. pylori* (Neshani et al. 2019).

1.5.1.3 Salivaricin

Salivaricin has been demonstrated to be beneficial in protecting oral health and skin and lung infection prevention and treatment. *St. salivarius* K12 synthesized bacteriocin, which is named salivaricin A2. This compound could inhibit *St. pneumoniae* isolates (a cause of bacteremia, sinus infection, pneumonia, meningitis, and ear infection) (Geng et al. 2018). Additionally, the research of Reglinski and Sriskandan (2015) investigated that salivaricin B derived from *St. salivarius* K12 provided an antimicrobial effect on *St. pyogenes* (a cause of several human infective diseases such as flesh-eating disease, scarlatina, cellulitis, pharyngotonsillitis, rheumatic fever, and erysipelas). In addition, *Corynebacterium* spp. caused several human disorders such as endocarditis, gastrointestinal tract infection, pharyngitis, and skin infection. Interestingly, salivaricin B could inhibit *Corynebacterium* spp. GH 17 (Reglinski and Sriskandan 2015). Besides, *St. Salivarius* 5M6c produced salivaricin D, which showed an inhibitory effect on *Cl. bifermentans* (a pathogen that causes the infection disorders) (Birri, Brede, and Nes 2012, Edagiz et al. 2015). Additionally, the antibiotic resistance *Ln. lactis* can infect and impair the host immune system. However, Patel et al. (2012) showed that salivaricin D showed antimicrobial activity to *Ln. lactis.* Salivaricin D was also applied in pneumonia prevention because this bacteriocin could inhibit *St. pneumoniae* TIGR4, R6, and D36 (Birri, Brede, and Nes 2012).

1.5.1.4 Fermencin

Fermencin SA715 is synthesized from *Lb. fermentum* GA715, which contained inhibitory effects on several pathogens such as *P. aeruginosa* PA7, *M. luteus* ATCC 10240, *E. coli* UT181, *S. aureus* RF122, and *Corynebacterium* spp. GH17 (Wayah and Philip 2018). In addition, *Lb. fermentum* SD11 produced fermencin SD11 (MW = 33.5 kDa) (Wannun, Piwat, and Teanpaisan 2016), which could suppress the growth of oral pathogens such as *Aggregatibacter actinomycetemcomitans* ATCC 33384, *Fusobacterium nucleatum* ATCC 25586, *St. sobrinus* ATCC 33478, *St. mutans* ATCC 25175, *Porphyromonas gingivalis* ATCC 33277, and fungal (i.e., *Can. albicans* ATCC 90028) (Periasamy and Kolenbrander 2009).

1.5.1.5 Enterocin

Enterocin A could suppress and protect against listeriosis in pregnant women, the elderly, newborns, and immunodeficiency patients, which are the groups most susceptible to infection (Mateus et al.

2013). *L. monocytogenes* EGDe and *E. coli* NCDC 135 were inhibited by 4.57 μg/ml and 112 μg/ml of enterocin A (Ghrairi and Hani 2015, Sheoran and Tiwari 2019). Also, the synergistic effects of thyme essential oil and enterocin A showed a tremendous inhibitory effect on *L. monocytogenes* EGDe (Ghrairi and Hani 2015). Furthermore, *En. hirae* 20C synthesized enterocin 20C. This novel bacteriocin E20c of molecular weight 6.5 kDa could against *Sa. enterica*, which may develop several disorders such as bacteremia, enteric fever, and gastroenteritis. Thus, enterocin 20C could be applied for foodborne illness prevention (Sharma, Rashid, and Kaur 2020). Additionally, enterocin LD3 was obtained from *En. hirae* LD3. Enterocin LD3 exhibited an inhibitory effect on *M. luteus* MTCC 106 (Sheoran and Tiwari 2019).

1.5.1.6 Other bacteriocins

Bacteriocin OR-7 was effective against *Cam. jejuni* NCTC 11168 (the human gastroenteritis pathogen) (Stern et al. 2006). Simova, Beshkova, and Dimitrov Zh (2009) reported that bulgaricin BB18 exhibited antimicrobial activity to Gram-positive and Gram-negative such as *Cl. difficile* TCN16, *L. monocytogenes* C12, *B. subtilis* 1A95, *Sa. typhimurium* FVK781, and *H. pylori* HPK78. In addition, several Gram-positive and Gram-negative pathogens are suppressed by Bac7293A and Bac7293B produced by *W. hellenica* BCC 7293 (Woraprayote et al. 2015). Besides, bacteriocin DY4-2 showed an antimicrobial effect against Gram-positive and Gram-negative pathogens (Lv et al. 2018). Similarly, several pathogens such as L. monocytogenes, *Cl. sporogenes*, *Mycobacterium phlei*, *S. aureus*, *En. faecalis*, and *B. subtilis* were suppressed by mutacin B-Ny266 synthesized by *St. mutans* (Mota-Meira et al. 1997). The study of Corr et al. (2007) reported that the administration of *Lb. salivarius* UCC118 which could synthesize bacteriocins Abp118. Consequently, mice were protected from *L. monocytogenes* infection. **Besides, *Lb. casei* LAFTI L26 (bacteriocin-producing strain) suppressed the growth of *L. monocytogenes* and enterohemorrhagic *E. coli* in mice (Su, Henriksson, and Mitchell 2007). *Pc. acidilactici* MM33 and *Lc. lactis* MM19 were isolated from intestinal origin, and they could inhibit the vancomycin-resistant *Enterococcus* in mice (Millette et al. 2008). The potential probiotic *Lb. salivarius* DPC 6005 (bacteriocin-producing strain) suppressed and prevented the growth and infection of *Sa. typhimurium* in pigs (Soltani et al. 2021). In conclusion, bacteriocins may be regarded as a practical approach to reducing the development of antibiotic resistance.

1.5.2 Antiviral Activity

Noticeably, various bacteriocins can be beneficially used as antiviral compounds to inhibit *Herpes simplex* virus (HSV-1 and HSV-2). Bacteriocin ST5Ha from *En. faecium* ST5Ha exhibited anti-HSV activity (Todorov et al. 2010). In addition, the enterocin CRL35 and ST4V inhibited HSV-1 and HSV-2. The modes of action of these compounds on HSV are to impair the viral-intracellular replication and interfere with the final phase of replication (Wachsman et al. 2003). Furthermore, PV-1 (poliovirus) was suppressed by bacteriocin GEn12 and GEn09 synthesized from *En. durans* GEn12 and GEn09, but also by bacteriocin GLc05 synthesized from *Lc. lactis* GLc05 (Cavicchioli et al. 2018). However, the nisin had no antiviral activity on HSV-1, influenza A H1N1, Newcastle disease virus Montana or Feline Herpesvirus KS285, and murine norovirus (Lange-Starke et al. 2014).

The mechanisms of bacteriocins against viruses were described by Wachsman et al. (2003). Bacteriocins may inhibit viral particle aggregation and impede receptor sites on host cells. In addition, bacteriocin may have a vital part in preventing the viral proliferation phase. Besides, bacteriocins containing antiviral activity have hydrophobic properties so that they may attach to the viral envelope. As a result, bacteriocin can interfere with the fusion of viral membranes and cellular membranes and provide its antiviral activity. Hence, the inactive bacteriocins against non-enveloped viruses may be due to structural variation from those discovered active, particularly the hydrophobic deficiency (Badani, Garry, and Wimley 2014).

1.5.3 Antiparasite Infection

The parasite–host interaction is complicated since it may promote or cause disease in the host (Rook 2007, Massey et al. 2004). *Plasmodium*, *Trypanosoma*, and *Entamoeba* are protozoa that may cause significant diseases in humans (e.g., malaria, sleeping sickness, and amoebic dysentery) (Pollitt et al. 2011, Esch and Petersen 2013). Numerous bacteriocins have been shown to have a suppressive impact on various parasites and aid in the treatment of parasitic disorders. For instance, enterocin AS-48 could successfully lower the *Trypanosoma cruzi* by mitochondrial membrane depolarization and reactive oxygen species generation, therefore relieving the symptoms of Chagas' disease (Martín-Escolano et al. 2020). Though studies on bacteriocins' ability to suppress parasites are currently limited, bacteriocins have the potential to be a useful therapeutic in the battle against parasite infection.

1.5.4 Treatment of Cancer

Cancer has become a substantial public health threat and a significant concern in the past half-century. According to current WHO records, there were 608,570 cancer mortality and nearly 1.9 million new cancer cases. For cancer mortality, lung cancer is the primary critical cause of cancer mortality among males (22%), followed by prostate (11%) and colorectal (9%) cancers. Among females, lung (22%), breast (15%), and colorectal (8%) cancers are the leading causes of cancer mortality.

For new cancer cases, prostate cancer is the most frequent cancer in males (26%), followed by lung (12%) and colorectal (8%) cancers; females are more likely to get breast (30%), lung (13%), and colorectal (8%) cancers (Siegel et al. 2021). Numerous studies suggest that bacteriocins exert antitumour cell activity. Bacteriocins against cancer cells are achieved by stimulating apoptosis and/or depolarizing the cell membrane, resulting in permeability alterations (Kaur and Kaur 2015), since bacteriocins are naturally and legally used in foods and may be suitable as a potential anticancer therapeutic approach.

For cancer treatment, Joo et al. (2012) discovered that nisin can suppress cancer growth in cancer treatment. Nisin at 40 and 80 µg/ml doses was used to treat three head and neck squamous cell carcinomas (HNSCC): 17B, HSC, and 14A. After 24 hours, nisin caused increased DNA shattering or apoptosis, an arrest in the cell cycle, and decreased cell proliferation in HNSCC. The anti-HNSCC activity of nisin was evaluated using the oral cancer xenograft mice model. For three weeks, a 150 mg/kg dosage of nisin was supplemented orally daily, and tumour sizes greatly decreased in nisin-treated mice. From this finding, it could be recommended that nisin is a safe and recent alternative medication for treating HNSCC. Besides, the 105.46 and 112.25 µM (IC_{50}) of nisin showed cytotoxic effects on MCF-7 (human breast adenocarcinoma cell line) and HepG2 (liver hepatocellular carcinoma), respectively (Paiva et al. 2012). Nisin exhibited an antiproliferative effect on the Jurkat cell line (T-lymphocyte cell). However, the same nisin concentrations could also inhibit normal human lymphocytes (Begde et al. 2011). Moreover, nisin could be against HT-29 and Caco-2 cell lines (colon and colorectum adenocarcinoma) with 89.9 µM and 115 µM (IC50), respectively (Maher and McClean 2006). Nisin A and Z revealed a significant dose-dependent reduction in the cell proliferation of HSC-3, UM-SCC-17B, and UM-SCC-14A at the concentration ranging from 800 to 100 µg/ml, as well as comparable similar dose-dependent effects were found on the enhancement in the number of apoptotic cells for two of the cell HSC-3 and UM-SCC-17B (Kamarajan et al. 2015). The synergistic effect of nisin A and doxorubicin could decrease the tumour size by 66.82% by specifically apoptosis induction compared to the control group. This finding could be concluded that nisin can enhance the effectiveness of chemotherapeutic medicines (i.e., doxorubicin) (Preet et al. 2015).

For pediocin, *Pc. acidilactici* PAC1.0 synthesized pediocin PA-1, which could suppress the proliferation of cell lines DLD-1 (human colon adenocarcinoma) and A-549 (human lung carcinoma) (Beaulieu 2005). Moreover, PA-1 was synthesized from *Pc. acidilactici* K2a2-3, bacteriocin PA-1 was shown to be cytotoxic to HeLa (cervical epithelial carcinoma) and HT-29 (colon adenocarcinoma) cell lines (Villarante et al. 2011). The proliferation of MCF7, HepG2, and HeLa cell lines was inhibited by 25 μg/ml of pediocin CP2, synthesized by *Pc. acidilactici* CP2 MTCC5101 (Reiss 2016). Furthermore, plantaricin A (2.4 kDa; class II bacteriocin) was synthesized by *Lb. plantarum* C11 (Diep, Håvarstein, and Nes 1996). It displays an antimicrobial effect at pH values ranging from 4.0 to 6.5 against closely related LAB species (Daeschel, McKenney, and McDonald 1990). *In vitro*, plantaricin A showed a cytotoxicity effect on the human T-cell leukaemia cell line (Jurkat) (Zhao et al. 2006).

Finally, non-pathogenic bacteriocin producers can be used to synthesize anticancerous bacteriocins in specific tumours. Additionally, the application of bacteriocin in cancer treatment may also reduce the risk of secondary infections. Nevertheless, it is crucial to highlight that most research linked to the anticancer activity of bacteriocins has been conducted *in vitro*, suggesting that *in vivo* studies are necessary.

1.5.5 Immunomodulatory Effects and Anti-inflammation Activity

The immune system is a complex network of cells, tissues, and organs that defends the body against diseases and harmful compounds (Hooper, Littman, and Macpherson 2012). Several disorders are associated with immune system problems (Varadé, Magadán, and González-Fernández 2021). Immunomodulatory effects and anti-inflammatory effects of bacteriocins are shown in Table 1.5. Bacteriocins can suppress the inflammatory condition caused by pathogen-associated molecular patterns (PAMPs) or other stimulants by cytokine modulation. This is characterized by an elevation in anti-inflammatory cytokines and a reduction in pro-inflammatory cytokines via controlling the activation of specific pathways such as mitogen-activated protein kinase (MAPK), nuclear factor kappa-B (NF-B), and Toll-like receptor (TLR). Bacteriocins can stimulate epithelial cells to synthesize antimicrobial compounds that have inhibitory effects on pro-inflammatory bacteria.

TABLE 1.5
Immunomodulatory effects and anti-inflammation activity.

Producer strains	Bacteriocins	Features	References
Lc. subsp. *lactis*	Nisin	Enhancing the level of IL-12 production in macrophages, regulating the levels of inflammatory factors in both directions, and promoting immune balance, reducing the levels of TNF-α, TNF-β, NF-κB, IL-1, and ROS in mice	(Moein, Fooladi, and Hosseini 2018, Małaczewska et al. 2019, Preet et al. 2019)
Lc. subsp. *lactis*	Nisin A	Suppressing the growth of bacteria in the wound and reducing the level of IL-6, IL-8, and TNF-α	Mouritzen et al. (2019)
Lc. subsp. *lactis*	Nisin Z	Mastitis therapy and anti-*St. agalactiae*, anti-*S. aureus*	Wu, Hu, and Cao (2007)
St. lactis SMN003	Nisin P	Modulating the cytokine level to alleviate uterine inflammation in rats	Jia et al. (2019)
Lb. gasseri LA39	Gassericin A	Binding to KRT19 hence enhances fluid absorption and reduces secretion early weaned piglets	Hu et al. (2018)
Lb. plantarum	Plantaricin EF	Suppressing lipid inflammation and lowering obesity	Heeney et al. (2019)
Lb. salivirius	Salivaricin LHM	Limiting inflammation caused by *P. aeruginosa*, with immune modulation in mice	Mahdi, Jabbar, and Auda (2019)

Furthermore, bacteriocin can also reduce pathogen migration and infection-induced inflammation by enhancing the expression of tight junction proteins (TJP), strengthening the intestinal barrier, and lowering the invasion of pro-inflammatory pathogens into the bloodstream and extra-intestinal organs.

In vivo and *in vitro* studies have shown that nisin has a remarkable anti-inflammatory effect. Nisin A can stimulate human keratinocytes HaCaT activity, suppress LPS-induced pro-inflammatory cytokine levels (i.e., TNF-α), limit bacterial growth, and promote the process of wound healing (Mouritzen et al. 2019). In cows with mastitis, nisin Z inhibited *St. agalactiae* and *S. aureus* and resulted in a significantly reduced milk somatic cell count, potentially treating the symptoms of mastitis (Wu, Hu, and Cao 2007). Besides, *St. lactis* SMN0003 produced nisin P, which could relieve the inflammation of the uterine in rats by modulation of the pro-inflammatory and anti-inflammatory cytokines levels (i.e., B7-2, IFN-γ , IL-2, and IL-8) and manipulating uterine neutrophils, hence repairing endometrial tissue (Jia et al. 2019). In obese mice, plantaricin EF could enhance the expression of TJP, improve the integrity of the intestinal barrier, lower the weight, and reduce lipid inflammation (Heeney et al. 2019). In addition, *Lb. salivarius* generated salivaricin LHM, which may inhibit the growth and biofilm formation of *P. aeruginosa* and reduce inflammation and prevent harm caused by *P. aeruginosa* infection. Therefore, it could be concluded that the salivaricin LHM contained an anti-inflammation effect both *in vivo* and *in vitro* (Mahdi, Jabbar, and Auda 2019).

It is suggested that nisin has potent immune-modulatory effects. In truth, whether the function is anti-infective, anti-inflammatory, or anti-tumour, it is inextricably linked to immune modulation. Under certain situations, nisin can suppress the level of pro-inflammatory factors to perform an anti-inflammatory activity, but it can also enhance the production of pro-inflammatory factors. For instance, nisin and Ag (nisin-Ag) nanoparticles stimulated the pro-inflammatory cytokine IL-12 production in macrophages (Moein, Fooladi, and Hosseini 2018). Surprisingly, nisin helped boost the peripheral blood mononuclear cell (PBMC) proliferation, enhance IL-1 and IL-6 production, and elevate the proportion of $CD4^+$ $CD8^+$ T cells. In contrast, when LPS triggered PBMC, nisin could suppress the secretion of the pro-inflammatory cytokine IL-6 (Małaczewska et al. 2019).

As mentioned previously, bacteriocins have a range of biological actions, including anti-infective agents and effective therapeutic agents in treating several immune-related diseases. They may well have potential cancer therapy applications.

1.6 THE EFFECT OF CRUCIAL FACTORS ON ANTIMICROBIAL ACTIVITY OF BACTERIOCINS

1.6.1 Temperature

The temperature plays an important role in the mass and bacteriocin production. Also, the temperature during fermentation can affect the antimicrobial activity of bacteriocin because the elevated temperature can degrade and denature the bacteriocin structure (Kim et al. 2006, Timothy, Iliyasu, and Anvikar 2021). Simova, Beshkova, and Dimitrov Zh (2009) investigated the heat sensitivity of an antimicrobial peptide synthesized by *Lb. bulgaricus* BB18. The antimicrobial peptide's antimicrobial activity was insignificantly diminished after heat treatments at 100 °C for 60 minutes and 115 °C for 15 minutes. The antimicrobial activity of bacteriocin DY4-2, which is synthesized from *Lb. plantarum* DY4-2, is stable to heating at 100 °C for 30 minutes (Lv et al. 2018). The antimicrobial activity of bacteriocin OR-7 was not affected after heat treatments at 37 °C for 2 h, 37 °C for 24 h, and 90 °C for 20 min (Stern et al. 2006). In addition, bacteriocin MN047, derived from *Lb. crustorum* MN047, is resistant to heating at 100 °C for 10, 20, and 30 min (Yi et al. 2016). All of the evidence suggests that antimicrobial peptides are stable to heating at 100 °C for 10–30 minutes. Hence, bacteriocins are suitable for high-temperature food processing.

1.6.2 pH

One of the most critical factors, pH affects antimicrobial activity, such as stability, the charge state of amino acid, substituent functional groups, peptide assembly/disassembly, and synergistic and antagonistic effects. Also, the pH of the environment/medium plays a vital role in bacterial generation and bacteriocin synthesis (Kanmani et al. 2011). As a result, studying pH is necessary to find and evaluate the optimally effective antimicrobial activity and bacteriocin production of targeted bacteriocin for further scientific research and practical application (Tian et al. 2015). Several research studies have focused on the influence of pH on the antipathogenic action of antimicrobial peptides. According to Woraprayote et al. (2015), the activity of purified bacteriocins Bac7293A and Bac7293B obtained from *W. hellenica* BCC 7293 was reduced when exposed to elevated pH (alkalinity condition). Bacteriocin DY4-2 had stability over a pH range from 2 to 9 and retained more than 70% activity at alkaline conditions (pH 10–11) (Yi et al. 2016). Bacteriocin OR-7 showed antimicrobial stability at pH 3–9, while the antimicrobial activity of OR-7 was hindered at pH 10 (Stern et al. 2006). Furthermore, Simova, Beshkova, and Dimitrov Zh (2009) discovered that the antimicrobial peptide derived from *Lb. bulgaricus* BB18 was stable to a wide pH range. From the point of view of bacteriocin production, the manipulation pH of fermentation can supportively enhance bacterial growth, bacteriocin synthesis, and bacteriocin charges (Anthony et al. 2009, Abbasiliasi et al. 2017).

1.6.3 Enzymes

Several enzymes, including lipases, amylase, and proteases, which are generally detected in foods and the gastrointestinal tract, were conducted to estimate the effect of these enzymes on the antimicrobial activity of AMPs. Proteases, in particular, have a significant impact on AMPs. AMP produced from *Lb. bulgaricus* BB18 was denatured by α-chymotrypsin, papain, pepsin, proteinase K, protease XI, protease XIV, and trypsin, but α-amylase and lipase did not affect antipathogenic activity (Simova, Beshkova, and Dimitrov Zh 2009). Purified bacteriocin Bac7293A and Bac7293B were completely degraded by proteolytic enzymes such as α-chymotrypsin, pepsin, trypsin, and proteinase K, whereas it could not be digested by α-amylase and lipase (Woraprayote et al. 2015). Bacteriocin OR-7 was susceptible to proteolytic enzymes such as β-chymotrypsin, proteinase K, and papain but resistant to lysozyme and lipase (Stern et al. 2006). Besides, the antipathogenic activity of bacteriocin DY4-2 was completely degraded by papain and pepsin but slightly inactivated by nutrase (Lv et al. 2018). Based on the finding, it is possible to conclude that practically all bacteriocins are susceptible to proteolytic enzymes such as proteases in the gastrointestinal tract and foods and so are non-toxic for human administration.

1.6.4 Food Matrices

Notwithstanding these achievements, it should be highlighted that the advantages of food products containing bacteriocin may be strictly limited by limitations, such as a too-narrow spectrum of inhibitory effects (Gharsallaoui et al. 2016) and/or a hydrophobic property, which may lead them to be partitioned in the organic lipid phase within a food matrix. Furthermore, the inappropriate distribution or inadequate solubility of bacteriocin molecules in foods might significantly impact the antimicrobial activity of these substances. To solve these limitations, bacteriocins may be combined with other preservation methods to enhance or maintain antimicrobial activity (Mills et al. 2011). For instance, plantaricin BM1 directly added to the ham surface is more potential to prevent *L. monocytogenes* than being blended into the cooked ham before homogenization. It was suggested that the depletion of activity of integrated bacteriocin was related to increased adsorption of bacteriocin molecules to meat components, inappropriate bacteriocin distribution, and diffusion in the food matrices (Zhou et al. 2015). In addition, because of interactions with its components or degradation by proteases, nisin can lose its activity in food products (Benech et al. 2002).

1.7 CONCLUSION AND FUTURE PROSPECTS

This chapter highlights the potential of bacteriocin in food preservation and as a novel therapeutic approach to pathogen infection, cancer, and the immune system. On the bacteriocins used for the food industries, there is adequate information. Bacteriocins are less likely than antibiotics to still be commercialized and approved by the FDA and WHO for use in preserving food and the therapy of infectious diseases. To achieve GRAS status and legal approval of bacteriocins for applications in the food industry, toxicological information and the fate of the molecule after ingestion are required. Furthermore, evidence is insufficient on human health *in vitro* and *in vivo*. Because some bacteriocins are sensitive to proteolytic enzymes, many investigations into bacteriocin activity have been confined to *in vitro* tests and have not been adequately investigated in animal studies. Some post-translationally modified bacteriocins have greater stability in the digestive system, but their effects *in vivo* are less well understood.

In the coming years, the bacteriocin method of delivery might play a crucial role in resolving bacteriocin destruction in the digestive tract. Furthermore, bioinformatic strategies predict an increasing number of bacteriocin biosynthesis clusters; however, achieving a bacteriocin-producing strain is difficult. Combining high-throughput sequencing and culture omics could lead to the discovery of novel bacteriocins and strains that synthesize them. Moreover, innovative techniques or technologies for incorporating bacteriocins, such as through antimicrobial or active packaging, are still challenging. The growing demand for safe, minimally processed, fresh foods presents significant challenges to the food packaging industry in terms of developing antimicrobial packaging techniques, specifically with the use of bacteriocin, to ensure the quality and safety of packaged products. Bacteriocins should conduct more research on their cytotoxicity, haemolytic activities, distribution, and metabolism until their benefits to human health can be demonstrated.

Additionally, the specific antimicrobial effect of bacteriocins compared to standard antibiotics makes them a promising alternative to antibiotics. Because of its low toxicity and proteinaceous composition, it has great potential to be employed as an antibiotic replacement in the future, possibly alleviating the problem of antibiotic resistance. Likewise, bacteriocin has a synergistic effect with antibiotics, which indicates that the combination of bacteriocin and antibiotics can significantly enhance the inhibitory or killing effects of antibiotics on pathogens and simultaneously decrease the dosage of antibiotics required to treat infections. More investigation into the role and mechanisms involved in bacteriocins will contribute to their practical application in infection prevention, antiproliferation of cancer cells, anti-inflammation, and immunomodulation.

REFERENCES

Abbasiliasi, Sahar, Joo Shun Tan, Tengku Azmi Tengku Ibrahim, Fatemeh Bashokouh, Nagasundara Ramanan Ramakrishnan, Shuhaimi Mustafa, and Arbakariya B. Ariff. 2017. "Fermentation factors influencing the production of bacteriocins by lactic acid bacteria: A review." *RSC Adv* 7(47):29395–29420. doi: 10.1039/C6RA24579J.

Acuña, Leonardo, Gianluca Picariello, Fernando Sesma, Roberto D. Morero, and Augusto Bellomio. 2012. "A new hybrid bacteriocin, Ent35–MccV, displays antimicrobial activity against pathogenic gram-positive and gram-negative bacteria." *FEBS Open Bio* 2:12–19. doi: 10.1016/j.fob.2012.01.002.

Agnes Mary, S., and V. R. Giri Dev. 2015. "Electrospun herbal nanofibrous wound dressings for skin tissue engineering." *J Text Inst* 106(8):886–895. doi: 10.1080/00405000.2014.951247.

Almeida-Santos, Ana C., Carla Novais, Luísa Peixe, and Ana R. Freitas. 2021. "*Enterococcus* spp. as a producer and target of bacteriocins: A double-edged sword in the antimicrobial resistance crisis context." *Antibiotics (Basel)* 10(10):1215.

Altveş, Safaa, Hatice Kübra Yildiz, and Hasibe Cingilli Vural. 2019. "Interaction of the microbiota with the human body in health and diseases." *Biosci Microbiota Food Health*. doi: 10.12938/bmfh.19-023.

Anand, S. K., R. A. Srinivasan, and L. K. Rao. 1984. "Antibacterial activity associated with *Bifidobacterium bifidum*." *Cult Dairy Prod J* 19(9): 6–8.

Ananou, Samir, Margarita Garriga, Anna Jofré, Teresa Aymerich, Antonio Gálvez, Mercedes Maqueda, Manuel Martínez-Bueno, and Eva Valdivia. 2010. “Combined effect of enterocin AS-48 and high hydrostatic pressure to control food-borne pathogens inoculated in low acid fermented sausages.” *Meat Sci* 84(4):594–600.

Anthony, Thangamani, Thangamani Rajesh, Nagarajan Kayalvizhi, and Paramasamy Gunasekaran. 2009. “Influence of medium components and fermentation conditions on the production of bacteriocin(s) by *Bacillus licheniformis* AnBa9.” *Bioresour Technol* 100(2):872–877. doi: 10.1016/j.biortech.2008.07.027.

Arbulu, Sara, Juan J. Jiménez, Loreto Gútiez, Cristina Campanero, Rosa Del Campo, Luis M. Cintas, Carmen Herranz, and Pablo E. Hernández. 2016. “Evaluation of bacteriocinogenic activity, safety traits and biotechnological potential of fecal lactic acid bacteria (LAB), isolated from griffon Vultures (*Gyps fulvus* subsp. *fulvus*).” *BMC Microbiol* 16(1):1–15.

Aspri, Maria, Paula M. O’Connor, Des Field, Paul D. Cotter, Paul Ross, Colin Hill, and Photis Papademas. 2017. “Application of bacteriocin-producing *Enterococcus faecium* isolated from donkey milk, in the bio-control of *Listeria monocytogenes* in fresh whey cheese.” *Int Dairy J* 73:1–9.

Ayers, Emily Alexandra. 2018. Evaluation of a flexographically applied antimicrobial coating for ready-to-eat turkey deli meat packaging. (Doctoral dissertation, Clemson University).

Ayivi, Raphael D., Rabin Gyawali, Albert Krastanov, Sulaiman O. Aljaloud, Mulumebet Worku, Reza Tahergorabi, Roberta Claro da Silva, and Salam A. Ibrahim. 2020. “Lactic acid bacteria: Food safety and human health applications.” *Dairy* 1(3):202–232.

Badani, H., R. F. Garry, and W. C. Wimley. 2014. “Peptide entry inhibitors of enveloped viruses: The importance of interfacial hydrophobicity.” *Biochim Biophys Acta* 1838(9):2180–2197. doi: 10.1016/j.bbamem.2014.04.015.

Balay, R. Danielle, Ramana V. Dangeti, Kamaljit Kaur, and Lynn M. McMullen. 2017. “Purification of leucocin A for use on wieners to inhibit *Listeria monocytogenes* in the presence of spoilage organisms.” *Int J Food Microbiol* 255:25–31.

Balay, Danielle R., Michael G. Gänzle, and Lynn M. McMullen. 2018. “The effect of carbohydrates and bacteriocins on the growth kinetics and resistance of *Listeria monocytogenes*.” *Front Microbiol* 9:347.

Balla, E., L. M. T. Dicks, M. Du Toit, M. J. Van Der Merwe, and W. H. Holzapfel. 2000. “Characterization and cloning of the genes encoding enterocin 1071A and enterocin 1071B, two antimicrobial peptides produced by *Enterococcus faecalis* BFE 1071.” *Appl Environ Microbiol* 66(4):1298–1304.

Barbiroli, A., A. Musatti, G. Capretti, S. Iametti, and M. Rollini. 2017. “Sakacin-A antimicrobial packaging for decreasing *Listeria* contamination in thin-cut meat: Preliminary assessment.” *J Sci Food Agric* 97(3):1042–1047.

Barbosa, Matheus S., Svetoslav D. Todorov, Cynthia H. Jurkiewicz, and Bernadette D. G. M. Franco. 2015. “Bacteriocin production by *Lactobacillus curvatus* MBSa2 entrapped in calcium alginate during ripening of salami for control of *Listeria monocytogenes*.” *Food Control* 47:147–153.

Beaulieu, Lucie. 2005. Production, purification et caracterisation de la pediocine PA-1 naturelle et de ses formes recombinantes: Contribution a la mise en evidence d'une nouvelle activite biologique (French and English text). [Ph.D. thesis], Universite Laval, Canada.

Begde, D., S. Bundale, P. Mashitha, J. Rudra, N. Nashikkar, and A. Upadhyay. 2011. “Immunomodulatory efficacy of nisin—A bacterial lantibiotic peptide.” *J Pept Sci* 17(6):438–444.

Ben Said, Laila, Hélène Gaudreau, Laurent Dallaire, Michèle Tessier, and Ismail Fliss. 2019. “Bioprotective culture: A new generation of food additives for the preservation of food quality and safety.” *Ind Biotechnol* 15(3):138–147. doi: 10.1089/ind.2019.29175.lbs.

Benech, R. O., E. E. Kheadr, C. Lacroix, and I. Fliss. 2002. “Antibacterial activities of nisin Z encapsulated in liposomes or produced *in situ* by mixed culture during cheddar cheese ripening.” *Appl Environ Microbiol* 68(11):5607–5619.

Bhalla, Tek Chand, Monika, Sheetal, and Savitri. 2019. “International laws and food-borne illness.” *Food Safety Human Health* 2019:319–371. doi: 10.1016/B978-0-12-816333-7.00012-6.

Birri, Dagim Jirata, Dag Anders Brede, and Ingolf F. Nes. 2012. “Salivaricin D, a novel intrinsically trypsin-resistant lantibiotic from *Streptococcus salivarius* 5M6c isolated from a healthy infant.” *Appl Environ Microbiol* 78(2):402–410. doi: 10.1128/AEM.06588-11.

Bosch, M., M. Rodriguez, F. Garcia, E. Fernández, M. C. Fuentes, and J. Cune. 2012. “Probiotic properties of *Lactobacillus plantarum* CECT 7315 and CECT 7316 isolated from faeces of healthy children.” *Lett Appl Microbiol* 54(3):240–246.

Budde, Birgitte Bjørn, Tina Hornbæk, Tomas Jacobsen, Vibeke Barkholt, and Anette Granly Koch. 2003. “*Leuconostoc carnosum* 4010 has the potential for use as a protective culture for vacuum-packed meats: culture isolation, bacteriocin identification, and meat application experiments.” *Int J Food Microbiol* 83(2):171–184.

Casaburi, Annalisa, Veronica Di Martino, Pasquale Ferranti, Luca Picariello, and Francesco Villani. 2016. "Technological properties and bacteriocins production by *Lactobacillus curvatus* 54M16 and its use as starter culture for fermented sausage manufacture." *Food Control* 59:31–45.

Casaus, Pilar, Trine Nilsen, Luis M. Cintas, Ingolf F. Nes, Pablo E. Hernández, and Helge Holo. 1997. "Enterocin B, a new bacteriocin from *Enterococcus faecium* T136 which can act synergistically with enterocin A." *Microbiology (Reading)* 143(7):2287–2294.

Castellano, P., and G. Vignolo. 2006. "Inhibition of *Listeria innocua* and *Brochothrix thermosphacta* in vacuum-packaged meat by addition of bacteriocinogenic *Lactobacillus curvatus* CRL705 and its bacteriocins." *Lett Appl Microbiol* 43(2):194–199.

Castellano, Patricia, Mariana Pérez Ibarreche, Mariana Blanco Massani, Cecilia Fontana, and Graciela M. Vignolo. 2017. "Strategies for pathogen biocontrol using lactic acid bacteria and their metabolites: A focus on meat ecosystems and industrial environments." *Microorganisms* 5(3):38.

Castro, Ana, Joana Silva, and Paula Teixeira. 2018. "*Staphylococcus aureus*, a food pathogen: Virulence factors and antibiotic resistance." In: *Foodborne Diseases*, edited by Alina Maria Holban and Alexandru Mihai Grumezescu, (15): 213–238. Elsevier.

Cavicchioli, V. Q., O. V. Carvalho, J. C. Paiva, S. D. Todorov, A. Silva Júnior, and L. A. Nero. 2018. "Inhibition of herpes simplex virus 1 (HSV-1) and poliovirus (PV-1) by bacteriocins from *Lactococcus lactis* subsp. *lactis* and *Enterococcus durans* strains isolated from goat milk." *Int J Antimicrob Agents* 51(1):33–37. doi: 10.1016/j.ijantimicag.2017.04.020.

Chausali, N., J. Saxena, and R. Prasad. 2022. "Recent trends in nanotechnology applications of bio-based packaging." *Journal of Agriculture and Food Research*, https://doi.org/10.1016/j.jafr.2021.100257.

Cheikhyoussef, Ahmad, Natascha Cheikhyoussef, Haiqin Chen, Jianxin Zhao, Jian Tang, Hao Zhang, and Wei Chen. 2010. "Bifidin I–A new bacteriocin produced by *Bifidobacterium infantis* BCRC 14602: Purification and partial amino acid sequence." *Food Control* 21(5):746–753.

Cheikhyoussef, Ahmad, Natascha Pogori, Haiqin Chen, Fengwei Tian, Wei Chen, Jian Tang, and Hao Zhang. 2009. "Antimicrobial activity and partial characterization of bacteriocin-like inhibitory substances (BLIS) produced by *Bifidobacterium infantis* BCRC 14602." *Food Control* 20(6):553–559.

Chi, Hai, and Helge Holo. 2018. "Synergistic antimicrobial activity between the broad spectrum bacteriocin garvicin KS and nisin, farnesol and polymyxin B against gram-positive and gram-negative bacteria." *Curr Microbiol* 75(3):272–277.

Choyam, Shilja, Alok Kumar Srivastava, Jae-Ho Shin, and Rajagopal Kammara. 2019. "Ocins for food safety." *Front Microbiol* 10:1736.

Cintas, Luis M., Pilar Casaus, Carmen Herranz, Leiv Sigve Håvarstein, Helge Holo, Pablo E. Hernández, and Ingolf F. Nes. 2000. "Biochemical and genetic evidence that *Enterococcus faecium* L50 produces enterocins L50A and L50B, the sec-dependent enterocin P, and a novel bacteriocin secreted without an N-terminal extension termed enterocin Q." *J Bacteriol* 182(23):6806–6814.

Corr, Sinéad C., Yin Li, Christian U. Riedel, Paul W. O'Toole, Colin Hill, and Cormac G. M. Gahan. 2007. "Bacteriocin production as a mechanism for the antiinfective activity of *Lactobacillus salivarius* UCC118." *Proc Natl Acad Sci* 104(18):7617–7621.

Cotter, P. D., C. Hill, and R. P. Ross. 2005. "Bacteriocins: Developing innate immunity for food." *Nat Rev Microbiol* 3(10):777–788. doi: 10.1038/nrmicro1273.

Cotter, Paul D., R. Paul Ross, and Colin Hill. 2013. "Bacteriocins — A viable alternative to antibiotics?" *Nat Rev Microbiol* 11(2):95–105. doi: 10.1038/nrmicro2937.

Cuozzo, Sergio A., Fernando Sesma, Jorge M. Palacios, Aída P. de Ruíz Holgado, and Raúl R. Raya. 2000. "Identification and nucleotide sequence of genes involved in the synthesis of lactocin 705, a two-peptide bacteriocin from *Lactobacillus casei* CRL 705." *FEMS Microbiol Lett* 185(2):157–161.

da Costa, Roger J., Flávia L. S. Voloski, Rafael G. Mondadori, Eduarda H. Duval, and Ângela M. Fiorentini. 2019. "Preservation of meat products with bacteriocins produced by lactic acid bacteria isolated from meat." *J Food Qual* 2019:4726510. doi: 10.1155/2019/4726510.

da Silva Sabo, Sabrina, Michele Vitolo, José Manuel Domínguez González, and Ricardo Pinheiro de Souza Oliveira. 2014. "Overview of *Lactobacillus plantarum* as a promising bacteriocin producer among lactic acid bacteria." *Food Res Int* 64:527–536.

Daba, H., S. Pandian, J. F. Gosselin, R. E. Simard, J. Huang, and C. Lacroix. 1991. "Detection and activity of a bacteriocin produced by *Leuconostoc mesenteroides*." *Appl Environ Microbiol* 57(12): 3450–3455.

Daeschel, M. A., M. C. McKenney, and L. C. McDonald. 1990. "Bacteriocidal activity of *Lactobacillus plantarum* C-11." *Food Microbiol* 7(2):91–98.

Davidson, P. M., and S. Zivanovic. 2003. "The use of natural antimicrobials." In: *Food Preservation Techniques*, edited by Peter Zeuthen and Leif Bogh-Sørensen, 5–30. Boca Raton Boston New York Washington, DC: CRC Press.

De Kwaadsteniet, M., T. Fraser, C. A. Van Reenen, and L. M. T. Dicks. 2006. "Bacteriocin T8, a novel class IIa sec-dependent bacteriocin produced by *Enterococcus faecium* T8, isolated from vaginal secretions of children infected with human immunodeficiency virus." *Appl Environ Microbiol* 72(7):4761–4766.

De Kwaadsteniet, M., K. Ten Doeschate, and L. M. T. Dicks. 2008. "Characterization of the structural gene encoding nisin F, a new lantibiotic produced by a *Lactococcus lactis* subsp. *lactis* isolate from freshwater catfish (Clarias gariepinus)." *Appl Environ Microbiol* 74(2):547–549.

Degli Esposti, Micaela, Maurizio Toselli, Carla Sabia, Patrizia Messi, Simona de Niederhäusern, Moreno Bondi, and Ramona Iseppi. 2018. "Effectiveness of polymeric coated films containing bacteriocin-producer living bacteria for *Listeria monocytogenes* control under simulated cold chain break." *Food Microbiol* 76:173–179.

Delves-Broughton, J. 2005. "Nisin as a food preservative." *Food Aust* 57(12):525–527.

Delves-Broughton, J., and G. Weber. 2011. "Nisin, natamycin and other commercial fermentates used in food biopreservation." In: *Protective cultures, antimicrobial metabolites and bacteriophages for food and beverage biopreservation*, edited by Christophe Lacroix, 63–99. Cambridge, UK: Woodhead publishing.

Dicks, L. M. T., F. D. Mellett, and L. C. Hoffman. 2004. "Use of bacteriocin-producing starter cultures of *Lactobacillus plantarum* and *Lactobacillus curvatus* in production of ostrich meat salami." *Meat Sci* 66(3):703–708.

Diep, Dzung B., and Ingolf F. Nes. 2002. "Ribosomally synthesized antibacterial peptides in gram positive bacteria." *Curr Drug Targets* 3(2):107–122.

Diep, Dzung Bao, Leiv Sigve Håvarstein, and Ingolf F. Nes. 1996. "Characterization of the locus responsible for the bacteriocin production in *Lactobacillus plantarum* C11." *J Bacteriol* 178(15):4472–4483.

Dillon, V. M. 2014. "Natural anti-microbial systems | Preservative effects during storage." In: *Encyclopedia of food microbiology (second edition)*, edited by Carl A. Batt and Mary Lou Tortorello, 941–947. Oxford: Academic Press.

Dos Santos, Carolina Alves, Gabriela Ribeiro dos Santos, Victória Soares Soeiro, Júlia Rodrigues dos Santos, Marcia de Araujo Rebelo, Marco Vinícius Chaud, Marli Gerenutti, Denise Grotto, Raksha Pandit, Mahendra Rai, and Angela Faustino Jozala. 2018. "Bacterial nanocellulose membranes combined with nisin: A strategy to prevent microbial growth." *Cellulose* 25(11):6681–6689.

Drolia, Rishi, and Arun K. Bhunia. 2019. "Crossing the intestinal barrier via *Listeria* adhesion protein and internalin A." *Trends Microbiol* 27(5):408–425. doi: 10.1016/j.tim.2018.12.007.

Drosinos, E. H., M. Mataragas, P. Nasis, M. Galiotou, and J. Metaxopoulos. 2005. "Growth and bacteriocin production kinetics of *Leuconostoc mesenteroides* E131." *J Appl Microbiol* 99(6):1314–1323.

Edagiz, S., P. Lagace-Wiens, J. Embil, J. Karlowsky, and A. Walkty. 2015. "Empyema caused by *Clostridium bifermentans*: A case report." *Can J Infect Dis Med Microbiol* 26(2):105–107. doi: 10.1155/2015/481076.

Eguchi, Tomoko, Kazuyo Kaminaka, Jun Shima, Shinichi Kawamoto, Katsumi Mori, Seong-Hyun Choi, Katsumi Doi, Sadahiro Ohmomo, and Seiya Ogata. 2001. "Isolation and characterization of enterocin SE-K4 produced by thermophilic enterococci, *Enterococcus faecalis* K-4." *Biosci Biotechnol Biochem* 65(2):247–253.

Ellis, Jenna-Claire, Reynolds P. Ross, and Colin Hill. 2019. "Nisin Z and lacticin 3147 improve efficacy of antibiotics against clinically significant bacteria." *Future Microbiol* 14(18):1573–1587.

Erhabor, C. R., J. O. Erhabor, and L. J. McGaw. 2019. "The potential of South African medicinal plants against microbial biofilm and quorum sensing of foodborne pathogens: A review." *S Afr J Bot* 126:214–231. doi: 10.1016/j.sajb.2019.07.024.

Esch, Kevin J., and Christine A. Petersen. 2013. "Transmission and epidemiology of zoonotic protozoal diseases of companion animals." *Clin Microbiol Rev* 26(1):58–85.

Feiner, Gerhard. 2006. "39 - The microbiology of specific bacteria." In: *Meat Products Handbook*, edited by Gerhard Feiner, 595–615. Woodhead Publishing.

Field, Des, Lisa Quigley, Paula M. O'Connor, Mary C. Rea, Karen Daly, Paul D. Cotter, Colin Hill, and R. Paul Ross. 2010. "Studies with bioengineered Nisin peptides highlight the broad-spectrum potency of Nisin V." *Microb Biotechnol* 3(4):473–486.

Fu, Yuxin, Dongdong Mu, Wanjin Qiao, Duolong Zhu, Xiangxiang Wang, Fulu Liu, Haijin Xu, Per Saris, Oscar P. Kuipers, and Mingqiang Qiao. 2018. "Co-expression of nisin Z and leucocin C as a basis for effective protection against *Listeria monocytogenes* in pasteurized milk." *Front Microbiol* 9:547.

Fujita, Koji, Shiro Ichimasa, Takeshi Zendo, Shoko Koga, Fuminori Yoneyama, Jiro Nakayama, and Kenji Sonomoto. 2007. "Structural analysis and characterization of lacticin Q, a novel bacteriocin belonging to a new family of unmodified bacteriocins of gram-positive bacteria." *Appl Environ Microbiol* 73(9):2871–2877.

Fusco, Vincenzina, Grazia M. Quero, Gyu-Sung Cho, Jan Kabisch, Diana Meske, Horst Neve, Wilhelm Bockelmann, and Charles M. A. P. Franz. 2015. "The genus *Weissella*: Taxonomy, ecology and biotechnological potential." *Front Microbiol* 6:155.

Gálvez, Antonio, Hikmate Abriouel, Rosario Lucas López, and Nabil Ben Omar. 2007. "Bacteriocin-based strategies for food biopreservation." *Int J Food Microbiol* 120(1):51–70. doi: 10.1016/j.ijfoodmicro.2007.06.001.

Gao, Yurong, Dapeng Li, and Xiaoyan Liu. 2014. "Bacteriocin-producing Lactobacillus sakei C2 as starter culture in fermented sausages." *Food Control* 35(1):1–6.

Garcia-Gutierrez, Enriqueta, Paula M. O'Connor, Gerhard Saalbach, Calum J. Walsh, James W. Hegarty, Caitriona M. Guinane, Melinda J. Mayer, Arjan Narbad, and Paul D. Cotter. 2020. "First evidence of production of the lantibiotic nisin P." *Sci Rep* 10(1):1–15.

García, Mª T., R. Lucas, H. Abriouel, N. Ben Omar, R. Pérez, Mª J. Grande, M. Martínez-Cañamero, and A. Gálvez. 2004. "Antimicrobial activity of enterocin EJ97 against '*Bacillus macroides/Bacillus maroccanus*' isolated from zucchini purée." *J Appl Microbiol* 97(4):731–737.

Garneau, Sylvie, Nathaniel I. Martin, and John C. Vederas. 2002. "Two-peptide bacteriocins produced by lactic acid bacteria." *Biochimie* 84(5–6):577–592.

Geng, Mengxin, Frank Austin, Ronald Shin, and Leif Smith. 2018. "Covalent structure and bioactivity of the type AII lantibiotic salivaricin A2." *Appl Environ Microbiol* 84(5):e02528-17. doi: 10.1128/AEM.02528-17.

Gharsallaoui, Adem, Nadia Oulahal, Catherine Joly, and Pascal Degraeve. 2016. "Nisin as a food preservative: Part 1: Physicochemical properties, antimicrobial activity, and main uses." *Crit Rev Food Sci Nutr* 56(8):1262–1274. doi: 10.1080/10408398.2013.763765.

Ghose, C., and C. P. Kelly. 2015. "The prospect for vaccines to prevent *Clostridium difficile* infection." *Infect Dis Clin North Am* 29(1):145–162. doi: 10.1016/j.idc.2014.11.013.

Ghrairi, Taoufik, and Khaled Hani. 2015. "Enhanced bactericidal effect of enterocin A in combination with thyme essential oils against *L. monocytogenes* and *E. coli* O157:H7." *J Food Sci Technol* 52(4):2148–2156. doi: 10.1007/s13197-013-1214-5.

Giraffa, G. 2014. "*Enterococcus*." In: *Encyclopedia of Food Microbiology (Second Edition)*, edited by Carl A. Batt and Mary Lou Tortorello, 674–679. Academic Press.

Grande Burgos, Maria Jose, Rubén Pérez Pulido, María Del Carmen López Aguayo, Antonio Gálvez, and Rosario Lucas. 2014. "The cyclic antibacterial peptide enterocin AS-48: Isolation, mode of action, and possible food applications." *Int J Mol Sci* 15(12):22706–22727.

Grinter, R., J. Milner, and D. Walker. 2012. "Bacteriocins active against plant pathogenic bacteria." *Biochem Soc Trans* 40(6):1498–1502. doi: 10.1042/bst20120206.

Gupta, Aabha, Santosh Kumar Tiwari, Victoria Netrebov, and Michael L. Chikindas. 2016. "Biochemical properties and mechanism of action of enterocin LD3 purified from *Enterococcus hirae* LD3." *Probiotics Antimicrob Proteins* 8(3):161–169.

Hanchi, Hasna, Walid Mottawea, Khaled Sebei, and Riadh Hammami. 2018. "The genus *Enterococcus*: between probiotic potential and safety concerns—An update." *Front Microbiol* 9:1791.

Hassan, A. H., and C. N. Cutter. 2020. "Development and evaluation of pullulan-based composite antimicrobial films (CAF) incorporated with nisin, thymol and lauric arginate to reduce foodborne pathogens associated with muscle foods." *Int J Food Microbiol* 320:108519.

Hata, Tomomi, Toyoki Sato, and Yasujiro Morimitsu. 2016. "Genetic characterization of *Enterococcus faecalis* N1-33 bacteriocin and obtained of the mutant strain that produce many bacteriocin by novobiocin agent effect, its merit as A food preservative in steamed rice model." *J Food Saf* 36(3):348–359.

Héchard, Yann, Benoit Dérijard, François Letellier, and Yves Cenatiempo. 1992. "Characterization and purification of mesentericin Y105, an anti-*Listeria* bacteriocin from *Leuconostoc mesenteroides*." *J Gen Microbiol* 138(12):2725–2731.

Héchard, Yann, and Hans-Georg Sahl. 2002. "Mode of action of modified and unmodified bacteriocins from gram-positive bacteria." *Biochimie* 84(5–6):545–557.

Heeney, Dustin D., Zhengyuan Zhai, Zach Bendiks, Javad Barouei, Alice Martinic, Carolyn Slupsky, and Maria L. Marco. 2019. "*Lactobacillus plantarum* bacteriocin is associated with intestinal and systemic improvements in diet-induced obese mice and maintains epithelial barrier integrity in vitro." *Gut Microbes* 10(3):382–397.

Heng, Nicholas C. K., Grace A. Burtenshaw, Ralph W. Jack, and John R. Tagg. 2007. "Ubericin A, a class IIa bacteriocin produced by *Streptococcus uberis*." *Appl Environ Microbiol* 73(23):7763–7766.

Hooper, Lora V., Dan R. Littman, and Andrew J. Macpherson. 2012. "Interactions between the microbiota and the immune system." *Science* 336(6086):1268–1273.

Hu, Jun, Libao Ma, Yangfan Nie, Jianwei Chen, Wenyong Zheng, Xinkai Wang, Chunlin Xie, Zilong Zheng, Zhichang Wang, and Tao Yang. 2018. "A microbiota-derived bacteriocin targets the host to confer diarrhea resistance in early-weaned piglets." *Cell Host Microbe* 24(6):817–832.

Huang, Fuqing, Kunling Teng, Yayong Liu, Yanhong Cao, Tianwei Wang, Cui Ma, Jie Zhang, and Jin Zhong. 2021. "Bacteriocins: Potential for human health." *Oxid Med Cell Longev* 2021.

Hwang, In-Chan, Ju Kyoung Oh, Sang Hoon Kim, Sejong Oh, and Dae-Kyung Kang. 2018. "Isolation and characterization of an anti-listerial bacteriocin from *Leuconostoc lactis* SD501." *Korean J Food Sci Anim Resour* 38(5):1008.

Ibarra-Sánchez, Luis A., Nancy El-Haddad, Darine Mahmoud, Michael J. Miller, and Layal Karam. 2020. "Invited review: Advances in nisin use for preservation of dairy products." *J Dairy Sci* 103(3):2041–2052.

Jack, Ralph W., and Günther Jung. 2000. "Lantibiotics and microcins: Polypeptides with unusual chemical diversity." *Curr Opin Chem Biol* 4(3):310–317.

Jia, Zhifeng, Meiling He, Chunjie Wang, Aorigele Chen, Xin Zhang, Jin Xu, He Fu, and Bo Liu. 2019. "Nisin reduces uterine inflammation in rats by modulating concentrations of pro-and anti-inflammatory cytokines." *Am J Reprod Immunol* 81(5):e13096.

Joerger, R. D. 2003. "Alternatives to antibiotics: Bacteriocins, antimicrobial peptides and bacteriophages." *Poult Sci* 82(4):640–647. doi: 10.1093/ps/82.4.640.

Johnsen, Line, Gunnar Fimland, and Jon Nissen-Meyer. 2005. "The C-terminal domain of pediocin-like antimicrobial peptides (class IIa bacteriocins) is involved in specific recognition of the C-terminal part of cognate immunity proteins and in determining the antimicrobial spectrum." *J Biol Chem* 280(10):9243–9250.

Johnson, E. M., D. Y. Jung, D. Y. Jin, D. R. Jayabalan, D. S. H. Yang, and J. W. Suh. 2018. "Bacteriocins as food preservatives: Challenges and emerging horizons." *Crit Rev Food Sci Nutr* 58(16):2743–2767. doi: 10.1080/10408398.2017.1340870.

Joo, Nam E., Kathryn Ritchie, Pachiyappan Kamarajan, Di Miao, and Yvonne L. Kapila. 2012. "Nisin, an apoptogenic bacteriocin and food preservative, attenuates HNSCC tumorigenesis via CHAC1." *Cancer Med* 1(3):295–305. doi: 10.1002/cam4.35.

Kamarajan, P., T. Hayami, B. Matte, Y. Liu, T. Danciu, A. Ramamoorthy, F. Worden, S. Kapila, and Y. Kapila. 2015. "Nisin ZP, a bacteriocin and food preservative, inhibits head and neck cancer tumorigenesis and prolongs survival." *PLOS ONE* 10(7):e0131008. doi: 10.1371/journal.pone.0131008.

Kang, Mi-Sun, Hoi-Soon Lim, Seon-Mi Kim, Hyun-Chul Lee, and Jong-Suk Oh. 2011. "Effect of *Weissella cibaria* on *Fusobacterium nucleatum*-induced interleukin-6 and interleukin-8 production in KB cells." *J Bacteriol Virol* 41(1):9–18.

Kanmani, P., R. S. Kumar, N. Yuvaraj, K. A. Paari, V. Pattukumar, and V. Arul. 2011. "Design and optimization of fermentation medium for enhanced bacteriocin production by probiotic bacterium *Enterococcus faecium* MC13." *Prep Biochem Biotechnol* 41(1):40–52. doi: 10.1080/10826068.2010.534221.

Karlo, J., R. Prasad, and S. P. Singh. 2023. "Biophotonics in food technology: Quo vadis?" *Journal of Agriculture and Food Research*, https://doi.org/10.1016/j.jafr.2022.100482.

Kaur, G., T. P. Singh, and R. K. Malik. 2013. "Antibacterial efficacy of Nisin, pediocin 34 and enterocin FH99 against *Listeria monocytogenes* and cross resistance of its bacteriocin resistant variants to common food preservatives." *Braz J Microbiol* 44(1):63–71.

Kaur, Sumanpreet, and Sukhraj Kaur. 2015. "Bacteriocins as potential anticancer agents." *Front Pharmacol* 6:272.

Kawai, Yasushi, Yasuyuki Ishii, Kensuke Arakawa, Koichiro Uemura, Boku Saitoh, Junko Nishimura, Haruki Kitazawa, Yukiko Yamazaki, Yoshio Tateno, and Takatoshi Itoh. 2004. "Structural and functional differences in two cyclic bacteriocins with the same sequences produced by lactobacilli." *Appl Environ Microbiol* 70(5):2906–2911.

Kawai, Yasushi, Rober Kemperman, Jan Kok, and Tadao Saito. 2004. "The circular bacteriocins gassericin A and circularin A." *Curr Protein Pept Sci* 5(5):393–398.

Khan, Haider, Steve Flint, and Yu Pak-Lam. 2010. "Enterocins in food preservation." *Int J Food Microbiol* 141(1–2):1–10.

Kim, Mi-Hee, Yoon-Jung Kong, Hong Baek, and Hyung-Hwan Hyun. 2006. "Optimization of culture conditions and medium composition for the production of micrococcin GO5 by *Micrococcus* sp. GO5." *J Biotechnol* 121(1):54–61. doi: 10.1016/j.jbiotec.2005.06.022.

Kingcha, Yutthana, Amonlaya Tosukhowong, Takeshi Zendo, Sittiruk Roytrakul, Plearnpis Luxananil, Kingeaw Chareonpornsook, Ruud Valyasevi, Kenji Sonomoto, and Wonnop Visessanguan. 2012. "Anti-*listeria* activity of *Pediococcus pentosaceus* BCC 3772 and application as starter culture for Nham, a traditional fermented pork sausage." *Food Control* 25(1):190–196.

Kitagawa, Norio, Takahito Otani, and Tetsuichiro Inai. 2019. "Nisin, a food preservative produced by *Lactococcus lactis*, affects the localization pattern of intermediate filament protein in HaCaT cells." *Anat Sci Int* 94(2):163–171. doi: 10.1007/s12565-018-0462-x.

Klaenhammer, Todd R. 1993. "Genetics of bacteriocins produced by lactic acid bacteria." *FEMS Microbiol Rev* 12(1–3):39–85.

Kuang, Dai, Jianmin Zhang, Xuebin Xu, Weimin Shi, Sheng Chen, Xiaowei Yang, Xudong Su, Xianming Shi, and Jianghong Meng. 2018. "Emerging high-level ciprofloxacin resistance and molecular basis of resistance in *Salmonella enterica* from humans, food and animals." *Int J Food Microbiol* 280:1–9. doi: 10.1016/j.ijfoodmicro.2018.05.001.

Lange-Starke, A., A. Petereit, U. Truyen, P. G. Braun, K. Fehlhaber, and T. Albert. 2014. "Antiviral potential of selected starter cultures, bacteriocins and D,L-lactic acid." *Food Environ Virol* 6(1):42–47. doi: 10.1007/s12560-013-9135-z.

Lavermicocca, Paola, Cristina Reguant, and Joaquin Bautista-Gallego. 2021. "Lactic acid bacteria within the food industry: What is new on their technological and functional role." *Front Microbiol* 12.

Le Lay, C., B. Akerey, I. Fliss, M. Subirade, and M. Rouabhia. 2008. "Nisin Z inhibits the growth of *Candida albicans* and its transition from blastospore to hyphal form." *J Appl Microbiol* 105(5):1630–1639. doi: 10.1111/j.1365-2672.2008.03908.x.

Le Lay, Christophe, Larbi Dridi, Michel G. Bergeron, and Marc Ouellette. 2016. "Nisin is an effective inhibitor of *Clostridium difficile* vegetative cells and spore germination." *J Med Microbiol* 65(2):169–175.

Lee, Ju-Hoon, Xiulan Li, and Daniel J. O'Sullivan. 2011. "Transcription analysis of a lantibiotic gene cluster from *Bifidobacterium longum* DJO10A." *Appl Environ Microbiol* 77(17):5879–5887.

Leisner, Jørgen J., Birgit Groth Laursen, Hervé Prévost, Djamel Drider, and Paw Dalgaard. 2007. "*Carnobacterium*: Positive and negative effects in the environment and in foods." *FEMS Microbiol Rev* 31(5):592–613.

Linares-Morales, José R., Néstor Gutiérrez-Méndez, Blanca E. Rivera-Chavira, Samuel B. Pérez-Vega, and Guadalupe V. Nevárez-Moorillón. 2018. "Biocontrol processes in fruits and fresh produce, the use of lactic acid bacteria as a sustainable option." *Front Sustain Food Syst* 2:50.

Linares, Daniel M., Carolina Gómez, Erica Renes, José M. Fresno, María E. Tornadijo, R. Paul Ross, and Catherine Stanton. 2017. "Lactic acid bacteria and bifidobacteria with potential to design natural biofunctional health-promoting dairy foods." *Front Microbiol* 8:846.

Luna-Guevara, J. J., M. M. P. Arenas-Hernandez, C. Martínez de la Peña, J. L. Silva, and M. L. Luna-Guevara. 2019. "The role of pathogenic *E. coli* in fresh vegetables: Behavior, contamination factors, and preventive measures." *Int J Microbiol* 2019:2894328. doi: 10.1155/2019/2894328.

Lv, X. R., H. H. Ma, M. T. Sun, Y. Lin, F. L. Bai, J. R. Li, and B. L. Zhang. 2018. "A novel bacteriocin DY4-2 produced by *Lactobacillus plantarum* from cutlassfish and its application as bio-preservative for the control of *Pseudomonas fluorescens* in fresh turbot (*Scophthalmus maximus*) fillets." *Food Control* 89:22–31. doi: 10.1016/j.foodcont.2018.02.002.

Madera, Carmen, Pilar García, Ana Rodríguez, Juan E. Suárez, and Beatriz Martínez. 2009. "Prophage induction in Lactococcus lactis by the bacteriocin lactococcin 972." *Int J Food Microbiol* 129(1):99–102.

Mahdi, Likaa Hamied, Hussein Satar Jabbar, and Ibtesam Ghadban Auda. 2019. "Antibacterial immunomodulatory and antibiofilm triple effect of Salivaricin LHM against *Pseudomonas aeruginosa* urinary tract infection model." *Int J Biol Macromol* 134:1132–1144.

Maher, Sam, and Siobhán McClean. 2006. "Investigation of the cytotoxicity of eukaryotic and prokaryotic antimicrobial peptides in intestinal epithelial cells in vitro." *Biochem Pharmacol* 71(9):1289–1298.

Małaczewska, Joanna, Edyta Kaczorek- Łukowska, Roman Wójcik, Wojciech Rękawek, and Andrzej K. Siwicki. 2019. "*In vitro* immunomodulatory effect of nisin on porcine leucocytes." *J Anim Physiol Anim Nutr (Berl)* 103(3):882–893.

Maldonado-Barragán, Antonio, Belén Caballero-Guerrero, Esther Jiménez, Rufino Jiménez-Díaz, José L. Ruiz-Barba, and Juan M. Rodríguez. 2009. "Enterocin C, a class IIb bacteriocin produced by *E. faecalis* C901, a strain isolated from human colostrum." *Int J Food Microbiol* 133(1–2):105–112.

Malhotra, Bhanu, Anu Keshwani, and Harsha Kharkwal. 2015. "Antimicrobial food packaging: Potential and pitfalls." *Front Microbiol* 6:611.

Maqueda, Mercedes, Antonio Gálvez, Manuel M. Bueno, Maria José Sanchez-Barrena, Carlos González, Armando Albert, Manuel Rico, and Eva Valdivia. 2004. "Peptide AS-48: Prototype of a new class of cyclic bacteriocins." *Curr Protein Pept Sci* 5(5):399–416.

Martín-Escolano, Rubén, Rubén Cebrián, Mercedes Maqueda, Desirée Romero, Maria José Rosales, Manuel Sánchez-Moreno, and Clotilde Marín. 2020. "Assessing the effectiveness of AS-48 in experimental mice models of Chagas' disease." *J Antimicrob Chemother* 75(6):1537–1545.

Martin-Visscher, Leah A., Marco J. van Belkum, Sylvie Garneau-Tsodikova, Randy M. Whittal, Jing Zheng, Lynn M. McMullen, and John C. Vederas. 2008. Isolation and characterization of carnocyclin A, a novel circular bacteriocin produced by *Carnobacterium maltaromaticum* UAL307. *Am Soc Microbiol* 74 (15): 4756–4763.

Martin-Visscher, Leah A., Sabesan Yoganathan, Clarissa S. Sit, Christopher T. Lohans, and John C. Vederas. 2011. "The activity of bacteriocins from *Carnobacterium maltaromaticum* UAL307 against gram-negative bacteria in combination with EDTA treatment." *FEMS Microbiol Lett* 317(2):152–159.

Martinez, Fabio Andres Castillo, Eduardo Marcos Balciunas, Attilio Converti, Paul D. Cotter, and Ricardo Pinheiro de Souza Oliveira. 2013. "Bacteriocin production by *Bifidobacterium* spp. A review." *Biotechnol Adv* 31(4):482–488.

Massani, M., V. Blanco, M. Molina, V. Renaud Sanchez, P. Eisenberg, and G. Vignolo. 2014. "Active polymers containing *Lactobacillus curvatus* CRL705 bacteriocins: Effectiveness assessment in Wieners." *Int J Food Microbiol* 178:7–12.

Massey, Ruth C., Angus Buckling, and Richard ffrench-Constant. 2004. "Interference competition and parasite virulence." *Proc R Soc B* 271(1541):785–788.

Masuda, Y., T. Zendo, N. Sawa, R. H. Perez, J. Nakayama, and K. Sonomoto. 2012. "Characterization and identification of weissellicin Y and weissellicin M, novel bacteriocins produced by *Weissella hellenica* QU 13." *J Appl Microbiol* 112(1):99–108.

Mateus, Teresa, Joana Silva, Rui L. Maia, and Paula Teixeira. 2013. "Listeriosis during pregnancy: A public health concern." *ISRN Obstet Gynecol* 2013:851712. doi: 10.1155/2013/851712.

Mauriello, G., D. Ercolini, A. La Storia, A. Casaburi, and F. Villani. 2004. "Development of polythene films for food packaging activated with an antilisterial bacteriocin from *Lactobacillus curvatus* 32Y." *J Appl Microbiol* 97(2):314–322.

Millette, Mathieu, Gilbert Cornut, Claude Dupont, François Shareck, Denis Archambault, and Monique Lacroix. 2008. "Capacity of human nisin-and pediocin-producing lactic acid bacteria to reduce intestinal colonization by vancomycin-resistant enterococci." *Appl Environ Microbiol* 74(7):1997–2003.

Mills, S., C. Stanton, C. Hill, and R. P. Ross. 2011. "New developments and applications of bacteriocins and peptides in foods." *Annu Rev Food Sci Technol* 2:299–329. doi: 10.1146/annurev-food-022510-133721.

Mirkovic, Nemanja, Natalija Polovic, Goran Vukotic, Branko Jovcic, Marija Miljkovic, Zorica Radulovic, Dzung B. Diep, and Milan Kojic. 2016. "*Lactococcus lactis* LMG2081 produces two bacteriocins, a nonlantibiotic and a novel lantibiotic." *Appl Environ Microbiol* 82(8):2555–2562. doi: 10.1128/AEM.03988-15.

Moein, Masood, Abbas Ali Imani Fooladi, and Hamideh Mahmoodzadeh Hosseini. 2018. "Determining the effects of green chemistry synthesized Ag-nisin nanoparticle on macrophage cells." *Microb Pathog* 114:414–419.

Montville, T. J., and Y. Chen. 1998. "Mechanistic action of pediocin and nisin: Recent progress and unresolved questions." *Appl Microbiol Biotechnol* 50(5):511–519.

Morgan, Sheila M., Paula M. O'Connor, Paul D. Cotter, R. Paul Ross, and Colin Hill. 2005. "Sequential actions of the two component peptides of the lantibiotic lacticin 3147 explain its antimicrobial activity at nanomolar concentrations." *Antimicrob Agents Chemother* 49(7):2606–2611. doi: 10.1128/AAC.49.7.2606-2611.2005.

Mota-Meira, Marilaine, Christophe Lacroix, Gisèle LaPointe, and Marc C. Lavoie. 1997. "Purification and structure of mutacin B-Ny266: A new lantibiotic produced by *Streptococcus mutans*." *FEBS Lett* 410(2–3):275–279.

Mouritzen, Michelle V., Athina Andrea, Katrine Qvist, Steen S. Poulsen, and Håvard Jenssen. 2019. "Immunomodulatory potential of Nisin A with application in wound healing." *Wound Repair Regen* 27(6):650–660.

Neetoo, Hudaa, Ye Mu, and Haiqiang Chen. 2008. "Potential antimicrobials to control *Listeria monocytogenes* in vacuum-packaged cold-smoked salmon pâté and fillets." *Int J Food Microbiol* 123(3):220–227.

Nes, Ingolf F., Dzung Bao Diep, Leiv Sigve Håvarstein, May Bente Brurberg, Vincent Eijsink, and Helge Holo. 1996. "Biosynthesis of bacteriocins in lactic acid bacteria." *Antonie Leeuwenhoek* 70(2):113–128.

Neshani, Alireza, Hosna Zare, Mohammad Reza Akbari Eidgahi, Amin Hooshyar Chichaklu, Aref Movaqar, and Kiarash Ghazvini. 2019. "Review of antimicrobial peptides with anti- *Helicobacter pylori* activity." *Helicobacter* 24(1):e12555.

Ng, Zhang Jin, Mazni Abu Zarin, Chee Keong Lee, and Joo Shun Tan. 2020. "Application of bacteriocins in food preservation and infectious disease treatment for humans and livestock: A review." *RSC Adv* 10(64):38937–38964.

Nieto-Lozano, J. C., J. I. Reguera-Useros, M. C. Peláez-Martínez, and A. Hardisson de la Torre. 2002. "Bacteriocinogenic activity from starter cultures used in Spanish meat industry." *Meat Sci* 62(2):237–243. doi: 10.1016/s0309-1740(01)00252-2.

Nieto-Lozano, Juan C., Juan I. Reguera-Useros, María del C. Peláez-Martínez, Gonzalo Sacristán-Pérez-Minayo, Ángel J. Gutiérrez-Fernández, Arturo Hardisson de la Torre. 2010. "The effect of the pediocin PA-1 produced by *Pediococcus acidilactici* against *Listeria monocytogenes* and *Clostridium perfringens* in Spanish dry-fermented sausages and frankfurters." *Food Control* 21(5):679–685.

Noonpakdee, W., C. Santivarangkna, P. Jumriangrit, K. Sonomoto, and S. Panyim. 2003. "Isolation of nisin-producing *Lactococcus lactis* WNC 20 strain from nham, a traditional Thai fermented sausage." *Int J Food Microbiol* 81(2):137–145.

O'Connor, E. B., B. O'Riordan, S. M. Morgan, H. Whelton, D. M. O'Mullane, R. P. Ross, and C. Hill. 2006. "A lacticin 3147 enriched food ingredient reduces *Streptococcus mutans* isolated from the human oral cavity in saliva." *J Appl Microbiol* 100(6):1251–1260.

O'Connor, Paula M., Eileen F. O'Shea, Caitriona M. Guinane, Orla O'Sullivan, Paul D. Cotter, R. Paul Ross, and Colin Hill. 2015. "Nisin H is a new nisin variant produced by the gut-derived strain *Streptococcus hyointestinalis* DPC6484." *Appl Environ Microbiol* 81(12):3953–3960.

Pag, Ulrike, and Hans-Georg Sahl. 2002. "Multiple activities in lantibiotics-models for the design of novel antibiotics?" *Curr Pharm Des* 8(9):815–833.

Paiva, Aline Dias, Michelle Dias de Oliveira, Sergio Oliveira de Paula, Maria Cristina Baracat-Pereira, Eefjan Breukink, and Hilario Cuquetto Mantovani. 2012. "Toxicity of bovicin HC5 against mammalian cell lines and the role of cholesterol in bacteriocin activity." *Microbiology (Reading)* 158(11):2851–2858.

Parente, Eugenio, Felicia Ciocia, Annamaria Ricciardi, Teresa Zotta, Giovanna E. Felis, and Sandra Torriani. 2010. "Diversity of stress tolerance in *Lactobacillus plantarum*, *Lactobacillus pentosus* and *Lactobacillus paraplantarum*: A multivariate screening study." *Int J Food Microbiol* 144(2): 270–279.

Patel, Trupti, Aoife Molloy, Robin Smith, and Indran Balakrishnan. 2012. "Successful treatment of *Leuconostoc* bacteremia in a neutropenic patient with tigecycline." *Infect Dis Rep* 4(2):e31–e31. doi: 10.4081/idr.2012.e31.

Pattanayaiying, Rinrada, Amporn Sane, Penchom Photjanataree, and Catherine N. Cutter. 2019. "Thermoplastic starch/polybutylene adipate terephthalate film coated with gelatin containing nisin Z and lauric arginate for control of foodborne pathogens associated with chilled and frozen seafood." *Int J Food Microbiol* 290:59–67.

Perez, Rodney H., Takeshi Zendo, and Kenji Sonomoto. 2014. "Novel bacteriocins from lactic acid bacteria (LAB): Various structures and applications." *Microb Cell Factories* 13(1) S3. doi: 10.1186/1475-2859-13-S1-S3.

Periasamy, Saravanan, and Paul E. Kolenbrander. 2009. "Mutualistic biofilm communities develop with *Porphyromonas gingivalis* and initial, early, and late colonizers of enamel." *J Bacteriol* 191(22):6804–6811.

Perna, Michele Christine. 2016. "Formulation and characterization of an antimicrobial coating containing nisin for large scale food package converting processes." (Doctoral dissertation, Clemson University, South Carolina).

Pilevar, Zahra, Hedayat Hosseini, Samira Beikzadeh, Elham Khanniri, and Adel M. Alizadeh. 2020. "Application of bacteriocins in meat and meat products: An update." *Curr Nutr Food Sci* 16(2):120–133.

Pollitt, Laura C., Paula MacGregor, Keith Matthews, and Sarah E. Reece. 2011. "Malaria and trypanosome transmission: Different parasites, same rules?" *Trends Parasitol* 27(5):197–203.

Preet, S., S. Bharati, A. Panjeta, R. Tewari, and P. Rishi. 2015. "Effect of nisin and doxorubicin on DMBA-induced skin carcinogenesis--A possible adjunct therapy." *Tumour Biol* 36(11):8301–8308. doi: 10.1007/s13277-015-3571-3.

Preet, Simran, Satish Kumar Pandey, Karamjit Kaur, Sonia Chauhan, and Avneet Saini. 2019. "Gold nanoparticles assisted co-delivery of nisin and doxorubicin against murine skin cancer." *J Drug Deliv Sci Technol* 53:101147.

Prudencio, C. V., M. T. dos Santos, and M. C. D. Vanetti. 2015. "Strategies for the use of bacteriocins in gram-negative bacteria: Relevance in food microbiology." *J Food Sci Technol Mysore* 52(9):5408–5417. doi: 10.1007/s13197-014-1666-2.

Qin, Yuxuan, Yao Wang, Yinghao He, Ying Zhang, Qianxuan She, Yunrong Chai, Pinglan Li, and Qingmao Shang. 2019. "Characterization of subtilin L-Q11, a novel class I bacteriocin synthesized by *Bacillus subtilis* L-Q11 isolated from orchard soil." *Front Microbiol* 10:484.

Quintavalla, Stefania, and Loredana Vicini. 2002. "Antimicrobial food packaging in meat industry." *Meat Sci* 62(3):373–380.

Raccach, M. 2014. "*Pediococcus*." In: *Encyclopedia of Food Microbiology (Second Edition)*, edited by Carl A. Batt and Mary Lou Tortorello, 1–5. Academic Press.

Raygada, Jose L., and Donald P. Levine. 2009. "Methicillin-resistant *Staphylococcus aureus*: A growing risk in the hospital and in the community." *Am Health Drug Benefits* 2(2):86–95.

Reddy, B.L., H. S. Jatav, V. D. Rajput, T. Minkina, A. Ranjan, A. Harikrishnan, V. K. Veena, A. Chauhan, S. Kumar, A. Prakash, and R. Prasad. 2022. "Nanomaterials based monitoring of food- and water-borne pathogens." *Journal of Nanomaterials*, Article ID 9543532, 13 pages, 2022. https://doi.org/10.1155/2022/9543532.

Reglinski, Mark, and Shiranee Sriskandan. 2015. "*Streptococcus pyogenes*." In: *Molecular Medical Microbiology*, edited by Yi-Wei Tang, Max Sussman, Dongyou Liu, Ian Poxton, and Joseph Schwartzman, 675–716. Academic Press Elsevier.

Reiss, J. 2016. "Molybdenum cofactor and sulfite oxidase deficiency." *Metabolomics (Los Angel)* 6(3):184.

Rivas, Franco P., Marcela P. Castro, Marisol Vallejo, Emilio Marguet, and Carmen A. Campos. 2014. "Sakacin Q produced by *Lactobacillus curvatus* ACU-1: Functionality characterization and antilisterial activity on cooked meat surface." *Meat Sci* 97(4):475–479.

Rook, Graham A. W. 2007. "The hygiene hypothesis and the increasing prevalence of chronic inflammatory disorders." *Trans R Soc Trop Med Hyg* 101(11):1072–1074.

Saleh, F. A., and E. M. El-Sayed. 2004. "Isolation and Characterization of Bacteriocins Produced by *Bifidobacterium lactis* BB-12 and *Bifidobacterium longum* BB-46." 2004.

Sánchez-Hidalgo, Marina, Mercedes Maqueda, Antonio Gálvez, Hikmate Abriouel, Eva Valdivia, and Manuel Martínez-Bueno. 2003. "The genes coding for enterocin EJ97 production by *Enterococcus faecalis* EJ97 are located on a conjugative plasmid." *Appl Environ Microbiol* 69(3):1633–1641.

Santiago-Silva, Paula, Nilda F. F. Soares, Juliana E. Nóbrega, Marcus A. W. Júnior, Kiriaque B. F. Barbosa, Ana Carolina P. Volp, Evelyn R. M. A. Zerdas, and Nédio J. Würlitzer. 2009. "Antimicrobial efficiency of film incorporated with pediocin (ALTA® 2351) on preservation of sliced ham." *Food Control* 20(1):85–89.

Santos, Johnson C. P., Rita C. S. Sousa, Caio G. Otoni, Allan R. F. Moraes, Victor G. L. Souza, Eber A. A. Medeiros, Paula J. P. Espitia, Ana C. S. Pires, Jane S. R. Coimbra, and Nilda F. F. Soares. 2018. "Nisin and other antimicrobial peptides: Production, mechanisms of action, and application in active food packaging." *Innov Food Sci Emerg Technol* 48:179–194. doi: 10.1016/j.ifset.2018.06.008.

Saraoui, Taous, Françoise Leroi, Frédérique Chevalier, Jean-Michel Cappelier, Delphine Passerini, and Marie-France Pilet. 2018. "Bioprotective effect of *Lactococcus piscium* CNCM I-4031 against *Listeria monocytogenes* growth and virulence." *Front Microbiol* 9:1564.

Sawa, N., K. Okamura, Takeshi Zendo, K. Himeno, Jiro Nakayama, and Kenji Sonomoto. 2010. "Identification and characterization of novel multiple bacteriocins produced by *Leuconostoc pseudomesenteroides* QU 15." *J Appl Microbiol* 109(1):282–291.

Sawa, Naruhiko, Takeshi Zendo, Junko Kiyofuji, Koji Fujita, Kohei Himeno, Jiro Nakayama, and Kenji Sonomoto. 2009. "Identification and characterization of lactocyclicin Q, a novel cyclic bacteriocin produced by *Lactococcus* sp. strain QU 12." *Appl Environ Microbiol* 75(6):1552–1558.

Schillinger, U., M. E. Stiles, and W. H. Holzapfel. 1993. "Bacteriocin production by *Carnobacterium piscicola* LV 61." *Int J Food Microbiol* 20(3):131–147.

Sharma, Preeti, Muzamil Rashid, and Sukhraj Kaur. 2020. "Novel enterocin E20c purified from *Enterococcus hirae* 20c synergised with ß-lactams and ciprofloxacin against *Salmonella enterica*." *Microb Cell Factories* 19(1):98. doi: 10.1186/s12934-020-01352-x.

Sheoran, P., and S. K. Tiwari. 2019. "Enterocin LD3 from *Enterococcus hirae* LD3 causing efflux of intracellular ions and UV-absorbing materials in gram-negative bacteria." *J Appl Microbiol* 126(4):1059–1069. doi: 10.1111/jam.14203.

Shi, Feng, Yanwei Wang, Yongfu Li, and Xiaoyuan Wang. 2016. "Mode of action of leucocin K7 produced by *Leuconostoc mesenteroides* K7 against *Listeria monocytogenes* and its potential in milk preservation." *Biotechnol Lett* 38(9):1551–1557.

Siegel, Rebecca L., Kimberly D. Miller, Hannah E. Fuchs, and Ahmedin Jemal. 2021. "Cancer statistics, 2021." *CA: A Cancer J Clin* 71(1):7–33. doi: 10.3322/caac.21654.

Silva, Célia C. G., Sofia P. M. Silva, and Susana C. Ribeiro. 2018. "Application of bacteriocins and protective cultures in dairy food preservation." *Front Microbiol* 9:594.

Simova, E. D., D. B. Beshkova, and P. Dimitrov Zh. 2009. "Characterization and antimicrobial spectrum of bacteriocins produced by lactic acid bacteria isolated from traditional Bulgarian dairy products." *J Appl Microbiol* 106(2):692–701. doi: 10.1111/j.1365-2672.2008.04052.x.

Singh, Veer Pal. 2018. "Recent approaches in food bio-preservation-a review." *Open Vet J* 8(1):104–111.

Soltani, Samira, Riadh Hammami, Paul D. Cotter, Sylvie Rebuffat, Laila Ben Said, Hélène Gaudreau, François Bédard, Eric Biron, Djamel Drider, and Ismail Fliss. 2021. "Bacteriocins as a new generation of antimicrobials: Toxicity aspects and regulations." *FEMS Microbiol Rev* 45(1):fuaa039.

Song, Adelene Ai-Lian, Lionel L. A. In, Swee Hua Erin Lim, and Raha Abdul Rahim. 2017. "A review on *Lactococcus lactis*: from food to factory." *Microb Cell Factories* 16(1):1–15.

Song, Han-Yi, Long Zhou, Dong-yan Liu, Xin-Jie Yao, and Yan Li. 2018. "What roles do probiotics play in the eradication of *Helicobacter pylori* Current knowledge and ongoing research." *Gastroenterol Res Pract* 2018:9379480. doi: 10.1155/2018/9379480.

Stern, N. J., E. A. Svetoch, B. V. Eruslanov, V. V. Perelygin, E. V. Mitsevich, I. P. Mitsevich, V. D. Pokhilenko, V. P. Levchuk, O. E. Svetoch, and B. S. Seal. 2006. "Isolation of a *Lactobacillus salivarius* strain and purification of its bacteriocin, which is inhibitory to *Campylobacter jejuni* in the chicken gastrointestinal system." *Antimicrob Agents Chemother* 50(9):3111–3116. doi: 10.1128/Aac.00259-06.

Su, P., A. Henriksson, and H. Mitchell. 2007. "Survival and retention of the probiotic *Lactobacillus casei* LAFTI® L26 in the gastrointestinal tract of the mouse." *Lett Appl Microbiol* 44(2):120–125.

Sun, Lin, Hui Song, and Wen Zheng. 2015. "Improvement of antimicrobial activity of pediocin PA-1 by site-directed mutagenesis in C-terminal domain." *Protein Pept Lett* 22(11):1007–1012.

Svetoch, Edward A., Boris V. Eruslanov, Vladimir V. Perelygin, Evgeni V. Mitsevich, Irina P. Mitsevich, Valery N. Borzenkov, Vladimir P. Levchuk, Olga E. Svetoch, Yuri N. Kovalev, and Yuri G. Stepanshin. 2008. "Diverse antimicrobial killing by *Enterococcus faecium* E 50–52 bacteriocin." *J Agric Food Chem* 56(6):1942–1948.

Terzić-Vidojević, Amarela, Katarina Veljović, Jelena Begović, Brankica Filipić, Dušanka Popović, Maja Tolinački, Marija Miljković, Milan Kojić, and Natasa Golić. 2015. "Diversity and antibiotic susceptibility of autochthonous dairy enterococci isolates: Are they safe candidates for autochthonous starter cultures?" *Front Microbiol* 6:954.

Thomas, L., M. Clarkson, and J. Delves-Broughton. 2000. *Natural Food Antimicrobial Systems*. CRC Press.

Tian, X. B., F. D. Sun, X. R. Zhou, S. Z. Luo, and L. Chen. 2015. "Role of peptide self-assembly in antimicrobial peptides." *J Pept Sci* 21(7):530–539. doi: 10.1002/psc.2788.

Timothy, Bamgbose, Atta Habiba Iliyasu, and Anupkumar R. Anvikar. 2021. "Bacteriocins of lactic acid bacteria and their industrial application." *Curr Top Lactic Acid Bacteria Probiotics* 7(1):1–13. doi: 10.35732/ctlabp.2021.7.1.1.

Tirloni, Erica, Vanessa Di Pietro, Giuseppe Rizzi, Francesco Pomilio, Patrizia Cattaneo, Cristian Bernardi, and Simone Stella. 2019. "Non-thermal inactivation of *Listeria* spp. in a typical dry-fermented sausage: 'Bergamasco salami'." *Ital J Food Saf* 8(3): 8112.

Titze, Isabel, and Volker Krömker. 2020. "Antimicrobial activity of a phage mixture and a lactic acid bacterium against *Staphylococcus aureus* from bovine mastitis." *Vet Sci* 7(1):31.

Todorov, S. D., M. Wachsman, E. Tomé, X. Dousset, M. T. Destro, L. M. Dicks, B. D. Franco, M. Vaz-Velho, and D. Drider. 2010. "Characterisation of an antiviral pediocin-like bacteriocin produced by *Enterococcus faecium*." *Food Microbiol* 27(7):869–879. doi: 10.1016/j.fm.2010.05.001.

Tong, Z., L. Dong, L. Zhou, R. Tao, and L. Ni. 2010. "Nisin inhibits dental caries-associated microorganism in vitro." *Peptides* 31(11):2003–2008. doi: 10.1016/j.peptides.2010.07.016.

Uriot, Ophélie, Sylvain Denis, Maira Junjua, Yvonne Roussel, Annie Dary-Mourot, and Stéphanie Blanquet-Diot. 2017. "*Streptococcus thermophilus*: from yogurt starter to a new promising probiotic candidate?" *J Funct Foods* 37:74–89.

Van Belkum, Marco J., and Michael E. Stiles. 2000. "Nonlantibiotic antibacterial peptides from lactic acid bacteria." *Nat Prod Rep* 17(4):323–335.

van Heel, Auke Johan, Manuel Montalban-Lopez, and Oscar P. Kuipers. 2011. "Evaluating the feasibility of lantibiotics as an alternative therapy against bacterial infections in humans." *Expert Opin Drug Metab Toxicol* 7(6):675–680.

Varadé, Jezabel, Susana Magadán, and África González-Fernández. 2021. "Human immunology and immunotherapy: Main achievements and challenges." *Cell Mol Immunol* 18(4):805–828.

Ventola, C. Lee. 2015. "The antibiotic resistance crisis: Part 1: Causes and threats." *Pharm Ther* 40(4):277.

Vidya, Chauhan. 2020. "Bacteriocin: A potent therapeutic weapon used as an alternative to antibiotics." *Arch Biomed Sci Eng* 039–040. doi: 10.17352/abse.000019.

Villalobos-Delgado, Luz H., G. V. Nevárez-Moorillon, Irma Caro, Emiliano J. Quinto, and Javier Mateo. 2019. "4 - Natural antimicrobial agents to improve foods shelf life." In: *Food Quality and Shelf Life*, edited by Charis M. Galanakis, 125–157. Academic Press.

Villarante, Karen I., Francisco B. Elegado, Shun Iwatani, Takeshi Zendo, Kenji Sonomoto, and Evelyn E. de Guzman. 2011. "Purification, characterization and *in vitro* cytotoxicity of the bacteriocin from *Pediococcus acidilactici* K2a2-3 against human colon adenocarcinoma (HT29) and human cervical carcinoma (HeLa) cells." *World J Microbiol Biotechnol* 27(4):975–980. doi: 10.1007/s11274-010-0541-1.

Vimont, Allison, Benoît Fernandez, Riadh Hammami, Ahlem Ababsa, Hocine Daba, and Ismaïl Fliss. 2017. "Bacteriocin-producing *Enterococcus faecium* LCW 44: A high potential probiotic candidate from raw camel milk." *Front Microbiol* 8:865.

Wachsman, M. B., V. Castilla, A. P. de Ruiz Holgado, R. A. de Torres, F. Sesma, and C. E. Coto. 2003. "Enterocin CRL35 inhibits late stages of HSV-1 and HSV-2 replication *in vitro*." *Antiviral Res* 58(1):17–24. doi: 10.1016/s0166-3542(02)00099-2.

Wannun, P., S. Piwat, and R. Teanpaisan. 2016. "Purification, characterization, and optimum conditions of fermencin SD11, a bacteriocin produced by human orally *Lactobacillus fermentum* SD11." *Appl Microbiol Biotechnol* 179(4):572–582. doi: 10.1007/s12010-016-2014-y.

Wayah, Samson Baranzan, and Koshy Philip. 2018. "Characterization, yield optimization, scale up and biopreservative potential of fermencin SA715, a novel bacteriocin from *Lactobacillus fermentum* GA715 of goat milk origin." *Microb Cell Factories* 17(1):125. doi: 10.1186/s12934-018-0972-1.

Whitford, M. F., M. A. McPherson, R. J. Forster, and R. M. Teather. 2001. "Identification of bacteriocin-like inhibitors from rumen *Streptococcus* spp. and isolation and characterization of bovicin 255." *Appl Environ Microbiol* 67(2):569–574.

Woraprayote, W., L. Pumpuang, A. Tosukhowong, S. Roytrakul, R. H. Perez, T. Zendo, K. Sonomoto, S. Benjakul, and W. Visessanguan. 2015. "Two putatively novel bacteriocins active against gram-negative food borne pathogens produced by *Weissella hellenica* BCC 7293." *Food Control* 55:176–184. doi: 10.1016/j.foodcont.2015.02.036.

Woraprayote, Weerapong, Yutthana Kingcha, Pannawit Amonphanpokin, Jittiporn Kruenate, Takeshi Zendo, Kenji Sonomoto, Soottawat Benjakul, and Wonnop Visessanguan. 2013. "Anti-listeria activity of poly (lactic acid)/sawdust particle biocomposite film impregnated with pediocin PA-1/AcH and its use in raw sliced pork." *Int J Food Microbiol* 167(2):229–235.

Woraprayote, Weerapong, Yuwares Malila, Supaluk Sorapukdee, Adisorn Swetwiwathana, Soottawat Benjakul, and Wonnop Visessanguan. 2016. "Bacteriocins from lactic acid bacteria and their applications in meat and meat products." *Meat Sci* 120:118–132.

Woraprayote, Weerapong, Laphaslada Pumpuang, Amonlaya Tosukhowong, Takeshi Zendo, Kenji Sonomoto, Soottawat Benjakul, and Wonnop Visessanguan. 2018. "Antimicrobial biodegradable food packaging impregnated with bacteriocin 7293 for control of pathogenic bacteria in Pangasius fish fillets." *LWT Food Sci Technol* 89:427–433.

World Health Organization. 2002. "Health and nutritional properties of probiotics in food including powder milk with live lactic acid bacteria, a joint FAO/WHO expert consultation." Cordoba, Argentina, 1–4 October 2001. http://www.who.int/foodsafety/publications/fs_management/probiotics/en/index.html.

Worobo, Randy W., Thomas Henkel, Miloslav Sailer, Kenneth L. Roy, John C. Vederas, and Michael E. Stiles. 1994. "Characteristics and genetic determinant of a hydrophobic peptide bacteriocin, carnobacteriocin A, produced by *Carnobacterium piscicola* LV17A." *Microbiology (Reading)* 140(3):517–526.

Wu, Junqiang, Songhua Hu, and Liting Cao. 2007. "Therapeutic effect of nisin Z on subclinical mastitis in lactating cows." *Antimicrob Agents Chemother* 51(9):3131–3135.

Xie, Lili, and Wilfred A. Van Der Donk. 2004. "Post-translational modifications during lantibiotic biosynthesis." *Curr Opin Chem Biol* 8(5):498–507.

Yerlikaya, Oktay. 2019. "Probiotic potential and biochemical and technological properties of *Lactococcus lactis* ssp. *lactis* strains isolated from raw milk and kefir grains." *J Dairy Sci* 102(1):124–134.

Yi, Lanhua, Ying Dang, Jingli Wu, Lihui Zhang, Liu Xiaojiao, Bianfang Liu, Yuan Zhou, and Xin Lu. 2016. "Purification and characterization of a novel bacteriocin produced by *Lactobacillus crustorum* MN047 isolated from koumiss from Xinjiang, China." *J Dairy Sci* 99(9):7002–7015.

Yildirim, Z., N. Öncül, M. Yildirim, and Ş. Karabiyikli. 2016. "Application of lactococcin BZ and enterocin KP against *Listeria monocytogenes* in milk as biopreservation agents." *Acta Aliment* 45(4):486–492.

Yildirim, Zeliha, and Michael G. Johnson. 1998. "Characterization and antimicrobial spectrum of bifidocin B, a bacteriocin produced by *Bifidobacterium bifidum* NCFB 1454." *J Food Prot* 61(1):47–51.

Yusuf, Mohd. 2018. "Natural antimicrobial agents for food biopreservation." In: *Food Packaging and Preservation*, edited by Alexandru Mihai Grumezescu and Alina Maria Holban, 409–438. UK: Academic Press.

Zanetti, Micheli, Thaís Karoline Carniel, Francieli Dalcanton, Raul Silva dos Anjos, Humberto Gracher Riella, Pedro H. H. de Araújo, Débora de Oliveira, and Marcio Antônio Fiori. 2018. "Use of encapsulated natural compounds as antimicrobial additives in food packaging: A brief review." *Trends Food Sci Technol* 81:51–60. doi: 10.1016/j.tifs.2018.09.003.

Zendo, Takeshi, Masanori Fukao, Kyoko Ueda, Tomoko Higuchi, Jiro Nakayama, and Kenji Sonomoto. 2003. "Identification of the lantibiotic nisin Q, a new natural nisin variant produced by *Lactococcus lactis* 61–14 isolated from a river in Japan." *BioSci Biotechnol Biochem* 67(7):1616–1619.

Zhang, Huiyun, Baohua Kong, Youling L. Xiong, and Xu Sun. 2009. "Antimicrobial activities of spice extracts against pathogenic and spoilage bacteria in modified atmosphere packaged fresh pork and vacuum packaged ham slices stored at 4°C." *Meat Sci* 81(4):686–692.

Zhang, Jinlan, Guorong Liu, Pinglan Li, and Yan Qu. 2010. "Pentocin 31–1, a novel meat-borne bacteriocin and its application as biopreservative in chill-stored tray-packaged pork meat." *Food Control* 21(2):198–202.

Zhao, Hongxia, Rohit Sood, Arimatti Jutila, Shambhunath Bose, Gunnar Fimland, Jon Nissen-Meyer, and Paavo K. J. Kinnunen. 2006. "Interaction of the antimicrobial peptide pheromone Plantaricin A with model membranes: Implications for a novel mechanism of action." *Biochim Biophys Acta Biomembranes* 1758(9):1461–1474.

Zhao, Xingchen, Ce Shi, Rizeng Meng, Zonghui Liu, Yanjun Huang, Ziwen Zhao, and Na Guo. 2016. "Effect of nisin and perilla oil combination against *Listeria monocytogenes* and *Staphylococcus aureus* in milk." *J Food Sci Technol* 53(6):2644–2653. doi: 10.1007/s13197-016-2236-6.

Zhou, Huimin, Yuanhong Xie, Hui Liu, Junhua Jin, Huixia Duan, and Hongxing Zhang. 2015. "Effects of two application methods of plantaricin BM-1 on control of *Listeria monocytogenes* and background spoilage bacteria in sliced vacuum-packaged cooked ham stored at 4°C." *J Food Prot* 78(10):1835–1841.

2 Fermented food microbes and their preservative byproducts

A. A. Amara

TABLE OF ABBREVIATIONS

AAB: acetic acid bacteria
BM: Beneficial microbes
GIS: Gastro intestinal system
GMP: Good manufacturing practices
GRAS: generally regarded as safe
LAB: Lactic acid bacteria
SCOBY: Symbiotic Culture of Bacterium and Yeast
FP: Fermentation process
SC: Seed culture
TFs: Traditional foods
TFFs: Traditional fermented foods

2.1 INTRODUCTION

Microbes present in fermented foods (FF) serve as external sources of oxidants, antioxidants, vitamins, hormones, minerals, amino acids, and other beneficial components. They also act as a safe reservoir for certain probiotic strains. Some microbial species have the ability to produce antimicrobial compounds that inhibit or eliminate pathogenic microbes. Furthermore, they have the capacity to generate other valuable compounds and macromolecules. When consumed along with food, these coexisting microbes can colonize the gastrointestinal system (GIS) and are commonly referred to as "probiotics." Beneficial microbes (BMs) form a larger group than probiotics and can be described as microbes capable of fermenting food and producing vital compounds and macromolecules. In addition to their antimicrobial properties, they employ various tactics and strategies to suppress the growth of pathogens, such as nutrient consumption, oxygen uptake, and colonization of available spaces in the gastrointestinal system.

Food can be broadly defined as any substance that can be metabolized by an organism to provide energy and contribute to tissue building. For humans, food encompasses any suitable source of nourishment. However, in literature, there is sometimes confusion between the terms "food" and "drink" (Toussaint-Samat 2009). Microbes and their spores are ubiquitous and can initiate fermentation or spoilage of our food (Figure 2.1). It is crucial to identify and investigate safe microbial strains that have the ability to preserve food and combat pathogens. These microbial alternatives offer potential solutions to replace chemical preservatives and antimicrobial compounds and drugs (Sybesma et al. 2006, Dugoua et al. 2009, Trapecar et al. 2011, Dinleyici et al. 2012, Xia, Zhu, and Zhu 2013, Jeong, Heo, and Lee 2017, Peivasteh-Roudsari et al. 2019, Zugic Petrovic et al. 2020). This tool is highly recommended for ensuring optimal performance and achieving desirable properties in finished food products. It facilitates better validation of qualities such as texture, sensory features, and more (Thierry et al. 2015, Terefe and Augustin 2020). Numerous literature sources have extensively described various traditional foods (TFs) that are prepared and consumed worldwide (Gok 2021).

 DOI: 10.1201/9781003268949-2

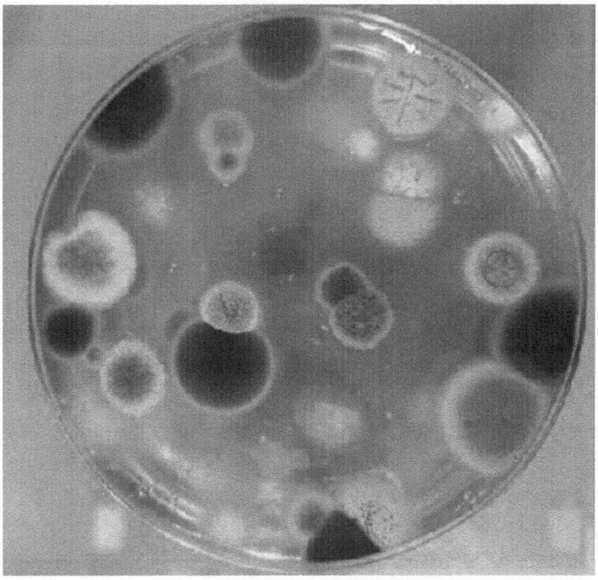

FIGURE 2.1 Microbes everywhere, non-autoclaved LB medium agar plat. (Prepared in this study)

Over time, different regulations and practices have been developed, and chronological documents have been published, highlighting the efforts of humans to devise new methods for food preservation. One of the most renowned stories related to food preservation comes from the Middle East and involves the Prophet Joseph (pronounced as "Yoosof" in Arabic). According to the story, he advised the ruler of Egypt to store the abundant grain crop from seven prosperous years to sustain the population during the following seven years of scarcity. How did he achieve this? By storing the seeds in their original protective coverings in well-ventilated and dry locations. This simple practice holds valuable lessons, as it effectively ensured the safety and security of Egypt and other nations.

Food preservation requires skill and involves managing various controlling factors, including inhibiting the food's own degrading enzymes, controlling coexisting microbes, preventing insect infestations, managing moisture and humidity levels, and removing excess water when necessary. The objective is to prevent any negative or harmful changes in the properties of the food. However, certain changes are accepted by humans and widely applied, leading to long-lasting preservation of the food. When these changes occur correctly, the food is not considered "spoiled" but rather "fermented." It's important to note that the concept of fermented food may vary from place to place and among different ethnic groups. What one ethnic group considers a fermented food, another might perceive as spoiled. In this chapter, a scientific approach will be used to avoid any conflicts.

Food fermentation and preservation can be seen as a battle between humans and microbes. Contaminated food can be a source of harmful microbes. However, some coexisting or intentionally added microbes are beneficial to humans. Collectively, there is a large number of microbes that align with humans and assist in fermenting and preserving their foods. In reality, even the microbes that cause food spoilage and degradation serve a crucial role by eliminating waste and completing natural food cycles in nature. This brings about differences in the perception of usefulness or harmfulness regarding food spoilage and preservation. If food is not consumed, it accumulates and disrupts the natural balance in nature. The microbial balance in the open environment should not be disturbed. On the other hand, microbes that come into contact with humans should be subjected to efficient control protocols that give humans the upper hand and enable the beneficial ones to assist us.

Meanwhile, humans have aqcuired significant knowledge about how to save foods from the spoilage process through different trials/errors, accidents, observations, and the like. The traditional use of salt (Solheim et al. 2014, Michelakou et al. 2021); vinegar (Díaz-Montes and Castro-Muñoz 2021, Pellerin, Bazinet, and Grenier 2021), and alcohol (Bartkiene et al. 2021) are examples. One could trace and relate some of the tactics. For example, historically desert travelers carried semi-cooked wheat or barley flour. They put this semi-cooked flour in small pouches that hung around their necks. Each meal they mixed some of the dry semi-cooked flour with water or milk and ate it, which allowed them to travel for up to two months in arid desert conditions by camel.

After the discovery of the microscope by Robert Hooks and Antoni Van Leeuwenhoek; microbes were isolated and identified, their growth conditions were specified, and, more recently, their types of activities were characterized using modern instrumentation. Some critical discoveries were achieved; like the discovery of antibiotics which are used widely for preserving food (20 years following the discovery of the microscope) (Mahalmani et al. 2019, Kasimanickam, Kasimanickam, and Kasimanickam 2021). Antibiotics are considered as one of the naturally produced microbial secondary metabolites but their unwise use has lead to catastrophic consequences. Initially, the explanation for the activity of fungi against *E. coli* was attributed to the consumption of the air (oxygen) by the fungus. The discovery of penicillin led many to believe that humans could eliminate undesirable microbes, including those causing food spoilage. However, such a misunderstanding of the power of nature gives a false signal about the possibility of complete eradication. The excessive use of antibiotics in various food chains, human treatment, agricultural phytopathogenic control, and food preservation has resulted in the emergence of antibiotic resistance (Devirgiliis, Barile, and Perozzi 2011, Founou, Founou, and Essack 2016, Kumar, Arnipalli, and Ziouzenkova 2020). The R-factor (resistance factor) was originally claimed as the responsible agent for the spread of antibiotic resistance. Today, most of the mechanisms involved in antibiotic resistance have been clarified.

Modern approaches for food preservation are the result of mixing old tactics and newly modified or invented ones. The most famous approaches for food preservation used in the last century include drying (Murphy et al. 2017), smoking, salting (Liu et al. 2018, Dias et al. 2020), sugaring (Zhao et al. 2021), canning, using spices (Naveed et al. 2013, Zhang et al. 2019, Sridhar et al. 2021), using chemical and natural preservatives (Yu, Chin, and Paik 2021), using radiation like UV (Lázaro, Monteiro, and Conte-Junior 2020), using antimicrobial agents (e.g., acids and antibiotics), using microbial—microbial antagonists (VidyaLaxme et al. 2014, Al-Madboly and Abdullah 2015, Uraipan and Hongpattarakere 2015, Lobos, Barrera, and Padilla 2017, Choi et al. 2018, Edalati et al. 2019, Kariyawasam et al. 2020, Dejene, Regasa Dadi, and Tadesse 2021), food fermenting (Zhu et al. 2021), pasteurizing (Sheoran and Tiwari 2019, Salazar et al. 2021), and the use of BMs and probiotics as preservatives (Vitetta, Hall, and Linnane 2014, Murphy et al. 2017, Choeisoongnern et al. 2021, Davachi et al. 2021, Pupa et al. 2021).

This chapter specifically focuses on tools and techniques related to food preservation that involve the usage of antimicrobial preservatives, fermented foods, probiotics, and microbial antagonists. It provides a detailed description of probiotic microbes and other beneficial microbes that possess probiotic properties, their byproducts, and safety concerning their use as bio-preservatives. Some types of traditional fermented food were included.

2.1.1 Use of Microbes

The individual who engages with microbes and attempts to to harness their benefits and avoid harmful or undesirable effects on food quality and production costs must be aware of the many issues concerning consumers, products, microbes, and the cost of production processes. The use of good microbes against bad ones will reduce efforts and costs, while protecting our health and improving the food industry. The existence of the BMs leads to the suppression of the harmful microbes' growth. In addition, their existence in food causes changes that farvour its consumption. They are able to produce compounds that preserve our food, facilitate its digestion and add additional positive values to the food. Some of them migrate to the GIS and colonize it. These are called 'probiotics' (Afrc 1989). Applying the best conditions for the growth of BMs in food will improve their impact on pathogens and at the same time will maintain the quality of final food product (Lilly and Stillwell 1965, O'Mahony et al. 2001, Kanmani et al. 2012, Siragusa et al. 2014). In many processes, both harmful microbes and beneficial ones coexist. Only production conditions can allow BMs to flourish and harmful ones to be supressed. This conversion can be achieved through different means such as influencing growth conditions to favor BMs. Enhancing their growth will increase their number and make them dominant in fermented food as well as in the GIS.

Other tactics could also be used such as reducing the number of harmful microbes that coexist with beneficial ones (Amara 2011, 2013a). By removing the substrates that induce their virulence and introducing simple sugary carbon sources, the behavior of microbes can be altered significantly. Sugar has the ability to bypass biochemical pathways and deactivate various processes that activate microbial virulence genes, including proteolytic and keratinolytic enzymes that promote other virulence-related activities (Amara 2013a). For instance, *Candida albicans* remains quiescent in the presence of sugar but exhibits virulence when protein is available, particularly when attacking the skin and tissues of immunocompromised patients (Amara 2013a).

The introduction of sugary components will promote the growth of beneficial microbes, particularly various yeast species, allowing them to thrive and become dominant. These beneficial microbes will outcompete harmful microbes by consuming essential growth factors present in the surrounding environment. Consequently, the number of harmful microbes will decrease significantly, limiting their ability to successfully colonize the area. Moreover, certain beneficial microbes have the capacity to produce antimicrobial products or byproducts that are capable of eliminating harmful microbes (Daeschel 1989, O'Mahony et al. 2001, Tharmaraj and Shah 2009, Suskovich et al. 2010, Denkova et al. 2017, Silva et al. 2020, Rodríguez-Sânchez et al. 2021, Yerlikaya, Saygili, and Akpinar 2021). For instance, many beneficial microbes have been found to produce antibiotics (Parker 1974). Additionally, these microbes can produce acids, such as lactic acid (Suskovich et al. 2010). Natural acids like lactic acid are safe for both food and consumers. Lactic acid, a simple acid, can effectively alter the pH of food, leading to the elimination of a large number of pathogens while also exerting control over others (Sturino, Rajendran, and Altermann 2013, Ohenhen et al. 2015).

2.1.1 Nature Against Nature

The power given to nature enables it to cycle/recycle its products. We need to control the output of new technology, including food products, according to the requirements of our environment (Amara 2011). When designing new food products, the safety of both humans and the entire natural ecosystem should be considered (Dugoua et al. 2009, Trapecar et al. 2011, Dinleyici et al. 2012, Jeong, Heo, and Lee 2017, Peivasteh-Roudsari et al. 2019, Zugic Petrovic et al. 2020). Fermentation biotechnology can play a significant role in maintaining this balance (Sarkar et al. 2010, Plessas et al. 2011, Jeong, Jung, et al. 2013), as it can accelerate the production of key natural products that have preservative properties and can be used in food preservation processes. In reality, the fermentation process is either a part or a controlled part of nature's duty to eliminate any accumulated products through microbial activity.

There is a real conflict between our unplanned demands and the safety of nature. In many cases, we have become our own enemy. In recent years, many microbes have acquired resistance or virulence genes. It is wise to allow microbes to live calmly in their natural habitats with their original natural distributions. Continuous use of antimicrobial agents (e.g., antibiotics or disinfectants) will create problems. An antimicrobial agent will kill most microbes but will leave behind a few that are either resistant or harbor hydrocarbon-degrading genes, enabling them to hydrolyze the antimicrobial agents. By killing the sensitive microbes (which are in large numbers) and allowing the survival of resistant ones (which are in smaller numbers), we create an environment where resistant microbes have no natural competition (not even competition for space). Over time, this changes the equation, and the resistant microbes, which are mostly opportunistic or pathogenic, become dominant. Using the same disinfectant repeatedly means that we increase the number of pathogens. This is exactly what has happened in hospitals.

Interestingly, most microbes are friendly, and even unfriendly ones can change their behavior and live in a friendly manner under specific conditions. This can be achieved through various mechanisms, such as

1. Forcing the microbe to lose mobile elements (plasmids) that carry virulence or resistance genes. This process, known as "plasmid curing," can be achieved by feeding the microbes simple substrates that do not require the functions encoded by the plasmid genes (e.g., using biopreservatives instead of antibiotics).
2. Modulating the microbial environment to prevent any stress that activates microbial virulence factors (e.g., giving sugar to *Candida albicans*).
3. Controlling key elements involved in pathogenesis or resistance processes. This can be accomplished by using chelating agents to bind divalent cations and deactivate the proteins and enzymes (Amara and Hussain 2006, Hussain and Amara 2006).
4. Microbes typically exist in communities and maintain a sort of balance. Surrounding virulent microbes with others in their microbial niche will induce competition that helps maintain a balance in their numbers.
5. Controlling the physical, chemical and biological factors that prevent the microbe from becoming virulent or turning into opportunistic pathogens. For example, in addition to being disinfectants, it is necessary to use drying procedures to control *P. aeruginosa* in hospitals (Amara 2011).
6. Using natural inexpensive tactics and products. For example, our oral cavity is well adapted to killing microbes that enter with food through saliva which contains lysozymes and other elements that can kill and lyse microbes (Huang et al. 2018, Perez et al. 2018). Chewing our food well can enhance this mechanism.

There is a need for further studies to evaluate the safety and side effects of antimicrobial agents used in our food. Sometimes, the pursuit of maximizing profit can lead to a deviation from focusing on certain safety issues (Magnuson et al. 2013, Koutsoumanis, Lianou, and Sofos 2014, Laulund et al. 2017).

By understanding various phenomena, we can anticipate potential outcomes. For instance, while literature describes probiotic strains that are able to resist different antibiotics as a positive trait, we must also consider the possibility of gene transfer. What if these genes are transferred to harmful microbes within our body during their battle with beneficial microbes? Such strains could establish themselves and colonize our gastrointestinal system. Therefore, many scientists recommend keeping probiotic strains and beneficial microbes in their natural state.

After the golden age of naturally and synthetically produced antibiotics, and recognizing the side effects of synthetic chemical compounds used in food preservation, scientists have turned to nature to extract substances that can control pathogens. Synthetic compounds are often imperfect, with traces of incomplete structures, analogs, and dextro-rotatory structures that can be harmful. The global war against microbes has not succeeded in eradicating them. Biopreservatives, on the other hand, have shown the ability to target specific microbes.

In addition to the search for natural biopreservatives, it is important to develop strategies to inhibit the growth of harmful microbes in food using food-like products. This can be achieved through the use of active short peptides, amino acids, organic structures, and natural acids that can preserve food, control or kill microbes, and keep us safe. This has led to the introduction of a new term in the scientific community and food industry: biopreservative.

The safety of the microbes used or their products should be thoroughly investigated. For example, some opportunistic microbes may not have reported side effects when present in a specific food. However, do consumers only consume that particular food?

Safety concerns should consider what can happen outside the package or can. Even if they may initially sound positive, natural preservatives still have activity (e.g., viable microbes). Therefore, it is important to provide safety instructions to control any processes. The presence of certain beneficial microbes like Lactobacillus in food can have desirable properties. However, consuming large quantities of yogurt, especially by children, can cause constipation (An et al. 2010, Guerra et al. 2011, Eswaran, Guentner, and Chey 2014, Kubota et al. 2020, Saviano et al. 2021).

BMs can be found in small numbers on the surface of fruits and vegetables (Faten et al. 2016, Daranas et al. 2018), so consuming fruits and vegetables is generally safe.

FFs contain a large number of microbes, and their safety is an essential consideration when consuming them. While fermented foods can be beneficial, they should be consumed in appropriate quantities and, in some cases, as appetizers. Traditionally produced fermented foods have been widely accepted by consumers and have demonstrated compatibility with modern food products and our modern lifestyle (Figures 2.2 and 2.3).

2.1.2 9.4 The Demand

The development of food preservation techniques has been driven by the need to keep food from spoiling over extended periods of time. However, this demand has varied depending on factors such as time, regions, and human groups. The preservation needs of a small family or individuals on trips in different environments like deserts, mountains, valleys, etc., differ from the needs of larger political entities like republics, kingdoms, or empires. Throughout history, practical and technical methods for preserving food, including the tools, practices, and techniques used, have varied across different places and time periods.

Regional factors have had a significant influence on food preservation methods. For instance, in hot regions, humans learned to use hot spices to preserve food. Hot spices contain antimicrobial compounds, such as volatile oils, which help inhibit the growth of microorganisms (Sridhar et al. 2021). In desert regions, humans learned to cut meat into thin, flat pieces and use the sun to dry them. In colder regions, meat was often covered with a thick layer of salt and left in a cold, well-ventilated area to dehydrate or subjected to safe fermentation processes (Li et al. 2019). Fermented foods have played a vital role in traditional food preservation methods. They represent a form of

a)

b)

FIGURE 2.2 a) and b) Two samples represent prehistoric Egyptian fermented food from two different locations in Menia governorate, Egypt. The product name is Kishk. Simply wheat seeds are mixed with Laben Rayeb and left to ferment for a suitable period of time. After fermentation, they are granulated and allowed to dry. There are two locations where this process takes place: a) from Galda village and b) from Malawy, both located in Menia governorate. This product is popular throughout Egypt, with a preference for those obtained from Upper Egypt. Upon closer inspection, one can observe the enlarged images of the wheat seeds. (Prepared in this study)

FIGURE 2.3 Fermented food in a modern city (Alexandria, Egypt) resists the modernization process. Apparently the first invented fermentor can be seen in the image. Observe the big plug that is similar to the one used nowadays for plugging flasks. (Prepared in this study)

acceptable food transformation. However, in the past, where survival was the most important factor, the primary focus of fermented foods was their preservation against spoilage factors. Nowadays, factors such as aroma, taste, texture, etc., have become additional important considerations in relation to fermented foods.

2.1.3 The Discovery of the Microbes That Claimed as Probiotics

Lilly and Stillwell (1965) were the first to use the term "probiotics." They defined probiotics as "substances secreted by one microorganism that stimulate the growth of another" (Lilly and Stillwell 1965). Probiotics are also referred to as "direct-fed microbial." The first discovered probiotic was a specific strain of Bacillus found in Bulgarian yogurt, which was named *Lactobacillus bulgaricus.* Bulgarian physician and microbiologist Stamen Grigorov made this discovery in 1905. The modern-day theory of probiotics is attributed to Nobel laureate Élie Metchnikoff, who postulated that the consumption of yogurt by Bulgarian peasants might be responsible for their long and healthy lives. In 1974, Parker proposed that probiotics are "organisms and substances that contribute to gastrointestinal balance." Metchnikoff suggested that probiotics could have an additional function by digesting unusual components, similar to the process in ruminant animals that eat rough, fibrous food. The unexpected effects of kefir led to its consideration as a treatment for various illnesses, as observed in Moscow hospitals at that time. The benefits of consuming fermented foods have been known for centuries, even before the observation of microorganisms. Metchnikoff theorized the concept of administering microorganisms to confer health benefits and observed a delay in senility after consuming bacteria found in yogurt (Mackowiak 2013, Fijan 2014). Parker (1974) considered probiotics as the other half of antibiotics, as certain species of probiotics can suppress the growth of pathogenic microbes (Parker 1974). Probiotics can produce active peptides and antimicrobial structures. Probiotics consist of a mono or mixed culture of live microorganisms and have a beneficial effect on the host by improving its intestinal microbial balance (Afrc 1989). Live probiotic cultures can be found in fermented dairy products, as well as in other types of fermented foods and probiotic-fortified foods such as yogurt (Mattila-Sandholm et al. 2002). According to the Food and Agriculture Organization/World Health Organization (FAO/WHO), probiotics are "live microorganisms that, when administered in adequate amounts, confer a health benefit on the host." The guidelines recommend developing methods (in vitro and in vivo) to evaluate the functionality and safety of probiotics (FAO/WHO 2002).

2.1.4 The Fermentation Process

Ancient documentation has described various types of fermented foods (FFs) and their beneficial effects. These foods include milk, bread, pickles, and wine. Throughout human history, different civilizations and cultures have developed important FF recipes. Some FFs contain remarkable beneficial microorganisms (BMs) that have made a significant impact. Examples of such FFs include kefir, yogurt, and kombucha tea. The mechanisms through which FFs benefit us have been investigated. In addition to their health benefits, FFs transform foods into newly popular products with unique properties such as tanginess, pungency, and aroma. From ancient times to the present day, we have intentionally or unintentionally consumed FFs, such as pickles, dairy products, fermented vegetables and fruits, and more. Consuming only canned foods, preserved foods, and boiled foods without including foods that contain BMs in our daily meals can lead to gastrointestinal troubles and health problems.

Fermentation describes different types of biochemical changes, including yeast fermentation, lactic acid fermentation, butyric acid fermentation, propionic acid fermentation, and acetic acid fermentation. The end products of fermentation often include CO_2 and organic compounds such as ethanol, organic acids, and other molecules.

Industrial fermentation involves the addition of specific starter cultures and the control of conditions such as temperature and pH (Surono 2016). In controlled fermentation, only desired microorganisms are inoculated. Pasteurization and ultra-pasteurization processes are commonly used, particularly with milk products. However, pasteurization also eliminates naturally coexisting microbes. To ensure successful FF, seed culture (SC) containing sufficient inoculum is used. The strains used are usually isolated from the same type of traditional fermented foods (TFFs), although additional beneficial strains may be added or certain strains may be omitted from the FF (Bokulich and Bamforth 2013). Controlled fermentation can also involve multiple steps, each supplied with one or more SCs. Each step may produce a specific product, such as in the production of vinegar, which involves two steps: ethanol production under anaerobic conditions, followed by conversion of ethanol to acetic acid under aerobic conditions. Disrupting the activity of one step may be crucial for the success of the subsequent step.

After the discovery of the health benefits of certain fermented products, such as yogurt and kefir, the idea spread to the rest of the world. Additionally, new fermented products have been designed and introduced to the market. In the early 1930s, Japanese scientist Shirota isolated strains of intestinal bacteria from healthy individuals that could survive and traverse the intestine. He used these strains to create fermented milk and investigated the effects of this milk on patients. The first product, named Shirota (later named L. *casei* Shirota), was released into the market by the Yakult Honsha Company. Many of the byproducts of BMs have yet to be fully investigated. This is likely due to our trust in them and their products, which stems from a long history of open validation throughout human civilization.

Probiotics distinguish themselves from other BMs by their ability to colonize the gastrointestinal system (GIS). They also perform other beneficial activities such as preventing pathogenic microbes from taking hold, fermenting food into a more digestible form, producing valuable health-promoting products, complementing missing genes in our bodies, and aiding in disease control and treatment (Rezac et al. 2018, Sivamaruthi, Kesika, and Chaiyasut 2019, Chaves-López et al. 2020).

2.2 REQUIREMENTS FOR PROBIOTIC STRAINS

Certain strains of *Lactobacilli* and *Bifidobacteria* are naturally present in the healthy human intestinal microflora. These strains have been included in the composition of probiotics and probiotic foods due to their well-established health benefits (Denkova et al. 2017)]. *Lactobacilli* and *Bifidobacteria* play a crucial role in maintaining the balance of the gastrointestinal microflora (Denkova et al. 2017).

2.2.1 Probiotic Common Properties

Probiotics that inhibit the GIS have some common properties:

- The ability to colonize the GIS as natural microflora in humans and animals.
- The ability to adhere themselves to the epithelial cells or cell lines or colonize the ileum (temporarily).
- The ability to tolerate conditions of the stomach and intestines (acidic pH in the stomach and to withstand the effects of bile).
- The ability to reproduce in the gastrointestinal tract.
- Possessing antimicrobial activity against conditionally pathogenic microorganisms, thereby displacing them from their biological niche.
- Modulating the immune response.
- Safe for children.
- Positively influencing the metabolic activity of the consumer.

2.2.2 Examples of Microbes Claimed as Probiotic

2.2.2.1 Bacterial strains

2.2.2.1.1 Bifidobacteria

The family Bifidobacteriaceae, including the genus *Bifidobacterium*, consists of various Gram-positive, non-motile, anaerobic bacteria. Bifidobacteria have a fermentative metabolism and produce lactic acid. They are commonly added as probiotics to food products or used during fermentation processes to provide health benefits. Their primary habitat is the intestinal tract, where their presence reduces the incidence of enteric infections and improves intestinal health. They are now the most commercially important products for stock culture companies. Bifidobacteria are endosymbiotic inhabitants of the gastrointestinal tract and vagina (Chen, Cai, and Feng 2007, Fijan 2014). They are known for their variety of resistance mechanisms against bile salts. Bifidobacteria naturally reside in the human colon and approximately thirty species have been identified, including *B. bifidum, B. adolescentis, B. longum, B. infantis, B. longum subsp. infantis, B. animalis subsp. animalis, B. animalis subsp. lactis, B. pseudolongum subsp. pseudolongum, B. thermophilum*, and *B. breve* (Xiao et al. 2010, Bourdichon et al. 2012, Anupma and Tamang 2020). In a healthy adult gut, they are nearly seven times more abundant than Lactobacilli. Bifidobacteria are found in large numbers in the intestines of babies, acquired through their mother's milk. These beneficial microorganisms prevent the colonization of virulent strains, particularly in the stomach. Studies have shown that Bifidobacteria can modulate the immune system. They can be found in foods such as yogurt, kefir, fermented vegetables like sauerkraut, and drinks like kombucha. For more information about specific microbial strains, please refer to Table 2.1 and the corresponding references.

2.2.2.1.2 Lactic acid bacteria

The genus *Lactobacillus* belongs to the family Lactobacillaceae and is considered the oldest documented probiotic. It comprises various facultative anaerobic or microaerophilic, rod-shaped Gram-positive bacteria. *Lactobacillus* has low cost and low energy requirements for food processing and preparation. Acid

TABLE 2.1
Different bacterial strains claimed as probiotics: their taxonomy, species names and applications

Systemetic classification	Most important species	Applications	References
Phylum: *Actinobacteria* Class: *Actinobacteria* *Order:* Bifidobacteriales *Family:* Bifidobacteriaceae *Genus: Bifidobacterium*	*B. bifidum, B. pseudologum* subsp *pseudologum, B. animalis, B. animalis* subsp *animalis, B. longum* subsp *infantis, B. infantis, B. theromophilum, B. adolescentis, B. breve, B. adolescentis, B. animalis* subsp *lactis.*	Dairy products and *B. breve* is used in addition to fermenting soy products	(Xiao et al. 2010, Tamang, Holzapfel, and Watanabe 2016, Voidarou et al. 2021a)
Phylum: Actinobacteria Class: Actinobacteria Order: Actinomycetales Suborder: Streptomycineae Family: Streptomycetaceae Genus: *Streptomyces*	*S. griseussubspgriseus*	Fermented meat products	(Neffe-Skoci´nsk, Wǒjciak, and Zielinska 2016, Voidarou et al. 2021a)
Phylum: Actinobacteria Class: Actinobacteria Subclass: Actinobacteridae VP Order: Actinomycetales Suborder: Propionibacterineae Family: Propionibacteriaceae AL Genus: *Propionibacterium*	*P. freudenreichii* subsp *shemani, P. freudenreichii* subsp *germanii, P. freudenreichii* subsp *freudenreichii, P. freudenreichii* subsp *globosum, P. thoeni, P. jensenii, P. acidipropionici.*	Dairy products and add to the buttermilk as a secondary cultured microflora	(Giraffa 2004, Voidarou et al. 2021a)
Phylum: Actinobacteria Class: Actinobacteria Subclass: Actinobacteridae VP Order: Actinomycetales Suborder: Corynebacterineae Family: Corynebacteriaceae AL Genus: *Corynebacterium*	*C. variabile, C. manihot, C. variabile, C. flavescens, C. casei, C. ammoniagenes, C. manihot*	Dairy products and *C. manihot* dominate the first stage of fermented cassava production (a traditional staple food)	(Bockelmann et al. 2005, Voidarou et al. 2021a)
Phylum: Actinobacteria Class: Actinobacteria Subclass: Actinobacteridae VP Order: Actinomycetales Suborder: Micrococcineae VP Family: Micrococcaceae AL Genus: *Micrococcus*	*M. lylae, M. luteus*	Cheese ripening and additionally could be used in the meat fermentation	(Bonnarme et al. 2001, O'Mahony et al. 2001, Voidarou et al. 2021a)
Phylum: Actinobacteria Class: Actinobacteria Subclass: Actinobacteridae VP Order: Actinomycetales Suborder: Micrococcineae VP Family: Micrococcaceae Genus: *Kocuria*	*K. varians, K. rhizophila,*	Traditional dairy and meat fermentation and as bioactive agent to suppress the growth of pathogens in chilled dairy products	(O'Mahony et al. 2001, Voidarou et al. 2021a)
Phylum: Actinobacteria Class: Actinobacteria Subclass: Actinobacteridae VP Order: Actinomycetales Suborder: Micrococcineae VP Family: Micrococcaceae AL Genus: *Arthrobacter*	*A. bergerei, A. arilatensis, A. ricotianae, A. globiformis*	Cheese production	(Haruta et al. 2006, Tamang, Holzapfel, and Watanabe 2016, Voidarou et al. 2021a)

(*Continued*)

TABLE 2.1 (CONTINUED)
Different bacterial strains claimed as probiotics: their taxonomy, species names and applications

Systemetic classification	Most important species	Applications	References
Phylum: Proteobacteria Class: Alphaproteobacteria Order: Rhodospirillales Family: Acetobacteraceae Genus: *Acetobacter*	*A. aceti* subsp*aceti*, *A. malorum*, *A. pasteurianus* (subsp*pasteurianus*), *A. oryzae*, *A. syzgii*, *A. fabarum*, *A. pomorum*, *A. orientalis*, *A. lovaniesis*, *A. tropicalis*, *A. polyxygenes*, *A. xylinum*,	Can be isolated from fresh vinegar but also can be isolated from some other fermented products such as fermented cocoa, coffee and tea	(Haruta et al. 2006, Sengun 2011, Voidarou et al. 2021a)
Phylum: Proteobacteria Class: Alphaproteobacteria Order: Rhodospirillales Family: Acetobacteraceae Genus: *Gluconacetobacter*	*G. azotocaptans*, *G. johannae*, *G. entanii*, *G. hansenii*, *G. diazotrophicus*, *G. oboediens*, *G. oxydans*, *G. europaeus*, *G. xylinus*		(Voidarou et al. 2021a)
Phylum: Proteobacteria Class: Gammaproteobacteria Order: Oceanospirillales Family: Halomonadaceae VP Genus: *Halomonas*	*H. elongate, Halomonas sp*	Meat fermentation, anticancer potential	(Hinrichsen, Montel, and Talon 1994, Tamang et al. 2015, El-Garawani et al. 2020, Voidarou et al. 2021a)
Phylum: Proteobacteria Class: Gammaproteobacteria Order: EnterobacterialesNP Family: Enterobacteriaceae Genus: *Hafnia*	*H. alvei*	Cocoa bean fermentation	(Papalexandratou et al. 2011, Voidarou et al. 2021a)
Phylum: Proteobacteria Class: Gammaproteobacteria Order: EnterobacterialesNP Family: Enterobacteriaceae Genus: *Tatumella*	*T. ptyseos*, *T. citrea*	Cocoa bean fermentation	(Voidarou et al. 2021a)
Phylum: Proteobacteria Class: Gammaproteobacteria Order: EnterobacterialesNP Family: Enterobacteriaceae AL Genus: *Escherichia*	*E. coli* Nissle 1917	Treat constipation (Chmielewska 2010) and inflammatory bowel disease (Behnsen et al. 2013). This strain could also relieve gastrointestinal disorder, ulcerative colitis, Crohn's disease, even colon cancer	(Plessas et al. 2011, Xia, Zhu, and Zhu 2013, Voidarou et al. 2021b)
Phylum: Proteobacteria NP 59 Class: Alphaproteobacteria Order: Sphingomonadales NP Family: Sphingomonadaceae Genus: *Zymomonas*	*Z. mobilis subspmobilis*	In alcoholic beverages particularly in different tropical areas of America, Africa and Asia	(Voidarou et al. 2021a)
Phylum: Proteobacteria Class: Gammaproteobacteria Order: Enterobacteriales Family: Enterobacteriaceae Genus: *Tatumella*	*T. citrea*, *T. ptyseos*	Cocoa bean fermentation	(Voidarou et al. 2021a)

(*Continued*)

TABLE 2.1 (CONTINUED)
Different bacterial strains claimed as probiotics: their taxonomy, species names and applications

Systemetic classification	Most important species	Applications	References
Phylum: Firmicutes Class: Bacilli Order: Lactobacillales Family: Leuconostocaceae" *Genus: Leuconostoc*	*L. lactis*, *L. mesenteroides* subsp. *mesenteroides*, *L. inhae*, *L. mesenteroides* subsp. *cremoris*,*L. pseudomesenteroides*, *L. fallax*, *L. mesenteroides* subsp. *dextranicum*, *L. holzapfelii*, *L. kimchi*, *L. citreum*, *L. carnosumm*, *L. palmae*, *L. gasicomitatum*, *L. gelidum*,	All fields of the fermented food industry	(Collins et al. 1993, Lee et al. 2011, Plessas et al. 2011, VidyaLaxme et al. 2014, de Paula et al. 2015, Liu et al. 2017, Voidarou et al. 2021b)
Phylum: Firmicutes Class: Bacilli Order: Lactobacillales Family: Leuconostocaceae Genus: *Oenococcus*	*Oenococcusoeni* (was known as *Leuconostoc oeni* until 1995)	Fermented grapes and apple juices, It can also be found as a detected bacterium during malolactic fermentation	(Voidarou et al. 2021a)
Phylum: Firmicutes Class: Bacilli Order: Bacillales Family: Staphylococcaceae Genus: *Staphylococcus* (Coagulase negative Staphylococci, CoNS)	*S. carnosus* subsp, *S. carnosus* subsp. *utilis* *S. sciuri* subsp. *sciuri*, *S. equorum* subsp. *equorum*, *S. carnosus*, *S. condimenti*, *S. succinus* subsp. *casei*, *S. cohnii*, *S. linens*, *S. fleuretiii*, *S. saprophyticus*, *S. piscifermentans*, *S. succinus* subsp. *succinus*, *S. epidermidis*, *S. kloosii*, *S. gallinarum*, *S. warneri*, *S. vitulinus*, *S. xylosus*	Part of the natural microbiota of fermented meat products as well as of hard and of soft cheeses and of fermented vegetables	(Plessas et al. 2011, Voidarou et al. 2021b)
Phylum: Firmicutes Class: Bacilli Order: Lactobacillales Family: Streptococcaceae Genus: *Lactococcus*	*Lactis macedonicus* subsp., *L. lactis* subsp subsp. *lactis*. *L. raffinolactis*	Yoghurt production and in some other dairy products as starter culture	
Phylum: Firmicutes Class Bacilli Order: Lactobacillales Family: Streptococcaceae AL Genus: *Streptococcus*	*thermophilus*, *S. gallolyticus* subsp. *Macedonicus*, *S. salivarius* subsp, *Streptococcus salivarius* NU10, *Streptococcus salivarius* 5M6c, *S. salivarius* subsp. *salivarius*,	Milk fermentation, cheeses production, production of Salivaricin 9, Salivaricin D Lantibiotic	(Corr et al. 2007, Birri, Brede, and Nes 2012, Barbour, Philip, and Muniandy 2013, Voidarou et al. 2021b)

(Continued)

TABLE 2.1 (CONTINUED)
Different bacterial strains claimed as probiotics: their taxonomy, species names and applications

Systemetic classification	Most important species	Applications	References
Phylum: Firmicutes Class: Bacilli Order: Lactobacillales Family: Lactobacillaceae AL Genus: *Lactobacillus*	Dairy product *L. lactis, L. salivarius, L. sakei, L. helveticus, L. nodensis, L. fermentum, L. plantarum/ paraplantarum, L. casei, L. paracasei, L. pentosus,* *L. salivarius, L. buchneri,L. rhamnosus, L. bulgaricus,L. curvatus, L. kefiri, L. kefiranofaciens, L. brevis, L. sakei, L. acidophilus, L. acidipiscis, L. reuteri, L. johnsonii, L. parakefiri, L. crispatus,* *L. delbrueckii, L. gasseri, L. coryniformis* subsp. *coryniformis* Fermented vegetables—fruits, *L. acidifarinae, L. acetotolerans, L. alimentarius,* *L. amolylovorus, L. cacaonum, L. buchneri, L. collinoides, L. farciminis, L. frumenti, L. kisonensis, L. diolvorans, L. rapi, L. vaccinostercus, L. portis, L. mali, L. spicheri, L. surkii, L. frumenti, L. crustorum,* *L. curvatus, L. kimchii, L. mucosae, L. nagelii, L. sakei* subsp. *carnosus, L. tuceti, L. similis*	Fermented dairy products Fermented vegetables—fruits, meat—fish and cocoa, sourdough, cereals, beverages	(Tamang et al. 2015) (Bourdichon et al. , Bourdichon et al. 2012, Tamang et al. 2015, Abdolalipour et al. 2020, Anupma and Tamang 2020, Assimos 2020, Babashahi et al. 2020, Dong et al. 2020, Hassan et al. 2020, Simoes da Silva et al. 2020) (Bourdichon 2012, Tamang et al. 2015, Tamang, Holzapfel, and Watanabe 2016) (Bourdichon et al. , Bourdichon et al. 2012, Tamang et al. 2015, Anupma and Tamang 2020)
Phylum: Firmicutes Class: Bacilli Order: Bacillales AL Family: Bacillaceae AL Genus: *Bacillus*	*B. coagulans, B. subtilis, B. subtilis var natto,* *B. amyloliquefaciens, B. licheniformism,* *B. circulans, B. firmus, B. megaterium,* *B. pumilus, B. thermoamylovorans,* *B. thuringiensis, B. macrerans* *B. coagulans, B. cereus. B. cereus* NVH 75/95, *B. cereus* NVH 75/95, *B. subtilis, B. coagulans, B. amyloliquefaciens, B. subtilis* var natto, *B. licheniformism, B. firmus, B. circulans, B. megaterium, B. thermoamylovorans, B. pumilus, B. macrerans, B. thuringiensis, B. subtilis* Marburg 168	Fermentation of cocoa, soy, "Pidan" (preserved alkali-treated fresh duck eggs, fish products, China) and industrial producers of γ—PGA in fermented soybean foods, respectively. B. cereus NVH 75/95 has also proven to be an efficient probiotic for animals. B. cereus NVH 75/95 has also proven to be an efficient probiotic for animals.	(Kubo et al. 2011, Tamang, Holzapfel, and Watanabe 2016, Tamang et al. 2016, Voidarou et al. 2021a) (Doron, Hibberd, and Gorbach 2008, Hempel et al. 2012) (Trapecar et al. 2011).

(Continued)

TABLE 2.1 (CONTINUED)
Different bacterial strains claimed as probiotics: their taxonomy, species names and applications

Systemetic classification	Most important species	Applications	References
Phylum: Firmicutes Class: Bacilli Order: Bacillales Family: Staphylococcaceae Genus: *Gemella*		Legume undergoing alkaline fermentation	(Diaz et al. 2019, Voidarou et al. 2021a)
Phylum: Firmicutes Class: Bacilli Order: Bacillales Family: Staphylococcaceae Genus: *Jeotgalicoccus*		Legume undergoing alkaline fermentation	(Diaz et al. 2019, Voidarou et al. 2021a)
Phylum: Firmicutes Class : Bacilli Order : Lactobacillales Family :Carnobacteriaceae Genus : *Carnobacterium*	*C. divergens*, *C. piscicola*, *C. maltaromanticum*.	fish fermentation and can be used in some cases in the in dairy products	(Tamang et al. 2015, Tamang, Holzapfel, and Watanabe 2016, Voidarou et al. 2021a)
Phylum: Firmicutes Class : Bacilli Order : Lactobacillales Family IV: Enterococcaceae Genus: *Enterococcus*	*E. faecalis*, *E. durans*, *E. cloacae*, *E. faecium*,	Fermentations of meat, milk, vegetables and cereals	(Foulquié Moreno et al. 2006, Voidarou et al. 2021a)
Phylum: "Firmicutes Class :II. "Bacilli Order : Lactobacillales Family: Enterococcaceae Genus: *Tetragenococcus*	*T. koreensis*, *T. halopilus*	Vegetable fermentation soybean paste	(Tamang et al. 2015, Anupma and Tamang 2020, Voidarou et al. 2021a) (Jeong, Heo, and Lee 2017)
Phylum: Firmicutes Class :. Bacilli Order: Lactobacillales Family: Enterococcaceae Genus: *Vagococcus*	*Vagococcus sp.*	Traditional fermented soybean food in India	(Voidarou et al. 2021a)
Phylum: Firmicutes Class : Bacilli Order : Lactobacillales Family : Leuconostocaceae Genus: *Weissella*	*W. confusa*, *W. beninensis*, *W. ghanensis*, *W. hellenica*, *W. fabaria*, *W. koreensis*, *W. paramesenteroides*, *W. ciburia*, *W. viridencens*, *W. thailandensis*	Fermented vegetables, meat, fish and cocoa. Some strains have remarkable antimicrobial activity against Gram-negative and Gram-positive pathogens	(Collins et al. 1993, Voidarou et al. 2021a)

fermentation combined with food salting remains the most common practical method of preservation. For example, it is used for preserving raw vegetables, cereal gruels, and milk-cereal mixtures. There are approximately 183 known species of these rod-shaped bacteria, and they are widely applied in different industrial food processes (König and Fröhlich 2009). They can prevent food spoilage and improve the nutritive value of the foods they inhabit. Several *Lactobacillus* species are commonly selected as probiotics due to their crucial properties, including high tolerance to acid and bile, capability to adhere to intestinal surfaces, and resistance to the low pH of gastric juice. These characteristics enable them to inhibit potentially pathogenic species through antimicrobial activity. *Lactobacillus rhamnosus* CRL1505, for example, has been shown to reduce viral-associated pulmonary damage by controlling immune-coagulative responses and clearing respiratory viruses (Zelaya et al. 2014). *Lactobacillus* strains are also safe and effective in preventing recurrent urinary tract infections in adult women (Grin et al. 2013). The complete genome sequences of some *Lactobacillus* species, such as *L. bulgaricus*, *Lactobacillus plantarum*, *L. johnsonii*, *L. acidophilus*, *L. sakei*, *L. salivarius*, have been published (Makarova et al. 2006, Fijan 2014). Certain *Lactobacillus* species are utilized in the production of yogurt, cheese, sauerkraut, pickles, sourdough, wine, and other fermented products (Bauer, du Toit, and Kossmann 2010, De Vuyst et al. 2014, Fijan 2014, Siragusa et al. 2014). In all cases, sugars are metabolized into lactic acid, creating a hostile environment for spoilage microorganisms and enabling food preservation. Commonly used *Lactobacillus* species in dairy products are listed in Table 2.1. Other fermented products that contain lactic acid bacteria (LAB) capable of fermenting vegetables include pickled vegetables (Breidt et al. 2013), kimchi (Oh, Oh, and Kim 2004, Breidt et al. 2013), pao cai (Pederson, Niketic, and Albury 1962), sauerkraut (Sybesma et al. 2006), tempeh (Moreno et al. 2002), miso, soy sauce (Tanasupawat et al. 2002), and non-dairy products such as bee pollen (Mohammad, Mahmud-Ab-Rashid, and Zawawi 2020). Some strains of LAB that ferment vegetables are included in Table 2.1.

Most *Lactobacilli* do not form stable and numerically significant populations in the human intestinal tract, especially in the small intestine where they are presumed to form epithelial associations (Walter 2008). Similarly, in the research conducted to modulate intestinal *bifidobacteria* and *lactobacilli*, after one-week supplementation with commercial probiotic food supplements, only a significant enlargement in the intestinal *Lactobacillus acidophilus* group was observed and even this enlargement disappeared after a few days (Fijan 2014, Taverniti et al. 2014). *Lactobacilli* are often described as indigenous inhabitants of the human intestinal tract (Walter 2008, Fijan 2014, Taverniti et al. 2014). The antimicrobial effect of LAB can be caused by some factors, such as producing lactic acid, aldehydes, bacteriocin antibiotics, and other compounds. Also, they are able to reduce the pH due to the production of lactic acid and acetic acid, hydrogen peroxide. *L. bulgaricus* strains were reported to have a 100% antimicrobial effect against foodborne pathogens and spoilage bacteria. All *L. bulgaricus* strains were reported to have antimicrobial activity against *Campylobacter jejuni* NCTC 11351 (Yerlikaya, Saygili, and Akpinar 2021).

The addition of viable cells of mesophilic *Lactococcus lactis*, some *Lactobacillus* species, and *Pediococcus* species in high numbers can control spoilage and pathogenic bacteria during the refrigerated storage of food at or below 5 °C (Ray 1992). In the existence of mesophilic LABs, the growth of psychotropic spoilage and pathogenic bacteria is reported to be controlled. Studies were conducted by adding LABs to fresh meat, seafood, liquid egg, and some processed meat products, such as bacon against *Clostridium botulinum*, *Salmonella* serovars, and *Staphylococcus aureus*. For more microbial strains refer to strains in Table 2.1 and the references within.

2.2.2.2 Fungi

The presence of mold growth in food is a sign of spoilage, and it is generally discarded by consumers. It is uncommon for laypeople to consume food with visible mold or fungus. However, there are certain foods in which visible fungal mycelium is intentionally part of the product. Numerous fungal genera are involved in the production of fermented foods. For more details, please refer to (McGovern 2007, Anupma and Tamang 2020, Ashaolu and Reale 2020, McGovern 2020) and the references cited within.

2.2.2.2.1 Yeasts (time-limited probiotic-like microbes)

The beneficial yeast includes hundreds of species, with the most famous being *S. cerevisiae*. It is used in the production of bread, beer, and wine. Yeast is typically employed for fermentation and leavening, utilizing sugars to produce alcohol (ethanol) and carbon dioxide in beer and wine production. Yeast can ferment its weight in glucose per hour. It is rich in protein and has long been considered a single-cell protein. *Saccharomyces bayanus* is used for winemaking, while *S. boulardii* is utilized in medicine as a probiotic. *S. boulardii* is often marketed in a lyophilized form to treat diarrhea. It has a well-established reputation for safety and can reduce the duration of diarrhea and hospital stays, regardless of the cause (Szajewska and Mrukowicz 2005, Goldenberg et al. 2015, Dinleyici et al. 2012, Vieira, Teixeira, and Martins 2012, Shan et al. 2013). *S. boulardii* administration has shown positive effects in patients with irritable bowel syndrome (Choi et al. 2011). It can also prevent and treat relapses of inflammatory bowel disease and be used to treat moderate symptoms of ulcerative colitis (Guslandi et al. 2000). Daily supplementation of *S. boulardii* together with standard antibiotics can reduce recurrent pseudomembranous colitis infection caused by *Clostridium difficile*. Safety reports of probiotic use during pregnancy have not reported any malformations (Dugoua et al. 2009).

Saccharomyces yeast also forms symbiotic relationships with bacteria in kefir (Witthuhn, Schoeman, and Britz 2005) and is a component of kombucha (Marsh et al. 2014). Yeast can be found and isolated from various environments, particularly those rich in sugars.

Yeast is rich in many vitamins, including B1, B2, niacin, and folic acid. Brewer's yeast can be consumed as a vitamin supplement. There are numerous fermented foods that contain yeast, and yeast cells and their metabolic products are known to be safe and widely used in different forms of food processing and preservation worldwide, particularly in baking and brewing products (Tamang et al. 2015, Anupma and Tamang 2020).

Currently, yeast and yeast products are integral to many important market products. They are utilized in single-cell protein production, foods, beverages, pharmaceutical products, and industrial enzymes. The applications of yeast products extend beyond the sectors described.Top of Form.

2.2.2.2.2 Filamentous fungi

Some species of filamentous fungi are utilized for the production of essential biotechnological products. These products include alcohol, starter dry cultures, and enzymes used in milk and cheese fermentation. Fungal fermented foods and products are common in Asia and Europe (Tamang and Samuel 2010, Bourdichon et al. 2012). Filamentous fungi have been traditionally used as starter cultures in Asia, serving various functions such as saccharification, liquefaction, and ethanol production. They are employed in the production of low-alcohol beverages and high-alcohol distilled liquors (Thapa and Tamang 2004, Tamang and Samuel 2010, Chen, Wu, and Xu 2014, Lee, Billmyre, et al. 2014, Suesse, Norton, and van Leeuwen 2016). In Europe, filamentous fungi are used in the development of different dairy products, particularly in the ripening processes of various types of cheese (e.g., Roquefort, Camembert), and for enzyme production (Spinnler and Gripon 2004, Chavez et al. 2011, Anupma and Tamang 2020). The genus *Aspergillus* includes important species such as *Aspergillus acidus, A. oryzae, A. niger, A. sojae, A. sydowii, A. versicolor,* and *A. flavus*. These species are used in the traditional production of fermented foods such as soy sauce, miso, sake, awamori liquors, Puerh tea, and more (Nout 2000, 2007, Mogensen et al. 2009, Tamang and Samuel 2010).

2.3 FERMENTED BYPRODUCTS

2.3.1 Fermented Food Extracts

Extracts derived from fermented byproducts of tomatoes, melons, and carrots have been found to possess biopreservative activity on food (Ricci et al. 2021). Kefir fermented beverages and Kombucha fermented beverages exhibit different chemical compositions and antimicrobial activities (Al-Mohammadi, Ibrahim, et al. 2021; Al-Mohammadi, Ismaiel, et al. 2021). Fermented Maillard reaction products, a novel milk-derived material made from whey protein, have also been studied and reported for their antimicrobial properties (Kim et al. 2021). Additionally, fermented white cheeses have been investigated for their antimicrobial activities (Toğay et al. 2016), among other fermented products.

Other microbial byproduct could be used in food preservation or as biopreservatives (Ricci et al. 2021).

2.3.2 Viable Cells or Cell Extracts

In many biotechnological processes, viable cells or their extracts are byproducts. The use of viable cells in biological control and in food preservation is an old technique. In fact, food fermentation is a process where viable cells preserve food and change it to an acceptable form at the same time. Probiotic byproducts increase the shelf life of the products (Voidarou et al. 2021b). *Lactobacillus plantarum* PM411 viable cells are detected on aerial plant surfaces. Monitoring new *Bacillus* species on plant surfaces could be applied in biological control applications and the sustainable management of crop production, as well as a candidate for food preservatives (Daranas et al. 2018). Viable cells and cell-free extracts of *Lactobacillus casei* were investigated and applied to elderberry juice, proving a significant increase in anthocyanin content, suggesting the possible use of *Lactobacillus* for increasing specific molecules, even for non-multiplying bacterial cells (Ricci et al. 2018). The process involves the addition of viable cells of mesophilic *Lactococcus lactis*, some *Lactobacillus* species, and *Pediococcus* species in high numbers to control spoilage and pathogenic bacteria during the refrigerated storage of food at or below 5 °C (Ray 1992).

2.3.3 Acids

Microbes can produce acids as antimicrobial agents (Figure 2.3). The most common acids produced as byproducts of fermentation processes include *lactic acid* (Carr, Chill, and Maida 2002; Leroy and De Vuyst 2004), *acetic acid* (Papalexandratou et al. 2011; Sengun and Karabiyikli 2011), and *propionic acid* (Thierry et al. 2015). Lipophilic acids such as lactic acid and acetic acid, in undissociated form, can penetrate the microbial cell membrane and, at higher intracellular pH, dissociate to produce hydrogen ions. These ions can interfere with microbial metabolite activity, such as substrate translocation and oxidative phosphorylation (Naidu, Bidlack, and Clemens 1999). Lactic acid permeabilizes Gram-negative bacteria by disrupting the outer membrane (Alakomi et al. 2000; Suskovich et al. 2010; Papalexandratou et al. 2011; Sengun and Karabiyikli 2011). The combined action of *Streptococcus* (Yerlikaya, Saygili, and Akpinar 2021) and *Leuconostoc* (Collins et al. 1993; Lee et al. 2011; de Paula et al. 2015) utilizes milk sugar (and other types of food sugars) to produce lactic acid, which precipitates the protein in milk (casein), as in the production of buttermilk and sour cream. During the fermentation of cool cabbage, lactic acid is naturally produced (Song et al. 2019). The acidity of lactic acid eradicates pathogenic bacteria and can keep the cabbage preserved for more than one year in the refrigerator. Commercial acids, such as lactic acid, are produced by some *Lactobacillus* species, such as *L. sakei* and *L. curvatus* (Zhang, Zhu, et al. 2018). Acetic acid is produced by *Acetobacter aceti*, and *propionic acid* is produced by dairy *Propionibacterium* species, such as *Propionibacterium freudenreichii*, *P. jensenii*, and *P. acidipropionici* (Woskow and Glatz 1991; Brede et al. 2007; Liu et al. 2015; Liu et al. 2016; Piwowarek et al. 2018). They are generally regarded as safe (GRAS). Acetic acid is known to have more antimicrobial activity than lactic acid. Acids produced from certain fermentation processes are preferred for their aroma (Bahuguna et al. 2020). Besides being used as antimicrobial agents, they are used as additives to enhance flavor and shelf life (Baird-Parker 1980; Ray and Sandine 1992; Geremew, Kebede, and Andualem 2015; Agriopoulou et al. 2020). LABs (lactic acid bacteria) are used as antibacterial agents to extend the shelf life of fresh and minimally processed fruits and vegetables (Agriopoulou et al. 2020). Vinegar is produced by a group of bacteria called *Acetobacter*, which converts alcohol into acetic acid or vinegar. Examples of acetic acid fermentation include apple cider vinegar, wine vinegar, coconut water vinegar, and African palm vinegar. Fermentation can be carried out by naturally coexisting microbes (de Castro et al.2015; Lee et al. 2017; Jia, Yu, and Zhang 2020; Luzón-Quintana, Castro, and Durán-Guerrero 2021; O'Reilly et al. 2021). Initially, yeast produces alcohol, which is then suppressed at certain levels due to the cessation of enzymatic activities, leading to the death of the yeast and cessation of alcoholic fermentation. By creating aerobic conditions, the alcohol is further metabolized to produce acetic acid. The vinegar is then subjected to heat to deactivate any remaining microbes (some strains may convert acetic acid further into CO_2). Vinegar is considered a fermented food product but not a source of probiotics, particularly due to the final microbial deactivation process.

P. freudenreichii subsp. showed a considerable inhibitory effect against spoilage bacteria in cheese-based dips. It also exhibited a notable inhibitory effect against *P. aeruginosa*, but only slight inhibition against *B. cereus*. Organic acids such as lactic, formic, acetic, propionic, benzoic, and phenyl lactic acids are produced, known to play a critical role in inhibiting pathogenic bacteria (Tharmaraj and Shah 2009).

2.3.4 Ethanol and Carbon Dioxide

Carbon dioxide (CO_2) has antimicrobial activity (Shah and Prajapati 2014; Blatchford et al. 2019). Fermentation processes typically involve the production of CO_2 along with one or more organic products such as ethanol, organic acids, and other organic molecules. Microbes utilize sugars to produce alcohol and CO_2. In commercial bread production, CO_2 is generated by the action of baker's yeast and other microbes during the fermentation process (Minervini et al. 2012; Katsi et al. 2021). However, certain types of bread processes, such as salt uprising and some rye bread, do not involve the use of yeast (Koistinen et al. 2018). Baker's yeast is composed of living cells of *Saccharomyces cerevisiae*. Typically, yeast is added to the dough at a level of around 2 percent of its weight. Baker's yeast ferments sugars such as glucose, fructose, maltose, and sucrose, but cannot utilize lactose (the predominant sugar in milk) and some other carbohydrates. The main products of fermentation are CO2, which acts as a leavening agent (Katsi et al. 2021). Some microbes, such as *Zymomonas mobilis*, can be used as alternatives to yeast in certain applications (Musatti et al. 2020). Ethanol, a byproduct of fermentation, is important for the aroma of freshly baked bread. Ethanol is also a well-known antimicrobial agent, capable of killing the microbes that produce it. Therefore, its production ceases at a critical point where no further enzymatic reactions occur. Ethanol also contributes to flavor development and changes the physical properties of the dough. Carbon dioxide is also produced by heterofermentative lactic acid bacteria (LABs) and exhibits antimicrobial activity against a wide range of microorganisms. It is a major end product of hexose fermentation (Voidarou et al. 2021b). LABs are capable of CO_2 production from malate and citrate. They can also metabolize arginine via the arginine deaminase pathway, leading to the formation of CO_2. Furthermore, decarboxylation of amino acids such as histidine and tyrosine can also result in CO2 formation (Naidu, Bidlack, and Clemens 1999). The antimicrobial activity of CO_2 can be attributed to its ability to create an anaerobic environment by displacing molecular oxygen. Additionally, CO_2 can lower the pH and have destructive effects on cell membranes (Naidu, Bidlack, and Clemens 1999).

2.3.5 Bacteriocin

A bacteriocin is a ribosomally synthesized peptide originally defined as a proteinaceous compound that affects the growth or viability of closely related organisms. Bacteriocins are similar to antibiotics but are composed of amino acids and/or modified amino acids. They come in various types (O'Shea et al. 2013; Furtado et al. 2015; Shyntum et al. 2019). Bacteriocins are applied in the control of foodborne pathogenic and spoilage bacteria. For more details, refer to (Mokoena 2017; Dicks et al. 2018; Nazari and Smith 2020; Pogány Simonová, Chrastinová, and Lauková 2020) and the references within. They are widely used in food preservation and are considered safe products for preserving food (Gautam and Sharma 2009). Bacteriocins possess unique properties that make them suitable as preservatives in food. These properties include

1. Heat stability: For example, Vijayendra et al. (2010) reported that bacteriocins did not lose their anti-listerial activity when treated at 100°C for 30 minutes or at 121°C for 15 minutes (Vijayendra, Rajashree, and Halami 2010). However, bacteriocins can lose their antimicrobial activity when treated with trypsin, proteinase-K, proteases, and peptidase (Vijayendra, Rajashree, and Halami 2010).
2. Ability to maintain activity in different environments: Bacteriocins have a wide pH tolerance and are thermostable, allowing them to remain active in various conditions.
3. Producer strains usually have proteolytic activity (Gautam and Sharma 2009).

For more details about bacteriocin refer to the following references (Mokoena 2017, Dicks et al. 2018, Nazari and Smith 2020, Pogány Simonová, Chrastinová, and Lauková 2020).

Due to its thermostability and pH tolerance, bacteriocin can withstand the heat and acidity/alkalinity of food during storage conditions (Gautam and Sharma 2009). Bacteriocin produced by *Enterococcus faecalis* has shown a broad spectrum of antibacterial activity against important food-borne pathogens such as *L. monocytogenes*, Salmonella, and Vibrio strains (Perumal et al. 2016). The use of bacteriocin in the food industry was reviewed by Bali et al. 2016. Different microbes that produce bacteriocins are involved in food production either as probiotics or through the use of their purified bacteriocins. This includes

1. Dairy industry: For example, the bacteriocin Nisin (Figure 2.4) and Pediocin are used in the preservation of raw milk from dairy farms (Kiliç et al. 1996, Cosentino et al. 2012, M and Sh 2018, Campanero et al. 2020, Kumar et al. 2020).
2. Bakery products and ingredients.
3. Alcoholic beverages: such as Boza, a Bulgarian cereal-based beverage (Valledor et al. 2020), M-Tuba and Tepache, two traditional fermented beverages in México (de la Fuente-Salcido et al. 2015).
4. Shrimp preservation (Iyapparaj et al. 2013, Feliatra et al. 2018).
5. Meat Products: such as calabresa, a fermented meat product (Castilho et al. 2019), Slovak fermented meat products (Lauková, Kandričáková, and Bino 2020), ready-to-eat meats (Vijayakumar and Muriana 2017), artisanal Tunisian meat "dried ossban" (Zommiti et al. 2018), and the like.
6. Protection of fresh fruit (Tenea and Suárez 2020), vegetables (Gautam and Sharma 2015), and sauces.
7. Bacteriocin can be used as alternative to disinfectant (Kim et al. 2020).
8. Aquaculture (Desriac et al. 2010).
9. Baby foods.
10. Livestock (Iseppi, Lauková, and Sabia 2021).
11. In clinical applications (Cirkovic et al. 2016, Cebrián et al. 2019, Velázquez-Suárez et al. 2021).
12. Other applications such as honeybee bee bread (Audisio, Terzolo, and Apella 2005, Negash and Tsehai 2020, Zimina et al. 2020).

Experimental design was used to optimize the production of bacteriocin (Kanmani et al. 2011, 2012). One isolate of *Lactobacillus plantarum* (1QB77) was chosen, considering its ability to produce a bacteriocin-like antimicrobial (Margalho et al. 2021).

The gastrointestinal tract contains different antimicrobial proteins such as colicins (Toshima et al. 2007, Trivedi, Jena, and Seshadri 2014, Bosák et al. 2018), defensins (Kohlgraf et al. 2010), cecropins (Zhai et al. 2019), and magainins (Zhang, Shan, et al. 2018). LABs bacteriocin (Naidu,

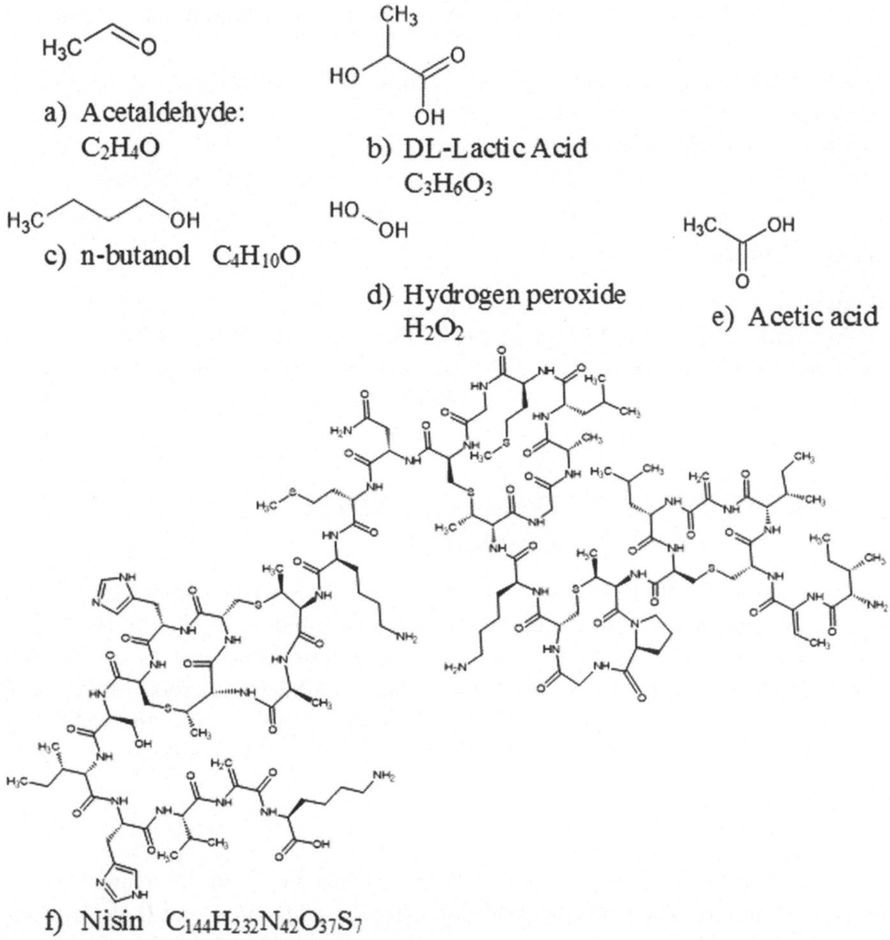

FIGURE 2.4 Some selected chemical structures for active compounds with antimicrobial activity used by probiotic and beneficial microbes: a) Acetaldehyde: C_2H_4O; b) DL-Lactic Acid $C_3H_6O_3$; c) n-butanol $C_4H_{10}O$; d) Hydrogen peroxide H_2O_2; e) Acetic acid; f) Nisin $C_{144}H_{232}N_{42}O_{37}S_7$. (Prepared in this study)

Bidlack, and Clemens 1999) are small antimicrobial peptides and are proteins in their nature that exhibit antimicrobial activity against the closely related Gram-positive bacteria, whereas producer cells are immune to their own bacteriocin (Suskovich et al. 2010). LABs bacteriocin inhibits Gram-negative bacteria (Yusuf and Thata 2013).

2.3.6 Hydrogen Peroxide (H_2O_2)

Some microbes produce H_2O_2 under aerobic conditions of growth. They release it into the environment to protect themselves (Daeschel 1989). Recently, H_2O_2 has been used in the evacuation of microbial cells and viruses (Amara, Salem-Bekhit, and Alanazim 2013; Amara 2015; Amara, Salem-Bekhit, and Alanazi 2015; Amara 2017; Menisy et al. 2017). H_2O_2 has been shown to have antimicrobial activity and the ability to degrade genomic elements in bacteria, fungi, yeast, and viruses, and it is utilized as part of the microbial evacuation process following the Sponge-Like-Protocol (Amara, Salem-Bekhit, and Alanazim 2013; Amara, Salem-Bekhit, and Alanazi 2015; Amara 2017). However, it can also cause mutations in unwanted genes, such as those responsible for exopolysaccharide production in *P. aeruginosa* (Hussain and Amara 2006; Amara 2011; Amara et al. 2011). *H2O2* is a strong oxidizing agent and can be used as an antimicrobial against bacteria, fungi, and viruses, including bacteriophages).

It is capable of degrading both RNA and DNA and can affect other macromolecules, as demonstrated by several studies on microbial cell and virus evacuation using the Sponge-Like-Protocol (Amara, Salem-Bekhit, and Alanazim 2013; Amara 2015). H_2O_2 exhibits antimicrobial activity against yeast, Gram-positive, and Gram-negative bacteria (Suskovich et al. 2010).

2.3.7 Other Antimicrobial Agents

There are many other antimicrobial byproducts that are produced by BMs and probiotics, such as acetaldehyde (Naidu, Bidlack, and Clemens 1999; Nieminen et al. 2014), cyclic dipeptides (Li et al. 2011; Liu et al. 2017; Jinendiran et al. 2020), and de-conjugated bile acids (Oelschlaeger 2010). These common microbial byproducts from the fermentation process support food preservation and resist spoilage (Figure 2.4). For more applications, please refer to the strains listed in Table 2.1

2.4 EXAMPLES OF FFS THAT CONTAIN MICROBES CLAIMED AS BMS AND/OR PROBIOTIC

There are different famous old fermented foods that have been reported to be used by various civilizations and trips. For example, the Romans had sauerkraut, Indians enjoyed Lassi (a pre-dinner yogurt drink), Bulgarians consumed fermented milk and kefir, Turks had Ariana drink, Asians pickled cabbage, turnips, eggplant, cucumbers, onion, squash, and carrots, Middle Easterners favored yogurt, and Arabs had Laban rayeb. Nowadays, the most popular fermented foods worldwide include sauerkraut, kimchi, pickles, yogurt, ariana, kefir, and laban rayeb.

2.4.1 Yogurt

The yogurt and other products of fermented milk are produced in a similar manner that uses different cultures of bacteria. The yogurt is produced using *L. acidophilus* and *L. delbrueckii* subsp. *bulgaricus* Delbruck and *Streptococcus thermophilus* bacteria. Besides, other *Lactobacilli* and *Bifidobacteria* are sometimes added during or after culturing yogurt. Many makers of yogurt add *L. acidophilus* to their bacterial cultures. *L. acidophilus* has possible health benefits in easing yeast infections and restoring the normal bacterial balance in the intestinal tract of humans after antibiotic treatment. Yogurt is the most purchased probiotic food. *L. bulgaricus*, *S. thermophilus*, and other *Lactobacilli* and *Bifidobacteria* are added to improve both taste and digestibility. However, yogurt made at home will guarantee the existence of viable bacteria.

It is better to use raw milk to preserve its beneficial bacteria and nutritional value, which will ultimately undergo fermentation. Yogurt is typically fermented for a short period of time. The longest fermentation, lasting twenty-four hours, results in the absence of lactose, making it ideal for lactose-intolerant individuals. Yogurt is also rich in protein, riboflavin, B vitamins, and calcium. The probiotics present in yogurt have been found to be beneficial in treating diarrhea caused by antibiotics. They can help replenish some of the beneficial bacteria in the gastrointestinal tract microflora that is often unintentionally destroyed by antibiotics.

2.4.2 Kefir

While yogurt is fermented by bacteria, kefir is fermented by both bacteria and yeast. Kefir is a fermented milk drink, similar to a thin yogurt or ayran, made with kefir grains and milk from *cows, goats, or sheep* (García Fontán et al. 2006, Altay et al. 2013, Ghane, Babaeekhou, and Ketabi 2020, Liu et al. 2020, Menezes et al. 2020, Plessas et al. 2020, Raveschot et al. 2020, Romero-Luna et al. 2020, Vasquez et al. 2020, Wang et al. 2020). The term "kefir" originated in the Caucasus region in the early 1900s. The kefir grains consist of colonies of bacteria, including *Lactobacilli*, and yeast. They coexist in a symbiotic relationship and ferment at room temperature. Traditionally, kefir was prepared in goatskin bags that were hung near doorways. The bag was knocked by anyone passing through to keep the milk and kefir grains well mixed. Similar practices were also observed among Bedouin travelers in the Middle East. The kefir grains are initially formed through auto-aggregations of *L. kefiranofaciens* and *Saccharomyces turicensis*, both of which are biofilm producers. They can adhere to surfaces and form three-dimensional microcolonies (Wyder, Meile, and Teuber 1999, Wang et al. 2012). Probiotic bacteria commonly found in kefir products include *L. acidophilus, Bifidobacterium bifidum, Streptococcus thermophilus, L. delbrueckii* subsp. *bulgaricus, L. helveticus, L. kefiranofaciens, Lactococcus lactis*, and *Leuconostoc* species (Guzel-Seydim et al. 2011, de Oliveira Leite et al. 2013). Yeast species found in kefir include *Candida kefyr* and *S. cerevisiae*, as well as *Kluyveromyces marxianus, Kluyveromyces lactis, Saccharomyces fragilis, Torulaspora delbrueckii*, and *Kazachstania unispora* (Guzel-Seydim et al. 2011). During the fermentation process, lactose is digested, making kefir suitable for individuals with lactose intolerance. Kefir is also rich in natural antibiotics and contains significant amounts of vitamins B1 and B12, calcium, folic acid, phosphorus, and vitamin K.

2.4.3 Kimchi

Kimchi, a vegetable pickle seasoned with garlic, red pepper, and ginger, is a national dish in Korea. It can also be spelled as "kimchee". On average, around 40 pounds (18 kg/person/year) of kimchi is consumed annually. Kimchi is believed to have played a role in maintaining a lower obesity rate due to its low-calorie content. It is primarily made using cabbage, while other vegetables like cucumbers and scallions are mixed in. Common additions include chilies, garlic, onions, and ginger. Kimchi is a lacto-fermented product, where the vegetables are submerged in brine water and allowed to ferment with the help of *Lactobacillus* bacteria. Vegetables commonly used in its preparation include celery cabbage, Chinese turnips, and cucumbers. The prepared vegetables are sliced and highly seasoned with red pepper, onion, and garlic. They are then fermented in brine in large earthenware jars. Additional seasonings such as dried and salted shrimp, anchovy paste, or oysters are sometimes used. The fermentation process typically takes about one month, depending on weather conditions. Kimchi jars are often stored (either partially or entirely) underground. Kimchi is well-known for its tangy, spicy, and acidic taste, and it is commonly consumed for breakfast, lunch, and dinner in Korea.

The microorganisms present in kimchi include *Bacillus mycoides, B. pseudomycoides, B. subtilis, L. brevis, L. curvatus, L. kimchii, L. parabrevis, L. pentosus, L. plantarum, L. sakei, L. spicheri, Lactococcus carnosum, Lc. gelidum, Lc. lactis, Leuconostoc carnosum, Ln. citreum, Ln. gasicomitatum, Ln. gelidum, Ln. holzapfelii, Ln. inhae, Ln. kimchii, Ln. lactis, Ln. mesenteroides, Serratia marcescens, Weissella cibaria, W. confusa, W. kandleri, W. kimchii, W. koreensis*, and *W. soli* (Lee et al. 2005, Hong 2013, Jeong, Jung, et al. 2013, Jeong, Lee, et al. 2013, Jung et al. 2013). Archaea and yeasts are also present in kimchi (Chang et al. 2008). Occasionally, undesirable white colonies may form in the product (Chang et al. 2008). Kimchi has been suggested to have the ability to improve mental functioning by increasing the expression of a gene responsible for regulating psychological stress and anxiety (Selhub, Logan, and Bested 2014).

2.4.4 Natto

Natto is mainly produced in Japan, although similar products are also produced in China, Thailand, and the Philippines. In Japan, approximately 1.2 kg/person/year of natto is consumed. Natto is made from small seeds of the soybean. The seeds are cooked and then inoculated with about 10^6–10^7 spores of *Bacillus subtilis* var. *natto* (Kubo et al. 2011). Natto is often served as a breakfast food. *B. subtilis* var. *natto* is resistant to lactic acid and imparts a strong, salty flavor and a pungent odor to the natto. The small size of soybeans allows the fermentation process to reach the center of the bean.

Nutritionally, natto is a rich source of protein, comparable to beef in terms of protein content. It is also high in vitamins B2 and K, as well as antioxidants, particularly selenium. Natto also contains nattokinase, which has shown anti-inflammatory activity and is used in the prevention and treatment of blood clots.

2.4.5 Tempeh or Tempe

Tempeh was first produced in Indonesia thousands of years ago and remains a staple in the country's cuisine, providing an inexpensive source of dietary protein. Over time, it has spread to other countries such as Malaysia and the Netherlands. Tempeh is made from partially cooked fermented soybeans and is particularly popular on the island of Java. In Indonesian cuisine, tempeh takes the form of soybean cakes and is produced through a natural culture and controlled fermentation process. The fermentation is carried out using *Rhizopus oligosporus*, known as tempeh starters (Supriyanto, Fujio, and Hayakawa 1994, Moreno et al. 2002). The main step in making tempeh is the fermentation of soybeans, which are inoculated with *Rhizopus* spp. Traditionally, a previously fermented tempeh is mixed in, containing the spores of *Rhizopus oligosporus* or *Rhizopus oryzae*. The mixture is then spread into a thin layer and allowed to ferment for around 24 hours at a temperature of approximately 30 °C. The soybeans need to cool down to allow spore germination. Typically, tempeh is ready to be harvested after 48 hours of fermentation. The final product has a distinguishable whitish color, firm texture, and nutty flavor. Fermented soybeans in tempeh contain vitamin B12, which is a byproduct of the fermentation process. Tempeh also enhances the body's absorption of isoflavones and removes naturally occurring enzyme inhibitors found in soybeans. It is high in fiber, beneficial bacteria, enzymes, and manganese. Tempeh serves as a good digestible protein source and a great meat substitute due to its high protein content. It is also a good source of calcium and contains all essential amino acids. Sometimes, tempeh is made from a blend of grains, beans, or other vegetables.

2.5 THE SAFE FERMENTED FOOD

The concept of safe fermented food refers to food that is free from danger or the risk of harm. In the past, before the discovery of the microscope and advancements in microbiology, safe food meant food that could be consumed without causing harm, even when consumed over a long period of time and possessing new accepted properties. This understanding of safety still holds true today. Some fermented foods have been observed for their potential medicinal properties and the comfort they provide when consumed. The concept of fermentation, as we understand it today, has been known for thousands of years. For example, in Asia, koji is used as a starter culture for various fermented foods. During the 19th century, yeast became available in the market, but this did not lead to the cessation of traditional fermented bread production. In some parts of the world that have not fully modernized, people still utilize old techniques to produce bread. These techniques include self-fermentation, using parts of certain plants that contain yeast, or continuing previous successful fermentation processes. In essence, scientists did not discover these practices but rather rediscovered them and placed them within a scientific framework that can be further developed and refined by anyone. Modern science has introduced an additional step to the fermentation process, which is ensuring the safety of the final product. This includes analyzing different fermentation processes and keeping Food Science under strict control to enhance quality and safety. However, it is also important to continue preserving and understanding traditional methods alongside these advancements until we have fully comprehended them, documented them, and potentially improved their safety if necessary.

2.5.1 First Scientific Trials to Analysis the Fermentation Phenomena

Food fermentation is an important spontaneous process that has continuously captured human attention. Successful fermentation processes (FPs) have been passed down from generation to generation, from home to home, and from one culture to another. Over time, these processes have reached both consumers and the industry. Initially, fermented foods (FFs) were primarily used for food preservation purposes, but today they have become tools for health promotion and the production of new products (Zugic Petrovic et al. 2020). Numerous economic studies have highlighted the significance of probiotics and FFs.

Understanding the conditions required for safe fermented food is a critical aspect. Microbes work in a simple and spontaneous manner. Prior to the discovery of the microscope, FFs and spoilage processes were attributed to the theory of spontaneous generation. The microbial activity was not well understood during the pre-Pasteurian era, mainly due to the lack of tangible evidence for the existence of microorganisms. Raw foods, even with a small number of microbes on their surfaces, do not spoil as long as they remain in their natural forms. The presence of a chitin layer on the surfaces of vegetables and fruits, for example, protects them from microbial activity. However, once suitable conditions arise, such as surface injuries or moisture, microbes initiate their activity. In most cases, FFs are more acceptable to consumers and are resistant to spoilage processes (Forssten et al. 2020, Pimentel et al. 2020).

Throughout history, on long journeys, dried or fermented foods were preferred. As humans gradually discovered the factors contributing to the success of these methods, the processes evolved into defined techniques rather than trial and error. Some of these techniques remained secrets for a long time, as in the case of kefir.

Certain civilizations appeared to be aware of microbial existence or, at the very least, their activity. For instance, food, animals, and body parts were preserved as part of the mummification ceremonies, including the mummification of human bodies, in Egypt. The use of rice paste (Koji) for the fermentation of microbes in Asia is another example. Modern food technology can draw valuable insights from these ancient practices. However, in conclusion, the key elements of the old methods involve removing or preventing moisture (dehydration), using natural preservatives like resins, and storing or fermenting the food in a dry environment. "Drying/preventing moisture" is one of the most successful traditional preservation methods.

Technology provides us with more opportunities for control and the invention of new methods based on efficient traditional practices. Modern technology enables the detection of even trace amounts of contaminants in food analysis, using sophisticated instruments such as real-time PCR (Gest 2004, Dowdell et al. 2020, Shehata and Newmaster 2020).

Foods that undergo fermentation naturally, without the addition of specific starter cultures, are known as self-fermented foods. Various terms are used to describe this process, including "wild" or "spontaneous" fermentation. These traditional methods involve a wide variety of microorganisms that originate from the local environment. In natural fermentation, active microbes coexist with the food, and the fermentation process begins whenever the required conditions are met, such as the presence of anaerobic conditions or the use of salt. Adding a portion of previously fermented food is often done to accelerate the fermentation process and achieve consistent quality. In fact, such practices are still actively followed in different parts of the world (Tamang et al. 2015).

2.5.2 Safety Considerations

Probiotics are generally considered safe microbes. However, it's important to recognize that they are still microbes, and as such, there is a possibility that they could acquire mobile resistance factors or undesired virulence factors (Amara 2011). Probiotics have the potential to acquire genes responsible for antibiotic resistance, hydrocarbon degradation, and other virulence factors, which are often carried on genetic elements called plasmids. Plasmids can be transferred to other microbes through various mechanisms. Furthermore, excessive exposure to certain compounds or nutrients that can promote mutation may lead to the mutagenesis of genes, resulting in the production of modified proteins with new properties (Venkatachalam et al. 1994).

As part of the defense mechanism of microbes, they may start to excrete harmful compounds that are not normally excreted (Hussain and Amara 2006, Menisy et al. 2017). For example, during a serial dilution experiment to determine the minimum inhibition concentration (MIC) of SDS on a microbial strain, the microbe did not yield accurate results due to the excretion of exopolysaccharides resulting from the specific type of medium used. Only after changing the medium, the correct MIC could be determined (Menisy et al. 2017).

The role of changing molecular structure on function can be illustrated through analogs, which have structures similar to certain compounds but with minor differences. Some analogs are more stable or even non-degradable, such as IPTG. Analog compounds can demonstrate that our biological system, despite its specificity, is governed by chemical and physical laws. Similar compounds to a specific substrate, like analogs, can interfere with related enzymes or activate bio-pathways incorrectly. While this phenomenon can be helpful in research, it may pose risks in real-life situations. The biological system functions mechanically and efficiently but without discretion (Amara 2013a).

Mutations in the β -lactamase gene can lead to multiple antibiotic resistance. β -lactamase is a resistance factor against β-lactam antibiotics. The discovery of new generations of β -lactam antibiotics has induced the emergence of new strains that are initially reported as sensitive. Venkatachalam et al. (1994) investigated TEM β -lactamase variants with amino acid substitutions in the active-site pocket of the enzyme. The experiments identified natural isolates with increased resistance to extended-spectrum cephalosporins like cefotaxime and ceftazidime. Mutants were selected for 100-fold greater resistance to ceftazidime compared to the wild-type. All mutants exhibited a serine substitution at position 238, lysine or arginine at position 240, and a small amino acid at position 241 (Venkatachalam et al. 1994). This example illustrates what can occur in a laboratory setting.

Nature is dynamic, and the probability of encountering similar or new phenomena is high. Microbes possess integrated systems to protect themselves against antimicrobial agents. Spore-forming bacteria, for instance, produce spores for protection. Exopolysaccharide formation, as mentioned earlier, provides mechanical protection by coating or immobilizing microbial cells and can act as a chelating agent for divalent ions (Amara and Hussain 2006, Hussain and Amara 2006, Amara 2011). Although some antimicrobial drugs can penetrate alginate biofilms and kill bacterial cells, they often cannot completely eradicate the entire vegetative population. The combination of dead cells and alginate forms a biological cover that protects a portion of the microbial population. The immense microbial biodiversity and the robustness of the biological system provide microbes under attack with alternative genetic material from other surviving microbes, which can supply them with genes responsible for survival (Amara 2011). Plasmid transduction and phage attacks are considered as means of transmitting resistant genes among microbes. Additionally, genes can be transmitted through genomic DNA. Dead cells with intact DNA serve as an unlimited source for transferring new genetic material (Blattner et al. 1997, Casadevall and Pirofski 1999, Amara 2011). Hybridization can also occur naturally.

An example highlighting the dynamic nature of gene transfer mechanisms is the finding that enterohemorrhagic *E. coli* isolates contain virulence plasmids and pathogenicity islands similar to those found in *Shigella* spp. (Blattner et al. 1997, Casadevall and Pirofski 1999). As many as 1387 new genes were acquired compared to the previously sequenced nonpathogenic laboratory strain *E. coli* K–12 (Makino et al. 1998). Geographical variations have been shown to play a role in the acquisition of resistance, pathogenicity, and virulence factors (Amara 2011). The biological system adapts to ensure its survival.

Unwise microbial treatment can amplify the harmful aspects of microbes. The concept of microbial treatment needs to be reevaluated, and knowledge gaps should be addressed. For instance, H_2O_2, which is produced during the natural defense mechanism, can lead to negative results. H2O2 mutates *P. aeruginosa* and induces increased alginate production. The reactions of biological systems are unpredictable. Amara and Hussain (2006), Hussain and Amara (2006), and other authors have recommended a case-by-case approach for evaluating each microbe and strain as a unique case, rather than applying general methods (Amara and Hussain 2006, Hussain and Amara 2006).

Recently, a new method for microbial cell evacuation has been described, involving the use of the MIC to eliminate microbes (Amara, Salem-Bekhit, and Alanazim 2013, Amara, Salem-Bekhit, and Alanazi 2015). This idea could be of interest in the field of food technology by determining the MIC of antimicrobial compounds and using it, or a slightly higher concentration, to ensure microbial control with minimal use of antimicrobial compounds (Amara, Salem-Bekhit, and Alanazim 2013, Amara 2015, Amara, Salem-Bekhit, and Alanazi 2015, Amara and Shib 2015, Amara 2017).

Geographic variation is an essential factor in the transmission of resistance factors. Food and geographic factors can directly influence antibiotic resistance genes through mechanisms proposed by Amara and Shibl (2015) (Amara and Shib 2015). This effect is thought to involve an epigenetic-like mechanism (Amara and Shib 2015). Other human activities, such as the use of antimicrobial agents at different stages of plant growth, can also affect our food. The type of antimicrobial agent used may contaminate plant tissues, affecting food quality and influencing microbial activity within specific formulas.

It is worth noting that certain conditions can divert the biochemical flux in microbes to produce indigestible products, such as polyhydroxyalkanoates and polyhydroxybutyrate, or inhibit the β -oxidation pathway. For more details on the role of medium components in redirecting microbial behavior, refer to Hoffman et al. (2002) and Rhem et al. (2002) (Hoffmann et al. 2002, Rehm et al. 2002).

Observations have shown that certain compounds exhibit different activities at different concentrations (Awny et al. 1997, Amara, El masry, and Bogdady 2008).

Microbes and biological cells are highly selective in their carbon source utilization. It is crucial to analyze the different types of carbon sources added to the medium (Amara 2013a). Foods containing cysteine are preferable to those containing cystine because cell membranes transport cysteine ten times faster than cystine. Understanding such properties enables us to select foods that benefit us (Amara 2013b, Amara 2013a).

he literature usually contains alarm signals that lead to better control for any overestimated issue. That's because one should not expect that a particular issue like the probiotic is safe forever. That's exactly what has happened after the discovery of the antibiotic. De Paula et al. (2015) warning that *Leuconostoc mesenteroides* strain has two faces. They present promising characteristics, such as probiotic potential and bacteriocin production. Nonetheless, rare strains exhibit pathogenic profile and cause spoilage in different food matrices. The two faces of *Lc. mesenteroides* show that further studies should be performed to better understand the behavior of this microorganism in different fields, and that there is a need to validate each bacterial species and each strain (de Paula et al. 2015). The great majority of cases of pathogenic *Lc. mesenteroides* is related to underlying diseases and immunocompromised patients. Immune system changes can expose the patients to different opportunistic pathogens such as *Lc. mesenteroides*, increasing the rates of opportunistic illness. Moreover, the intake of dairy products containing this microorganism apparently is not related to any disease (Lee et al. 2011, de Paula et al. 2015).

There is mounting evidence on the biological efficacy of probiotics; however, the indication of these microorganisms for clinical use should be seen with caution (*de Melo Pereira et al.* 2018, *Silva et al.* 2020). Several pathogenic bacterial strains, such as some strains of *S. typhimurium* and *E. coli* 0157:H7, were found to be relatively resistant to low pH due to their ability to overproduce some proteins induced by the acid environment. The presence of such strains in food, which is freshly acidified at a 1–2% level to control microorganisms, may cause a problem.

On the other hand, one must not neglect that in immunocompromised individuals or other patients, *S. boulardii* can cause fungemia or localized infections (Hennequin et al. 2000, Thygesen, Glerup, and Tarp 2012).

In 2002, the United Nations Food and Agriculture Organization/World Health Organization (FAO/WHO) published the "Guidelines for the evaluation of probiotics in food." These guidelines established safety and efficacy standards for probiotics, systematizing their discovery and selection (2002). Thus, the FAO/WHO guidelines suggest different criteria that should be evaluated for the selection of probiotics, namely resistance to unfavorable conditions in the human body, ability to adhere to epithelial tissues, antimicrobial activity, and safety for use (de Melo Pereira et al. 2018).

The safety of a probiotic strain is defined by its origin and lack of association with pathogenic cultures (Markowiak and Śliżewska 2017). Peivasteh-Roudsari et al. (2019) reported that the use of live bacteria called probiotics in food and dietary supplements requires accurate safety assessment. The bacteria used in probiotic products should be completely safe. Many research tools, animal models, and laboratory-based techniques are available. Nevertheless, probiotic safety should usually be confirmed by human studies (Peivasteh-Roudsari et al. 2019). There are some important points that could guarantee better control over the various steps concerning the use of probiotics and the beneficial microorganisms (BMs), which could include some necessarily needed activities:

- A need for standardized antimicrobial bioassays.
- A need for specifying the practices that prevent the spreading/elevation of microbial resistance.
- A need to specify the unique properties for a microbe to be claimed/and used as probiotic.
- A need to follow up the positive/negative effect of the products that contains probiotics in the human/animal health.
- Public education about how to use probiotics for health protection and diseases/pathogens control and treatment.
- Using international microbial codes and suitable culture collections for preserving probiotic microbes.
- Good manufacturing practices (GMP) roles must be followed and controlled by the quality control lab. (QC) and quality assurance lab. (QA), shelf life conditions, labeling, minimum/maximum dosage, and demonstrable health claims.
- The accepted amount that could be consumed concerning sex/age/weight and the like.
- Consumer feedback concerning any elevated health problems.
- International homogenization for the GMP, QC, QA, market practices, feedback, consumer service, etc.
- Restricted governmental lows for practices, regulation, and control.

The full FAO/WHO report of 2001 is available at the FAO website (ftp://ftp.fao.org/es/esn/food/probio_report_en.pdf) (FAO/WHO 2002). The report provided FAO, WHO, and their member states with scientific recommendations in relation to the nutritional and safety assessment of probiotics, in particular powdered milk with life LABs; provided general guidance for the assessment of these probiotics; and addressed specific questions arising in relation to their toxigenicity, pathogenicity, allergenicity, or other specific features relevant to their nutritional properties or safety (FAO/WHO 2002, Pineiro and Stanton 2021).

2.6 CONCLUSION

BMs provide us with many benefits and can be used as biopreservatives. Most fermentation processes contribute to food preservation, as they result in the production of active compounds by microbes that can be safely used as preservatives. This is a preferable option compared to chemical preservatives or antibiotics that can be harmful to our health. Some TFs are popular and contain BMs. Among these, specific microbes known as probiotics can colonize our gastrointestinal system. Consuming fermented foods that contain BMs or probiotics can support our health, prevent various diseases, and aid in their treatment. While beneficial microbes, such as probiotics, and microbial biopreservatives are valuable, they require regulation and management. Establishing successful protocols for production, application, regulation, and control is crucial to support the growing trend of using biopreservatives in food preservation.

REFERENCES

2002. Guidelines for the evaluation of probiotics in food guidelines for the evaluation of probiotics in food:1–11.

Abdolalipour, E., M. Mahooti, A. Gorji, and A. Ghaemi. 2020. Synergistic therapeutic effects of probiotic Lactobacillus casei TD-2 consumption on GM-CSF-induced immune responses in a murine model of cervical cancer. *Nutr Cancer* 74(1):372–382s. doi: 10.1080/01635581.2020.1865419.

Agriopoulou, S., E. Stamatelopoulou, M. Sachadyn-Król, and T. Varzakas. 2020. Lactic acid bacteria as antibacterial agents to extend the shelf life of fresh and minimally processed fruits and vegetables: Quality and safety aspects. *Microorganisms* 8(6):952. doi: 10.3390/microorganisms8060952.

Al-Madboly, L. A., and A. K. Abdullah. 2015. Potent antagonistic activity of Egyptian *Lactobacillus plantarum* against multiresistant and virulent food-associated pathogens. *Front Microbiol* 6:347. doi: 10.3389/fmicb.2015.00347.

Al-Mohammadi, A. R., R. A. Ibrahim, A. H. Moustafa, A. A. Ismaiel, A. Abou Zeid, and G. Enan. 2021. Chemical constitution and antimicrobial activity of kefir fermented beverage. *Molecules* 26(16):5026. doi: 10.3390/molecules26092635.

Al-Mohammadi, A. R., A. A. Ismaiel, R. A. Ibrahim, A. H. Moustafa, A. Abou Zeid, and G. Enan. 2021. Chemical constitution and antimicrobial activity of kombucha fermented beverage. *Molecules* 26(16). doi: 10.3390/molecules26165026.

Alakomi, H. L., E. Skyttä, M. Saarela, T. Mattila-Sandholm, K. Latva-Kala, and I. M. Helander. 2000. Lactic acid permeabilizes gram-negative bacteria by disrupting the outer membrane. *Appl Environ Microbiol* 66(5):2001–2005.

Altay, F., F. Karbancıoglu-Güler, C. Daskaya-Dikmen, and D. Heperkan. 2013. A review on traditional Turkish fermented non-alcoholic beverages: Microbiota, fermentation process and quality characteristics. *Int J Food Microbiol* 167(1):44–56. doi: 10.1016/j.ijfoodmicro.2013.06.016.

Amara, A. A. 2011. Opportunistic pathogens and their biofilm "food for thought". In Vilas, M. (Ed.), *Science Against Microbial Pathogens: Communicating Current Research and Technology Advances. FORMATEX Microbiology Series No 3* 2(3):813–824.

Amara, A. A. 2013a. The inevitability of balanced lives: Genes and, foods in action and, interactions. *IIOABJ* 3(2):1–27.

Amara, A. A. 2013b. The inevitability of balanced lives: Genes and, foods in action and, interactions. *IIOAB J* 4(2):1–27.

Amara, A. A. 2015. Kostenlose viral ghosts, bacterial ghosts, microbial ghosts and more. *Schüling Verlage Germany*: 130. Isbn: 9783865232601.

Amara, A. A. 2017. Cell evacology: Could it be a new science. *Cell Mol Med* 3(2):9.

Amara, A. A., M. H. El-Masry, and H. H. Bogdady. 2008. Plant crude extracts could be the solution: Extracts showing in vivo antitumorigenic activity. *Pak J Pharm Sci* 21(2):159–171.

Amara, A. A., M. A. Hassan, A. T. Abulhamd, and B. M. Haroun. 2011. algs genes: H_2O_2 map their viability and validity in *Pseudomonas aeruginosa*. *IJBB* 7(3): 321–330.

Amara, A. A., and M. Z. Hussain. 2006. Offline controlling of *Pseudomonas aeruginosa* resistant to protein inhibitor antibiotics using combination of EDTA and Na-citrate or disinfectant(s). *Microbiology (China)* 33(33):95–100.

Amara, A. A., M. M. Salem-Bekhit, and F. K. Alanazi. 2015. Plackett-Burman randomization method for bacterial ghosts preparation form *E. coli* JM109. *Saudi Pharm J* 22(3):273–279.

Amara, A. A., M. M. Salem-Bekhit, and F. K. Alanazim. 2013. Sponge-like: A new protocol for preparing bacterial ghosts. *TSWJ* 1013:545741.

Amara, A. A., and A. Shib. 2015. A hypothesis about the variation within β -lactamases: An epigenetic-like mechanism. *IIOABJ* 6(1):1–11.

An, H. M., E. H. Baek, S. Jang, D. K. Lee, M. J. Kim, J. R. Kim, K. O. Lee, J. G. Park, and N. J. Ha. 2010. Efficacy of Lactic Acid Bacteria (LAB) supplement in management of constipation among nursing home residents. *Nutr J* 9:5. doi: 10.1186/1475-2891-9-5.

Anupma, A., and J. P. Tamang. 2020. Diversity of filamentous fungi isolated from some amylase and alcohol-producing starters of India. *Front Microbiol* 11:905. doi: 10.3389/fmicb.2020.00905.

Ashaolu, T. J., and A. Reale. 2020. A holistic review on euro-Asian lactic acid bacteria fermented cereals and vegetables. *Microorganisms* 8(8):1176. doi: 10.3390/microorganisms8081176.

Assimos, D. G. 2020. Re: Metabolomic profiling of oxalate-degrading probiotic *Lactobacillus acidophilus* and *Lactobacillus gasseri*. *J Urol* 203(2):247–248. doi: 10.1097/01.JU.0000612972.99650.f8.

Audisio, M. C., H. R. Terzolo, and M. C. Apella. 2005. Bacteriocin from honeybee beebread Enterococcusavium, active against Listeriamonocytogenes. *Appl Environ Microbiol* 71(6):3373–3375. doi: 10.1128/aem.71.6.3373-3375.2005.

Awny, H. A., A. A. Amara, M. H. El-Masry, and H. H. Baghdadi. 1997. Effect of 12 plant extract on Hepatic microsomal penzo(a) pyrene hydroxylation and hydrogen peroxide production in mice treated with lendane. *J Environ Nutr Interact* 1:129–142.

Babashahi, M., M. Mirlohi, R. Ghiasvand, L. Azadbakht, L. Mosharaf, and S. Torki-Baghbadorani. 2020. Effects of probiotic soy milk fermented by lactobacillus plantarum A7 (KC 355240) added with Cuminum cyminum essential oil on fasting blood glucose levels, serum lipid profile and body weight in diabetic Wistar rats. *Int J Prev Med* 11:8. doi: 10.4103/ijpvm.IJPVM_541_17.

Bahuguna, A., I. G. Jo, J. S. Lee, and M. Kim. 2020. Effects of radishes, apples, and pears on the lactic acid bacteria and nutritional and functional qualities of flavored soy sauce. *Foods* 9(11):1562. doi: 10.3390/foods9111562.

Baird-Parker, A. 1980. Organic acids. In Silliker, J. H. (Ed.), *Microbial Ecology of Foods*, Vol. I. New York: Academic Press, pp. 126–135.

Bali, V., P. S. Panesar, M. B. Bera, and J. F. Kennedy. 2016. Bacteriocins: Recent trends and potential applications. *Crit Rev Food Sci Nutr* 56(5):817–834.

Barbour, A., K. Philip, and S. Muniandy. 2013. Enhanced production, purification, characterization and mechanism of action of salivaricin 9 lantibiotic produced by *Streptococcus salivarius* NU10. *PLOS ONE* 8(10):e77751. doi: 10.1371/journal.pone.0077751.

Bartkiene, E., E. Zokaityte, P. Zavistanaviciute, E. Mockus, D. Cernauskas, M. Ruzauskas, E. Tolpeznikaite, and R. P. F. Guiné. 2021. Nutraceutical chewing candy formulations based on acetic, alcoholic, and lactofermented apple juice products. *Foods* 10(10):2329. doi: 10.3390/foods10102329.

Bauer, R., M. du Toit, and J. Kossmann. 2010. Influence of environmental parameters on production of the acrolein precursor 3-hydroxypropionaldehyde by Lactobacillus reuteri DSMZ 20016 and its accumulation by wine lactobacilli. *Int J Food Microbiol* 137(1):28–31. doi: 10.1016/j.ijfoodmicro.2009.10.012.

Behnsen, J., E. Deriu, M. Sassone-Corsi, and M. Raffatellu. 2013. Probiotics: Properties, examples, and specific applications. *Cold Spring Harb Perspect Med* 3(3):a010074–a010074. doi: 10.1101/cshperspect.a010074.

Birri, D. J., D. A. Brede, and I. F. Nes. 2012. Salivaricin D. a Novel Intrinsically trypsin-Resistant lantibiotic from *Streptococcus salivarius* 5M6c Isolated from a Healthy Infant. *Appl Environ Microbiol* 78(2):402–410. doi: 10.1128/aem.06588-11.

Blatchford, P. A., S. G. Parkar, W. Hopkins, J. R. Ingram, and K. H. Sutton. 2019. Dose-dependent alterations to in vitro human microbiota composition and butyrate inhibition by a supercritical carbon dioxide hops extract. *Biomolecules* 9(9). doi: 10.3390/biom9090390.

Blattner, F. R., G. Plunkett, C. A. Bloch, N. T. Perna, V. Burland, M. Riley, J. Collado-Vides, J. D. Glasner, C. K. Rode, G. F. Mayhew, J. D. N. Gregor, H. A. Kirkpatrick, M. A. Goeden, D. J. Rose, B. Mau, and Y. Shao. 1997. The complete genome sequence of Escherichia coli K-12. *Science* 277(5331):1453–1462.

Bockelmann, W., K. P. Willems, H. Neve, and K. H. Heller. 2005. Cultures for the ripening of smear cheeses. *Int Dairy J* 15(6–9):719–732.

Bokulich, N. A., and C. W. Bamforth. 2013. The microbiology of malting and brewing. *Microbiol Mol Biol Rev* 77(2):157–172. doi: 10.1128/mmbr.00060-12.

Bonnarme, P., C. Lapadatescu, M. Yvon, and H. E. Spinnler. 2001. L-methionine degradation potentialities of cheese-ripening microorganisms. *J Dairy Res* 68(4):663–674.

Bosák, J., L. Micenková, M. Hrala, K. Pomorská, M. Kunova Bosakova, P. Krejci, E. Göpfert, M. Faldyna, and D. Šmajs. 2018. Colicin FY inhibits pathogenic *Yersinia enterocolitica* in mice. *Sci Rep* 8(1):12242. doi: 10.1038/s41598-018-30729-7.

Bourdichon, F., S. Casaregola, C. Farrokh, J. C. Frisvad, M. L. Gerds, W. P. Hammes, J. Harnett, G. Huys, S. Laulund, A. Ouwehand, I. B. Powell, J. B. Prajapati, Y. Seto, E. Ter Schure, A. Van Boven, V. Vankerckhoven, A. Zgoda, S. Tuijtelaars, and E. Bech Hansen. 2012. Food fermentations: Microorganisms with technological beneficial use. *Int J Food Microbiol* 154(3):87–97. doi: 10.1016/j.ijfoodmicro.2011.12.030.

Brede, D. A., S. Lothe, Z. Salehian, T. Faye, and I. F. Nes. 2007. Identification of the Propionicin F bacteriocin immunity gene (pcfI) and development of a food-grade cloning system for *Propionibacterium freudenreichii. Appl Environ Microbiol* 73(23):7542–7547. doi: 10.1128/aem.01023-07.

Breidt, F., R. F. McFeeters, I. Perez-Diaz, and C.-H. Lee. 2013. Fermented vegetables. In Doyle M. P. and Buchanan R. L. (Eds.), Chapter 33. *Food Microbiology: Fundamentals and Frontiers*, 4th Ed. ASM Press.

Campanero, C., E. Muñoz-Atienza, D. B. Diep, J. Feito, S. Arbulu, R. del Campo, I. F. Nes, P. E. Hernández, C. Herranz, and L. M. Cintas. 2020. Biochemical, genetic and transcriptional characterization of multibacteriocin production by the anti-pneumococcal dairy strain Streptococcus infantarius LP90. *PLOS ONE* 15(3):e0229417. doi: 10.1371/journal.pone.0229417.

Carr, F. J., D. Chill, and N. Maida. 2002. The lactic acid bacteria: A literature survey. *Crit Rev Microbiol* 28(4):281–370.

Casadevall, A., and L. A. Pirofski. 1999. Host-pathogen interactions: Redefining the basic concepts of virulence and pathogenicity. *Infect Immun* 67(8):3703–3713.

Castilho, N. P. A., M. Colombo, L. Ld. Oliveira, S. D. Todorov, and L. A. Nero. 2019. *Lactobacillus curvatus* UFV-NPAC1 and other lactic acid bacteria isolated from calabresa, a fermented meat product, present high bacteriocinogenic activity against Listeria monocytogenes. *BMC Microbiol* 19(1):63. doi: 10.1186/s12866-019-1436-4.

Cebrián, R., M. E. Rodríguez-Cabezas, R. Martín-Escolano, S. Rubiño, M. Garrido-Barros, M. Montalbán-López, M. J. Rosales, M. Sánchez-Moreno, E. Valdivia, M. Martínez-Bueno, C. Marín, J. Gálvez, and M. Maqueda. 2019. Preclinical studies of toxicity and safety of the AS-48 bacteriocin. *J Adv Res* 20:129–139. doi: 10.1016/j.jare.2019.06.003.

Chang, H., K. Kim, Y. Nam, S. Roh, M. Kim, C. Jeon, H. Oh, and J. Bae. 2008. Analysis of yeast and archaeal population dynamics in kimchi using denaturing gradient gel electrophoresis. *Int J Food Microbiol* 126(1–2):159–166. doi: 10.1016/j.ijfoodmicro.2008.05.013.

Chaves-López, C., C. Rossi, F. Maggio, A. Paparella, and A. Serio. 2020. Changes occurring in spontaneous maize fermentation: An overview. *Fermentation* 6(1):36. doi: 10.3390/fermentation6010036.

Chavez, R., F. Fierro, R. O. G. Rico, and F. Laich. 2011. Mold-fermented foods: *Penicillium spp.* as ripening agents in the elaboration of cheese and meat products. In Leitão, A. L. (Ed.), *Mycofactories*. Emirate of Sharjah: Bentham Science Publishers, pp. 73–98.

Chen, B., Q. Wu, and Y. Xu. 2014. Filamentous fungal diversity and community structure associated with the solid state fermentation of Chinese Maotai-flavor liquor. *Int J Food Microbiol* 179:80–84. doi: 10.1016/j.ijfoodmicro.2014.03.011.

Chen, J., W. Cai, and Y. Feng. 2007. Development of intestinal bifidobacteria and lactobacilli in breast-fed neonates. *Clin Nutr* 26(5):559–566. doi: 10.1016/j.clnu.2007.03.003.

Chmielewska, A., and H. Szajewska. 2010. Systematic review of randomised controlled trials: Probiotics for functional constipation. *World J Gastroenterol* 16(1):69–75.

Choeisoongnern, T., S. Sirilun, R. Waditee-Sirisattha, K. Pintha, S. Peerajan, and C. Chaiyasut. 2021. Potential probiotic Enterococcus faecium OV3-6 and its bioactive peptide as alternative bio-preservation. *Foods (Basel, Switzerland)* 10(10):2264. doi: 10.3390/foods10102264.

Choi, A. R., J. K. Patra, W. J. Kim, and S. S. Kang. 2018. Antagonistic activities and probiotic potential of lactic acid bacteria derived from a plant-based fermented food. *Front Microbiol* 9:1963. doi: 10.3389/fmicb.2018.01963.

Choi, C. H., S. Y. Jo, H. J. Park, S. K. Chang, J.-S. Byeon, and S.-J. Myung. 2011. A randomized, double-blind, placebo-controlled multicenter trial of Saccharomyces boulardii in irritable bowel syndrome. *J Clin Gastroenterol* 45(8):679–683. doi: 10.1097/mcg.0b013e318204593e.

Cirkovic, I., D. D. Bozic, V. Draganic, J. Lozo, T. Beric, M. Kojic, B. Arsic, E. Garalejic, S. Djukic, and S. Stankovic. 2016. Licheniocin 50.2 and bacteriocins from *Lactococcus lactis* subsp. lactis biovar. diacetylactis BGBU1-4 Inhibit biofilms of coagulase Negative *Staphylococci* and *Listeria monocytogenes* clinical isolates. *PLOS ONE* 11(12):e0167995. doi: 10.1371/journal.pone.0167995.

Collins, M. D., J. Samelis, J. Metaxopoulos, and S. Wallbanks. 1993. Taxonomic studies on some leuconostoc-like organisms from fermented sausages: Description of a new genus Weissellafor the *Leuconostoc paramesenteroides* group of species. *J Appl Bacteriol* 75(6):595–603.

Corr, S. C., Y. Li, C. U. Riedel, P. W. O'Toole, C. Hill, and C. G. M. Gahan. 2007. Bacteriocin production as a mechanism for the antiinfective activity of *Lactobacillus salivarius* UCC118. *Proc Natl Acad Sci U S A* 104(18):7617–7621. doi: 10.1073/pnas.0700440104.

Cosentino, S., M. E. Fadda, M. Deplano, R. Melis, R. Pomata, and M. B. Pisano. 2012. Antilisterial activity of nisin-like bacteriocin-producing Lactococcus lactis subsp. lactis isolated from traditional Sardinian dairy products. *J Biomed Biotechnol* 2012:376428. doi: 10.1155/2012/376428.

Daeschel, M. A. 1989. Antimicrobial substances from lactic acid bacteria for use as food preservatives. *Food Technol* 43(1):164s.

Daranas, N., A. Bonaterra, J. Francés, J. Cabrefiga, E. Montesinos, and E. Badosa. 2018. Monitoring viable cells of the biological control agent *Lactobacillus plantarum* PM411 in aerial plant surfaces by means of a strain-specific viability quantitative PCR method. *Appl Environ Microbiol* 84(10):e00107. doi: 10.1128/aem.00107-18.

Davachi, S. M., N. Pottackal, H. Torabi, and A. Abbaspourrad. 2021. Development and characterization of probiotic mucilage based edible films for the preservation of fruits and vegetables. *Sci Rep* 11(1):16608. doi: 10.1038/s41598-021-95994-5.

de Castro, R. D., A. Mota, E. de Oliveira Lima, A. U. D. Batista, J. de Araújo Oliveira, and A. L. Cavalcanti. 2015. Use of alcohol vinegar in the inhibition of Candida spp. and its effect on the physical properties of acrylic resins. *BMC Oral Health* 15:52–52. doi: 10.1186/s12903-015-0035-5.

de la Fuente-Salcido, N. M., J. C. Castañeda-Ramírez, B. E. García-Almendárez, D. K. Bideshi, R. Salcedo-Hernández, and J. E. Barboza-Corona. 2015. Isolation and characterization of bacteriocinogenic lactic bacteria from M-Tuba and Tepache, two traditional fermented beverages in México. *Food Sci Nutr* 3(5):434–442. doi: 10.1002/fsn3.236.

de Melo Pereira, G. V., B. de Oliveira Coelho, A. I. Magalhães Júnior, V. Thomaz-Soccol, and C. R. Soccol. 2018. How to select a probiotic? A review and update of methods and criteria. *Biotechnol Adv* 36(8):2060–2076.

de Oliveira Leite, A. M., M. A. Miguel, R. S. Peixoto, A. S. Rosado, J. T. Silva, and V. M. F. Paschoalin. 2013. Microbiological, technological and therapeutic properties of kefir: A natural probiotic beverage. *Braz J Microbiol* 44(2):341–349. doi: 10.1590/s1517-83822013000200001.

de Paula, A. T., A. B. Jeronymo-Ceneviva, S. D. Todorov, and A. L. B. Penna. 2015. The two faces of *Leuconostoc mesenteroides* in food systems. *Food Rev Int* 31(2):147–171.

De Vuyst, L., S. Van Kerrebroeck, H. Harth, G. Huys, H. M. Daniel, and S. Weckx. 2014. Microbial ecology of sourdough fermentations: Diverse or uniform? *Food Microbiol* 37:11–29. doi: 10.1016/j.fm.2013.06.002.

Dejene, F., B. Regasa Dadi, and D. Tadesse. 2021. *In vitro* antagonistic effect of lactic acid bacteria isolated from fermented beverage and finfish on pathogenic and foodborne pathogenic microorganism in Ethiopia. *Int J Microbiol* 2021:5370556. doi: 10.1155/2021/5370556.

Denkova, R., B. Goranov, D. Teneva, Z. Denkova, and G. Kostov. 2017. Antimicrobial activity of probiotic microorganisms: Mechanisms of interaction and methods of examination. *Antimicrobial Research: Novel Bioknowledge and Educational Programs* 1: 201–212.

Desriac, F., D. Defer, N. Bourgougnon, B. Brillet, P. Le Chevalier, and Y. Fleury. 2010. Bacteriocin as weapons in the marine animal-associated bacteria warfare: Inventory and potential applications as an aquaculture probiotic. *Mar Drugs* 8(4):1153–1177. doi: 10.3390/md8041153.

Devirgiliis, C., S. Barile, and G. Perozzi. 2011. Antibiotic resistance determinants in the interplay between food and gut microbiota. *Genes Nutr* 6(3):275–284. doi: 10.1007/s12263-011-0226-x.

Dias, I., M. Laranjo, M. E. Potes, A. C. Agulheiro-Santos, S. Ricardo-Rodrigues, A. R. Fialho, J. Véstia, M. J. Fraqueza, M. Oliveira, and M. Elias. 2020. Autochthonous starter cultures are able to reduce biogenic amines in a traditional Portuguese smoked fermented sausage. *Microorganisms* 8(5):686. doi: 10.3390/microorganisms8050686.

Díaz-Montes, E., and R. Castro-Muñoz. 2021. Edible films and coatings as food-quality preservers: An overview. *Foods* 10(2):249. doi: 10.3390/foods10020249.

Diaz, M., L. Kellingray, N. Akinyemi, O. O. Adefiranye, A. B. Olaonipekun, G. R. Bayili, J. Ibezim, A. S. du Plessis, M. Houngbédji, D. Kamya, I. M. Mukisa, G. Mulaw, J. S. Manthi, C. W. Onyango, A. Atter, E. Agbemafle, T. Annan, A. N. Bernice, E. M. Buys, H. D. Joseph, C. Muyanja, J. Nakavuma, D. A. Odeny, L. H. Sawadogo, T. A. Tesfaye, A. W. Amoa, M. Obodai, M. J. Mayer, F. A. Oguntoyinbo, and A. Narbad. 2019. Comparison of the microbial composition of African fermented foods using amplicon sequencing. *Sci Rep* 9(1):13863.

Dicks, L. M. T., L. Dreyer, C. Smith, and A. D. van Staden. 2018. A review: The fate of bacteriocins in the human gastro-intestinal tract: Do they cross the gut–blood barrier? *Front Microbiol* 9:2297. doi: 10.3389/fmicb.2018.02297.

Dinleyici, E. C., M. Eren, M. Ozen, Z. A. Yargic, and Y. Vandenplas. 2012. Effectiveness and safety of *Saccharomyces boulardiifor* acute infectious diarrhea. *Expert Opin Biol Ther* 12(4):395–410. doi: 10.1517/14712598.2012.664129.

Dong, Y., J. Zhu, M. Zhang, S. Ge, and L. Zhao. 2020. Probiotic *Lactobacillus salivarius* Ren prevent dimethylhydrazine-induced colorectal cancer through protein kinase B inhibition. *Appl Microbiol Biotechnol* 104(17):7377–7389. doi: 10.1007/s00253-020-10775-w.

Doron, S. I., P. L. Hibberd, and S. L. Gorbach. 2008. Probiotics for prevention of antibiotic-associated diarrhea. *J Clin Gastroenterol* 42(Supplement 2):S58–S63. doi: 10.1097/mcg.0b013e3181618ab7.

Dowdell, P., S. Chankhamhaengdecha, W. Panbangred, T. Janvilisri, and A. Aroonnual. 2020. Probiotic activity of Enterococcus faecium and Lactococcus lactis isolated from Thai fermented sausages and their protective effect against Clostridium difficile. *Probiotics Antimicrob Proteins* 12(2):641–648. doi: 10.1007/s12602-019-09536-7.

Dugoua, J.-J., M. Machado, X. Zhu, X. Chen, G. Koren, and T. R. Einarson. 2009. Probiotic safety in pregnancy: A systematic review and meta-analysis of randomized controlled trials of Lactobacillus, Bifidobacterium, and Saccharomyces spp. *J Obstet Gynaecol Can* 31(6):542–552. doi: 10.1016/s1701-2163(16)34218-9.

Edalati, E., B. Saneei, M. Alizadeh, S. Hosseini, A. Zahedi Bialvaei, and K. Taheri. 2019. Isolation of probiotic bacteria from raw camel's milk and their antagonistic effects on two bacteria causing food poisoning. *New Microbes New Infect* 27:64–68. doi: 10.1016/j.nmni.2018.11.008.

El-Garawani, I. M., S. M. El-Sabbagh, N. H. Abbas, H. S. Ahmed, O. A. Eissa, D. M. Abo-Atya, S. A. M. Khalifa, and H. R. El-Seedi. 2020. A newly isolated strain of Halomonas sp. (HA1) exerts anticancer potential via induction of apoptosis and G2/M arrest in hepatocellular carcinoma (HepG2) cell line. *Sci Rep* 10(1):14076. doi: 10.1038/s41598-020-70945-8.

Eswaran, S., A. Guentner, and W. D. Chey. 2014. Emerging pharmacologic therapies for constipation-predominant irritable bowel syndrome and chronic constipation. *J Neurogastroenterol Motil* 20(2):141–151. doi: 10.5056/jnm.2014.20.2.141.

FAO/WHO. 2002. Joint FAO/WHO working group report on drafting guidelines for the evaluation of probiotics in food London, Ontario, Canada, April 30 and May 1, 2002. 1:1–11.

Faten, K., K. Hamida, E. A. Soumya, I. S. K. Saad, M. Hasna, L. Hassan, and H. Moktar. 2016. Lactobacillus plantarum: Effect of a protective biofilm on the surface of olives during storage. *Braz J Microbiol* 47(1):202–209. doi: 10.1016/j.bjm.2015.11.028.

Feliatra, F., Z. A. Muchlisin, H. Y. Teruna, W. R. Utamy, N. Nursyirwani, and A. Dahliaty. 2018. Potential of bacteriocins produced by probiotic bacteria isolated from tiger shrimp and prawns as antibacterial to Vibrio, Pseudomonas, and Aeromonas species on fish. *F1000Res* 7:415. doi: 10.12688/f1000research.13958.1.

Fijan, S. 2014. Microorganisms with claimed probiotic properties: An overview of recent literature. *Int J Environ Res Public Health* 11(5):4745–4767. doi: 10.3390/ijerph110504745.

Forssten, S. D., A. Laitila, J. Maukonen, and A. C. Ouwehand. 2020. Probiotic triangle of success; strain production, clinical studies and product development. *FEMS Microbiol Lett* 367(19):fnaa167. doi: 10.1093/femsle/fnaa167.

Foulquié Moreno, M. R., P. Sarantinopoulos, E. Tsakalidou, and L. De Vuyst. 2006. The role and application of enterococci in food and health. *Int J Food Microbiol* 106(1):1–24. doi: 10.1016/j.ijfoodmicro.2005.06.026.

Founou, L. L., R. C. Founou, and S. Y. Essack. 2016. Antibiotic resistance in the food chain: A developing country-perspective. *Front Microbiol* 7:1881. doi: 10.3389/fmicb.2016.01881.

Fuller, R. 1989. Probiotics in man and animals. *J Appl Bacteriol* 66(5):365–378. doi: 10.1111/j.1365-2672.1989.tb05105.x.

Furtado, D. N., S. D. Todorov, M. Landgraf, M. T. Destro, and B. D. Franco. 2015. Bacteriocinogenic Lactococcus lactis subsp. lactis DF04Mi isolated from goat milk: Application in the control of Listeria monocytogenes in fresh Minas-type goat cheese. *Braz J Microbiol* 46(1):201–206. doi: 10.1590/s1517-838246120130761.

García Fontán, M. C., S. Martínez, I. Franco, and J. Carballo. 2006. Microbiological and chemical changes during the manufacture of Kefir made from cows' milk, using a commercial starter culture. *Int Dairy J* 16(7):762–767. doi: 10.1016/j.idairyj.2005.07.004.

Gautam, N., and N. Sharma. 2009. Bacteriocin: Safest approach to preserve food products. *Indian J Microbiol* 49(3):204–211. doi: 10.1007/s12088-009-0048-3.

Gautam, N., and N. Sharma. 2015. A study on characterization of new bacteriocin produced from a novel strain of Lactobacillus spicheri G2 isolated from Gundruk- a fermented vegetable product of North East India: A novel bacteriocin production from *Lactobacillus spicheri* G2. *J Food Sci Technol* 52(9):5808–5816. doi: 10.1007/s13197-015-1710-x.

Geremew, T., A. Kebede, and B. Andualem. 2015. The role of spices and lactic acid bacteria as antimicrobial agent to extend the shelf life of metata ayib (traditional Ethiopian spiced fermented cottage cheese). *J Food Sci Technol* 52(9):5661–5670. doi: 10.1007/s13197-014-1694-y.

Gest, H. 2004. The discovery of microorganisms by Robert Hooke and Antoni van Leeuwenhoek, fellows of the royal society. *Notes Rec R Soc Lond* 58(2):187–201. doi: 10.1098/rsnr.2004.0055.

Ghane, M., L. Babaeekhou, and S. S. Ketabi. 2020. Antibiofilm activity of kefir probiotic lactobacilli against uropathogenic *Escherichia coli* (UPEC). *Avicenna J Med Biotechnol* 12(4):221–229. doi: AJMB-12-221 [pii].

Giraffa, G. 2004. Studying the dynamics of microbial populations during food fermentation. *FEMS Microbiol Rev* 28(2):251–260.

Gok, I. 2021. Functional potential of several Turkish fermented traditional foods: Biotic properties, bioactive compounds, and health benefits. *Food Rev Int* 00(00):1–26.

Grin, P. M., P. M. Kowalewska, W. Alhazzan, A. E. Fox-Robichaud, P. M. Kowalewska, W. Alhazzan, and A. E. Fox-Robichaud. 2013. Lactobacillus for preventing recurrent urinary tract infections in women: Meta-analysis. *Can J Urol* 20(1):6607–6614.

Guerra, P. V. P., L. N. Lima, T. C. Souza, V. Mazochi, F. J. Penna, A. M. Silva, J. R. Nicoli, and E. V. Guimarães. 2011. Pediatric functional constipation treatment with Bifidobacterium-containing yogurt: A crossover, double-blind, controlled trial. *World J Gastroenterol* 17(34):3916–3921. doi: 10.3748/wjg.v17.i34.3916.

Guslandi, M., G. Mezzi, M. Sorghi, and P. A. Testoni. 2000. *Saccharomyces boulardii* in maintenance treatment of Crohn's disease. *Dig Dis Sci* 45(7):1462–1464. doi: 10.1023/a:1005588911207.

Guzel-Seydim, Z. B., T. Kok-Tas, A. K. Greene, and A. C. Seydim. 2011. Review: Functional properties of kefir. *Crit Rev Food Sci Nutr* 51(3):261–268. doi: 10.1080/10408390903579029.

Haruta, S., S. Ueno, I. Egawa, K. Hashiguchi, A. Fujii, M. Nagano, M. Ishii, and Y. Igarashi. 2006. Succession of bacterial and fungal communities during a traditional pot fermentation of rice vinegar assessed by PCR-mediated denaturing gradient gel electrophoresis. *Int J Food Microbiol* 109(1–2):79–87.

Hassan, M. U., H. Nayab, F. Shafique, M. P. Williamson, T. S. Almansouri, N. Asim, N. Shafi, S. Attacha, M. Khalid, N. Ali, and N. Akbar. 2020. Probiotic properties of Lactobacillus helveticus and *Lactobacillus plantarum* isolated from traditional Pakistani yoghurt. *BioMed Res Int* 2020:8889198. doi: 10.1155/2020/8889198.

Hempel, S., S. J. Newberry, A. R. Maher, Z. Wang, J. N. Miles, R. Shanman, B. Johnsen, and P. G. Shekelle. 2012. Probiotics for the prevention and treatment of antibiotic- associated diarrhea: A systematic review and meta-analysis. *JAMA* 307(18):1959–1969. doi: 10.1001/jama.2012.3507.

Hennequin, C., C. Kauffmann-Lacroix, A. Jobert, J. P. Viard, C. Ricour, J. L. Jacquemin, and P. Berche. 2000. Possible role of catheters in *Saccharomyces boulardii* fungemia. *Eur J Clin Microbiol Infect Dis* 19(1):16–20. doi: 10.1007/s100960050003.

Hinrichsen, L. L., M. B. Montel, and R. Talon. 1994. Proteolytic and lipolytic activities of *Micrococcus roseus* (65), *Halomonas elongate* (16) and *Vibrio spp.* (168) isolated from Danish bacon curing brines. *Int. J. Food Microbiol* 22(2–3):115–126.

Hoffmann, N., A. A. Amara, B. B. Beermann, Q. Qi, H. J. Hinz, and B. H. A. Rehm. 2002. Biochemical characterization of the *Pseudomonas putida* 3-hydroxyacyl ACP:CoA transacylase which diverts intermediates of fatty acid *de novo* biosynthesis. *J Biol Chem* 277(45):42926–42936.

Hong, Y. 2013. Comparison of bacterial community changes in fermenting kimchi at two different temperatures using a denaturing gradient gel electrophoresis analysis. *J Microbiol Biotechnol* 23(1):76–84. doi: 10.4014/jmb.1210.10002.

Huang, G., X. Li, D. Lu, S. Liu, X. Suo, Q. Li, and N. Li. 2018. Lysozyme improves gut performance and protects against enterotoxigenic Escherichia coli infection in neonatal piglets. *Vet Res* 49(1):20. doi: 10.1186/s13567-018-0511-4.

Hussain, M. Z., and A. A. Amara. 2006. Case-by-case study using antibiotic-EDTA combination to control *Pseudomonas aeruginosa. Pak J Pharm Sci* 19(3):236–243.

Iseppi, R., A. Lauková, and C. Sabia. 2021. Editorial: Bacteriocin-producing probiotic bacteria: A natural solution for increasing efficiency and safety of livestock food production. *Front Microbiol* 12:675483. doi: 10.3389/fmicb.2021.675483.

Iyapparaj, P., T. Maruthiah, R. Ramasubburayan, S. Prakash, C. Kumar, G. Immanuel, and A. Palavesam. 2013. Optimization of bacteriocin production by Lactobacillus sp. MSU3IR against shrimp bacterial pathogens. *Aquat Biosyst* 9(1):12. doi: 10.1186/2046-9063-9-12.

Jeong, D.-W., S. Heo, and J.-H. Lee. 2017. Safety assessment of *Tetragenococcus halophilus* isolates from doenjang, a Korean high-salt-fermented soybean paste. *Food Microbiol* 62:92–98. doi: 10.1016/j.fm.2016.10.012.

Jeong, S. H., J. Y. Jung, S. H. Lee, H. M. Jin, and C. O. Jeon. 2013. Microbial succession and metabolite changes during fermentation of dongchimi, traditional Korean watery kimchi. *Int J Food Microbiol* 164(1):46–53. doi: 10.1016/j.ijfoodmicro.2013.03.016.

Jeong, S. H., S. H. Lee, J. Y. Jung, E. J. Choi, and C. O. Jeon. 2013. Microbial succession and metabolite changes during long-term storage of kimchi. *J Food Sci* 78(5):M763–M769. doi: 10.1111/1750-3841.12095.

Jia, C. F., W. N. Yu, and B. L. Zhang. 2020. Manufacture and antibacterial characteristics of *Eucommia ulmoides* leaves vinegar. *Food Sci Biotechnol* 29(5):657–665. doi: 10.1007/s10068-019-00712-7.

Jinendiran, S., W. Teng, H. U. Dahms, W. Liu, V. K. Ponnusamy, C. C. C. Chiu, B. S. D. Kumar, and N. Sivakumar. 2020. Induction of mitochondria-mediated apoptosis and suppression of tumor growth in zebrafish xenograft model by cyclic dipeptides identified from *Exiguobacterium acetylicum. Sci Rep* 10(1):13721. doi: 10.1038/s41598-020-70516-x.

Goldenberg, J. Z., L. Lytvyn, J. Steurich, P. Parkin, S. Mahant, and B. C. Johnston. 2015. Probiotics for the prevention of pediatric antibiotic-associated diarrhea: Reviews. *Cochrane Database of Systematic Reviews* (12). Art. No.: CD004827.

Jung, J. Y., S. H. Lee, H. M. Jin, Y. Hahn, E. L. Madsen, and C. O. Jeon. 2013. Metatranscriptomic analysis of lactic acid bacterial gene expression during kimchi fermentation. *Int J Food Microbiol* 163(2–3):171–179. doi: 10.1016/j.ijfoodmicro.2013.02.022.

Kanmani, P., R. S. Kumar, N. Yuvaraj, K. A. Paari, V. Pattukumar, and V. Arul. 2011. Design and optimization of fermentation medium for enhanced bacteriocin production by probiotic bacterium *Enterococcus faecium* MC13. *Prep Biochem Biotechnol* 41(1):40–52.

Kanmani, P., R. S. Kumar, N. Yuvaraj, K. A. Paari, V. Pattukumar, and V. Arul. 2012. Application of response surface methodology in the optimisation of a growth medium for enhanced natural preservative bacteriocin production by a probiotic bacterium. *Nat Prod Res* 26(16):1539–1543.

Kariyawasam, K., S. J. Yang, N. K. Lee, and H. D. Paik. 2020. Probiotic properties of Lactobacillus brevis KU200019 and synergistic activity with fructooligosaccharides in antagonistic activity against foodborne pathogens. *Food Sci Anim Resour* 40(2):297–310. doi: 10.5851/kosfa.2020.e15.

Kasimanickam, V., M. Kasimanickam, and R. Kasimanickam. 2021. Antibiotics use in food animal production: Escalation of antimicrobial resistance: Where are we now in combating AMR? *Med Sci (Basel)* 9(1):14. doi: 10.3390/medsci9010014.

Katsi, P., I. S. Kosma, S. Michailidou, A. Argiriou, A. V. Badeka, and M. G. Kontominas. 2021. Characterization of artisanal spontaneous sourdough wheat bread from central Greece: Evaluation of physico-chemical, microbiological, and sensory properties in relation to conventional yeast leavened wheat bread. *Foods* 10(3):635. doi: 10.3390/foods10030635.

Kiliç, A. O., S. I. Pavlova, W. G. Ma, and L. Tao. 1996. Analysis of Lactobacillus phages and bacteriocins in American dairy products and characterization of a phage isolated from yogurt. *Appl Environ Microbiol* 62(6):2111–2116.

Kim, S. W., S. I. Kang, D. H. Shin, S. Y. Oh, C. W. Lee, Y. Yang, Y. K. Son, H. S. Yang, B. H. Lee, H. J. An, I. S. Jeong, and W. Y. Bang. 2020. Potential of cell-free supernatant from *Lactobacillus plantarum* NIBR97, including novel bacteriocins, as a natural alternative to chemical disinfectants. *Pharmaceuticals (Basel)* 13(10):266. doi: 10.3390/ph13100266.

Kim, Y., S. Kim, S. Lee, J. Ha, J. Lee, Y. Choi, H. Oh, Y. Lee, Ns. Oh, Y. Yoon, and H. Lee. 2021. Antimicrobial activity of fermented Maillard reaction products, novel milk-derived material, made by whey protein and *Lactobacillus rhamnosus* and *Lactobacillus gasseri* on *Clostridium perfringens*. *Anim Biosci* 34(9):1525–1531. doi: 10.5713/ab.20.0290.

Kohlgraf, K. G., L. C. Pingel, D. E. Dietrich, and K. A. Brogden. 2010. Defensins as anti-inflammatory compounds and mucosal adjuvants. *Future Microbiol* 5(1):99–113. doi: 10.2217/fmb.09.104.

Koistinen, V. M., O. Mattila, K. Katina, K. Poutanen, A. M. Aura, and K. Hanhineva. 2018. Metabolic profiling of sourdough fermented wheat and rye bread. *Sci Rep* 8(1):5684. doi: 10.1038/s41598-018-24149-w.

König, H., and J. Fröhlich. 2009. Lactic acid bacteria. In König, H., Unden, G., and Fröhlich, J. (Eds.), *Biology of Microorganisms on Grapes, in Must and in Wine*. Berlin Heidelberg: Springer, pp. 3–29.

Koutsoumanis, K., A. Lianou, and J. Sofos. 2014. Food safety: Emerging pathogens. In *Encyclopedia of Agriculture and Food Systems*, 250–272. doi: 10.1016/b978-0-444-52512-3.00049-8.

Kubo, Y., A. P. Rooney, Y. Tsukakoshi, R. Nakagawa, H. Hasegawa, and K. Kimura. 2011. Phylogenetic analysis of *Bacillus subtilis* Strains applicable to Natto (fermented soybean) production. *Appl Environ Microbiol* 77(18):6463–6469. doi: 10.1128/aem.00448-11.

Kubota, M., K. Ito, K. Tomimoto, M. Kanazaki, K. Tsukiyama, A. Kubota, H. Kuroki, M. Fujita, and Y. Vandenplas. 2020. *Lactobacillus reuteri* DSM 17938 and magnesium oxide in children with functional chronic constipation: A double-blind and randomized clinical trial. *Nutrients* 12(1):225. doi: 10.3390/nu12010225.

Kumar, N., V. Kumar, S. M. Waheed, and D. Pradhan. 2020. Efficacy of reuterin and bacteriocins nisin and pediocin in the preservation of raw milk from dairy farms. *Food Technol Biotechnol* 58(4):359–369. doi: 10.17113/ftb.58.04.20.6728.

Kumar, S. B., S. R. Arnipalli, and O. Ziouzenkova. 2020. Antibiotics in food chain: The consequences for antibiotic resistance. *Antibiotics (Basel)* 9(10):688. doi: 10.3390/antibiotics9100688.

Lauková, A., A. Kandričáková, and E. Bino. 2020. Susceptibility to enterocins and lantibiotic bacteriocins of biofilm-forming enterococci isolated from Slovak fermented meat products available on the market. *Int J Environ Res Public Health* 17(24):9586. doi: 10.3390/ijerph17249586.

Laulund, S., A. Wind, P. M. F. Derkx, and V. Zuliani. 2017. Regulatory and safety requirements for food cultures. *Microorganisms* 5(2):28. doi: 10.3390/microorganisms5020028.

Lázaro, C. A., M. L. G. Monteiro, and C. A. Conte-Junior. 2020. Combined effect of modified atmosphere packaging and UV-C radiation on pathogens reduction, biogenic amines, and shelf life of refrigerated tilapia (*Oreochromis niloticus*) fillets. *Molecules* 25(14):3222. doi: 10.3390/molecules25143222.

Lee, J.-S., G.-Y. Heo, J. W. Lee, Y.-J. Oh, J. A. Park, Y.-H. Park, Y.-R. Pyun, and J. S. Ahn. 2005. Analysis of kimchi microflora using denaturing gradient gel electrophoresis. *Int J Food Microbiol* 102(2):143–150. doi: 10.1016/j.ijfoodmicro.2004.12.010.

Lee, M. R., Y. T. Huang, P. I. Lee, C. H. Liao, C. C. Lai, L. N. Lee, and P. R. Hsueh. 2011. Health care associated bacteraemia caused by *Leuconostoc species* at a university hospital in Taiwan between 1995 and 2008. *J Hosp Infect* 78(1):45–49.

Lee, S., J. A. Lee, G. G. Park, J. K. Jang, and Y. S. Park. 2017. Semi-continuous fermentation of onion vinegar and its functional properties. *Molecules* 22(8):1313. doi: 10.3390/molecules22081313.

Lee, S. C., R. B. Billmyre, A. Li, S. Carson, S. M. Sykes, E. Y. Huh, P. Mieczkowski, D. C. Ko, C. A. Cuomo, and J. Heitman. 2014. Analysis of a food-borne fungal pathogen outbreak: Virulence and genome of a Mucor circinelloides isolate from yogurt. *mBio* 5(4):e01390. doi: 10.1128/mbio.01390-14.

Leroy, F., and L. De Vuyst. 2004. Lactic acid bacteria as functional starter cultures for the food fermentation industry. *Trends Food Sci Technol* 15(2):67–78.

Li, J., W. Wang, S. X. Xu, N. A. Magarvey, and J. K. McCormick. 2011. *Lactobacillus reuteri*-produced cyclic dipeptides quench agr-mediated expression of toxic shock syndrome toxin-1 in staphylococci. *Proc Natl Acad Sci U S A* 108(8):3360–3365. doi: 10.1073/pnas.1017431108.

Li, Y., L. Shan, C. Zhang, Z. Lei, and Y. Shang. 2019. Isolation and antibiotic resistant research of *Tetragenococcus halophilus* from Xuanwei ham, A China high-salt-fermented meat products. *Antibiotics (Basel)* 8(3):151. doi: 10.3390/antibiotics8030151.

Lilly, D. M., and R. H. Stillwell. 1965. Probiotics: Growth-promoting factors produced by microorganisms. *Science* 147(3659):747–748.

Liu, L., X. Zhuge, H. Shin, R. R. Chen, J. Li, G. Du, and J. Chen. 2015. Improved production of propionic acid in *Propionibacterium jensenii* via combinational overexpression of glycerol dehydrogenase and malate dehydrogenase from *Klebsiella pneumoniae*. *Appl Environ Microbiol* 81(7):2256–2264. doi: 10.1128/aem.03572-14.

Liu, L., N. Guan, G. Zhu, J. Li, H. Shin, G. Du, and J. Chen. 2016. Pathway engineering of *Propionibacterium jensenii* for improved production of propionic acid. *Sci Rep* 6:19963. doi: 10.1038/srep19963.

Liu, N., Q. Zhu, X. Zeng, B. Yang, M. Liang, L. Deng, L. He, C. Liang, R. Zhang, and J. Zhou. 2018. Optimization of processing conditions of Chinese smoke-cured bacon (Larou) with a new natural coating solution during storage period. *Korean J Food Sci Anim Resour* 38(3):636–652. doi: 10.5851/kosfa.2018.38.3.636.

Liu, R., A. H. Kim, M. K. Kwak, and S. O. Kang. 2017. Proline-based cyclic dipeptides from Korean fermented vegetable kimchi and from *Leuconostoc mesenteroides* LBP-K06 have activities against multidrug-resistant bacteria. *Front Microbiol* 8:761. doi: 10.3389/fmicb.2017.00761.

Liu, S., Y. Ma, Y. Zheng, W. Zhao, X. Zhao, T. Luo, J. Zhang, and Z. Yang. 2020. Cold-stress response of probiotic *Lactobacillus plantarum* K25 by iTRAQ proteomic analysis. *J Microbiol Biotechnol* 30(2):187–195. doi: 10.4014/jmb.1909.09021.

Lobos, O., A. Barrera, and C. Padilla. 2017. Microorganisms of the intestinal microbiota of Oncorhynchus mykiss produce antagonistic substances against bacteria contaminating food and causing disease in humans. *Ital J Food Saf* 6(2):6240. doi: 10.4081/ijfs.2017.6240.

Luzón-Quintana, L. M., R. Castro, and E. Durán-Guerrero. 2021. Biotechnological processes in fruit vinegar production. *Foods* 10(5):945. doi: 10.3390/foods10050945.

M, K., and S. N. Sh. 2018. Isolation and molecular identification of bacteriocin-producing enterococci with broad antibacterial activity from traditional dairy products in Kerman Province of Iran. *Korean J Food Sci Anim Resour* 38(1):172–179. doi: 10.5851/kosfa.2018.38.1.172.

Menisy, M., A. Hussein, A. A. E. Ghazy, S. Sheweita, and A. A. A. F. Amara. 2017. *Klebsiella pneumoniae* ghosts as vaccine using sponge like reduced protocol. *Cell Mol Med* 3(2):11.

Mackowiak, P. A. 2013. Recycling Metchnikoff: Probiotics, the intestinal microbiome and the quest for long life. *Front Public Health* 1:52. doi: 10.3389/fpubh.2013.00052.

Magnuson, B., I. Munro, P. Abbot, N. Baldwin, R. Lopez-Garcia, K. Ly, L. McGirr, A. Roberts, and S. Socolovsky. 2013. Review of the regulation and safety assessment of food substances in various countries and jurisdictions. *Food Addit Contam Part A Chem Anal Control Expo Risk Assess* 30(7):1147–1220. doi: 10.1080/19440049.2013.795293.

Mahalmani, V. M., P. Sarma, A. Prakash, and B. Medhi. 2019. Positive list of antibiotics and food products: Current perspective in India and across the globe. *Indian J Pharmacol* 51(4):231–235. doi: 10.4103/ijp.IJP_548_19.

Makarova, K., A. Slesarev, Y. Wolf, A. Sorokin, B. Mirkin, E. Koonin, A. Pavlov, N. Pavlova, V. Karamychev, N. Polouchine, V. Shakhova, I. Grigoriev, Y. Lou, D. Rohksar, S. Lucas, K. Huang, D. M. Goodstein, T. Hawkins, V. Plengvidhya, D. Welker, J. Hughes, Y. Goh, A. Benson, K. Baldwin, J. H. Lee, I.

Diaz-Muniz, B. Dosti, V. Smeianov, W. Wechter, R. Barabote, G. Lorca, E. Altermann, R. Barrangou, B. Ganesan, Y. Xie, H. Rawsthorne, D. Tamir, C. Parker, F. Breidt, J. Broadbent, R. Hutkins, D. O'Sullivan, J. Steele, G. Unlu, M. Saier, T. Klaenhammer, P. Richardson, S. Kozyavkin, B. Weimer, and D. Mills. 2006. Comparative genomics of the lactic acid bacteria. *Proc Natl Acad Sci* 103(42):15611–15616. doi: 10.1073/pnas.0607117103.

Makino, K., K. Ishii, T. Yasunaga, M. Hattori, K. Yokoyama, C. H. Yutsudo, Y. Kubota, Y. Yamaichi, T. Iida, K. Yamamoto, T. Honda, C. G. Han, E. Ohtsubo, M. Kasamatsu, T. Hayashi, S. Kuhara, and H. Shinagawa. 1998. Complete nucleotide sequences of 93-kb and 3.3-kb plasmids of an enterohemorrhagic Escherichia coli O157:H7 derived from Sakai outbreak. *DNA Res* 5(1):1–9.

Margalho, L. P., G. P. Jorge, D. A. P. Noleto, C. E. Silva, J. S. Abreu, M. V. F. Piran, M. Brocchi, and A. S. Sant'Ana. 2021. Biopreservation and probiotic potential of a large set of lactic acid bacteria isolated from Brazilian artisanal cheeses: From screening to in product approach. *Microbiol Res* 242:126622.

Markowiak, P., and K. Śliżewska. 2017. Effects of probiotics, prebiotics, and Synbiotics on human health. *Nutrients* 9(9):E1021.

Marsh, A. J., O. O'Sullivan, C. Hill, R. Paul Ross, and P. D. Cotter. 2014. Sequence-based analysis of the bacterial and fungal compositions of multiple kombucha (tea fungus) samples. *Food Microbiol* 38:171–178. doi: 10.1016/j.fm.2013.09.003.

Mattila-Sandholm, T., P. Myllärinen, R. Crittenden, G. Mogensen, R. Fondén, and M. Saarela. 2002. Technological challenges for future probiotic foods. *Int Dairy J* 12(2–3):173–182. doi: 10.1016/s0958-6946(01)00099-1.

McGovern, P. E. 2007. *Ancient Wine: The Search for the Origins of Viniculture*. Princeton, NJ: Princeton University Press.

McGovern, P. E. 2020. *Ancient Wine: The Search for the Origins of Viniculture*. Princeton, NJ: Princeton University Press.

Menezes, A. G. T., C. L. Ramos, G. Cenzi, D. S. Melo, D. R. Dias, and R. F. Schwan. 2020. Probiotic potential, antioxidant activity, and phytase production of indigenous yeasts isolated from indigenous fermented foods. *Probiotics Antimicrob Proteins* 12(1):280–288. doi: 10.1007/s12602-019-9518-z.

Michelakou, E., E. Giaouris, D. Doultsos, C. Nasopoulou, and P. Skandamis. 2021. Evaluation of the microbial stability and shelf life of 50% NaCl-reduced traditional Greek pork meat product "Syglino of Monemvasia" stored under vacuum at different temperatures. *Heliyon* 7(11):e08296. doi: 10.1016/j.heliyon.2021.e08296.

Minervini, F., R. Di Cagno, A. Lattanzi, M. De Angelis, L. Antonielli, G. Cardinali, S. Cappelle, and M. Gobbetti. 2012. Lactic acid bacterium and yeast microbiotas of 19 sourdoughs used for traditional/typical Italian breads: Interactions between ingredients and microbial species diversity. *Appl Environ Microbiol* 78(4):1251–1264. doi: 10.1128/aem.07721-11.

Mogensen, J. M., J. Varga, U. Thrane, and J. C. Frisvad. 2009. Aspergillus acidus from Puerh tea and black tea does not produce ochratoxin A and fumonisin B2. *Int J Food Microbiol* 132(2–3):141–144. doi: 10.1016/j.ijfoodmicro.2009.04.011.

Mohammad, S. M., N.-K. Mahmud-Ab-Rashid, and N. Zawawi. 2020. Probiotic properties of bacteria isolated from bee bread of stingless bee *Heterotrigona itama*. *J Apic Res* 60(1):172–187. doi: 10.1080/00218839.2020.1801152.

Mokoena, M. P. 2017. Lactic acid bacteria and their bacteriocins: Classification, biosynthesis and applications against uropathogens: A mini-review. *Molecules* 22(8):1255. doi: 10.3390/molecules22081255.

Moreno, M. R. F., J. J. Leisner, L. K. Tee, C. Ley, S. Radu, G. Rusul, M. Vancanneyt, and L. De Vuyst. 2002. Microbial analysis of Malaysian tempeh, and characterization of two bacteriocins produced by isolates of Enterococcus faecium. *J Appl Microbiol* 92(1):147–157. doi: 10.1046/j.1365-2672.2002.01509.x.

Murphy, D., A. Ricci, Z. Auce, J. G. Beechinor, H. Bergendahl, R. Breathnach, J. Bureš, J. P. Duarte Da Silva, J. Hederová, P. Hekman, C. Ibrahim, E. Kozhuharov, G. Kulcsár, E. L. Persson, J. M. Lenhardsson, P. Mačiulskis, I. Malemis, L. Markus-Cizelj, A. Michaelidou-Patsia, M. Nevalainen, P. Pasquali, J. Rouby, J. Schefferlie, W. Schlumbohm, M. Schmit, S. Spiteri, S. Srčič, L. Taban, T. Tiirats, B. Urbain, E. Vestergaard, A. Wachnik- Święcicka, J. Weeks, B. Zemann, A. Allende, D. Bolton, M. Chemaly, P. S. F. Escamez, R. Girones, L. Herman, K. Koutsoumanis, R. Lindqvist, B. Nørrung, L. Robertson, G. Ru, M. Sanaa, M. Simmons, P. Skandamis, E. Snary, N. Speybroeck, B. Ter Kuile, H. Wahlström, K. Baptiste, B. Catry, P. S. Cocconcelli, R. Davies, C. Ducrot, C. Friis, G. Jungersen, S. More, C. Muñoz Madero, P. Sanders, M. Bos, Z. Kunsagi, J. Torren Edo, R. Brozzi, D. Candiani, B. Guerra, E. Liebana, P. Stella, J. Threlfall, and H. Jukes. 2017. EMA and EFSA Joint Scientific Opinion on measures to reduce the need to use antimicrobial agents in animal husbandry in the European Union, and the resulting impacts on food safety (RONAFA). *EFSA J* 15(1):4666. doi: 10.2903/j.efsa.2017.4666.

Musatti, A., C. Cappa, C. Mapelli, C. Alamprese, and M. Rollini. 2020. *Zymomonas mobilis* in bread dough: Characterization of dough leavening performance in presence of sucrose. *Foods (Basel, Switzerland)* 9(1):89. doi: 10.3390/foods9010089.

Naidu, A. S., W. R. Bidlack, and R. A. Clemens. 1999. Probiotic spectra of Lactic Acid Bacteria (LAB). Crit Rev Food Sci Nutr. 39(1):13–126. doi: 10.1080/10408699991279187. PMID: 10028126.

Naveed, R., I. Hussain, A. Tawab, M. Tariq, M. Rahman, S. Hameed, M. S. Mahmood, A. B. Siddique, and M. Iqbal. 2013. Antimicrobial activity of the bioactive components of essential oils from Pakistani spices against Salmonella and other multi-drug resistant bacteria. *BMC Complement Altern Med* 13:265. doi: 10.1186/1472-6882-13-265.

Nazari, M., and D. L. Smith. 2020. A PGPR-produced bacteriocin for sustainable agriculture: A review of thuricin 17 characteristics and applications. *Front Plant Sci* 11:916. doi: 10.3389/fpls.2020.00916.

Neffe-Skoci´nsk, K., K. Wójciak, and D. Zielinska. 2016. Probiotic microorganisms in dry fermented meat products. V. Rao, and Rao, L.G. (Ed.). London: InTech.

Negash, A. W., and B. A. Tsehai. 2020. Current applications of bacteriocin. *Int J Microbiol* 2020:4374891. doi: 10.1155/2020/4374891.

Nieminen, M. T., L. Novak-Frazer, V. Rautemaa, R. Rajendran, T. Sorsa, G. Ramage, P. Bowyer, and R. Rautemaa. 2014. A novel antifungal is active against *Candida albicans* biofilms and inhibits mutagenic acetaldehyde production in vitro. *PLOS ONE* 9(5):e97864. doi: 10.1371/journal.pone.0097864.

Nout, R. 2000. Useful role of fungi in food processing. In Samson, R. A., Hoekstra, E. S., Frisvad, J. C., and Filtenborg, O. (Eds.), *Introduction to Food- and Airborne Fungi*, 6th ed. Utrecht: Centraal Bureau Voor Schimmel Cultures, pp. 295–303.

Nout, R. 2007. The colonizing fungus as a food provider. In Dijksterhuis, J., and Samson, R. A. (Eds.), *Food Mycology: A Multifaceted Approach to Fungi and Food*. Boca Raton, FL: CRC Press, pp. 335–352.

O'Shea, E. F., P. M. O'Connor, O. O'Sullivan, P. D. Cotter, R. P. Ross, and C. Hill. 2013. Bactofencin A, a new type of cationic bacteriocin with unusual immunity. *mBio* 4(6):e00498. doi: 10.1128/mBio.00498-13.

O'Mahony, T., N. Rekhif, C. Cavadini, and G. F. Fitzgerald. 2001. The application of a fermented food ingredient containing 'variacin', a novel antimicrobial produced by Kocuriavarians, to control the growth of *Bacillus cereus* in chilled dairy products. *J Appl Microbiol* 90(1):106–114.

O'Reilly, G. C., Y. Huo, S. J. Meale, and A. V. Chaves. 2021. Dose response of biochar and wood vinegar on in vitro batch culture ruminal fermentation using contrasting feed substrates. *Transl Anim Sci* 5(3):txab107. doi: 10.1093/tas/txab107.

Oelschlaeger, T. A. 2010. Mechanisms of probiotic actions: A review. *Int J Med Microbiol* 300(1):57–62.

Oh, C.-K., M.-C. Oh, and S.-H. Kim. 2004. The depletion of sodium nitrite by lactic acid bacteria isolated from kimchi. *J Med Food* 7(1):38–44. doi: 10.1089/109662004322984680.

Ohenhen, R. E., J. O. Isibor, G. Emonfonmwan, and S. A. Enabulele. 2015. Effects of pH and storage temperatures on antibacterial activity of bacteriocin produced by lactic acid bacteria isolated from OGI. *Br Microbiol Res J* 9(3):1–9.

Papalexandratou, Z., G. Vrancken, K. De Bruyne, P. Vandamme, and L. De Vuyst. 2011. Spontaneous organic cocoa bean box fermentations in Brazil are characterized by a restricted species diversity of lactic acid bacteria and acetic acid bacteria. *Food Microbiol* 28(7):1326–1338.

Parker, R. B. 1974. Probiotics, the other half of the antibiotic story. *Anim Nutr Health* 29:4–8.

Pederson, C. S., G. Niketic, and M. N. Albury. 1962. Fermentation of the Yugoslavian pickled cabbage. *Appl Microbiol* 10(1):86–89. doi: 10.1128/aem.10.1.86-89.1962.

Peivasteh-Roudsari, L., M. Pirhadi, H. Karami, B. Tajdar-oranj, E. Molaee-Aghaee, and P. Sadighara. 2019. Probiotics and food safety: An evidence-based review. *J Food Safe & Hyg* 5(1):1–9.

Pellerin, G., L. Bazinet, and D. Grenier. 2021. Deacidification of cranberry juice reduces its antibacterial properties against oral streptococci but preserves barrier function and attenuates the inflammatory response of oral epithelial cells. *Foods* 10(7):1634. doi: 10.3390/foods10071634.

Perez, H. A., A. Y. Bustos, M. P. Taranto, M. D. L. A. Frías, and A. E. Ledesma. 2018. Effects of lysozyme on the activity of ionic of fluoroquinolone species. *Molecules* 23(4):741. doi: 10.3390/molecules230 40741.

Perumal, V., A. Repally, A. Dasari, and A. Venkatesan. 2016. Partial purification and characterization of bacteriocin produced by *Enterococcus faecalis* DU10 and its probiotic attributes. *Prep Biochem Biotechnol* 46(7):686–694.

Pimentel, T. C., W. Costa, C. E. Barao, M. Rosset, and M. Magnani. 2020. Vegan probiotic products: A modern tendency or the newest challenge in functional foods. *Food Res Int* 140:110033. doi: 10.1016/j. foodres.2020.110033.

Pineiro, M., and C. Stanton. 2021. Probiotic bacteria: Legislative framework-requirements to evidence basis. *J Nutr* 137(3):851S–853S.
Piwowarek, K., E. Lipińska, E. Hać-Szymańczuk, M. Kieliszek, and I. Ścibisz. 2018. Propionibacterium spp.—Source of propionic acid, vitamin B12, and other metabolites important for the industry. *Appl Microbiol Biotechnol* 102(2):515–538. doi: 10.1007/s00253-017-8616-7.
Plessas, S., A. Alexopoulos, C. Voidarou, E. Stavropoulou, and E. Bezirtzoglou. 2011. Microbial ecology and quality assurance in food fermentation systems: The case of kefir grains application. *Anaerobe* 17(6):483–485. doi: 10.1016/j.anaerobe.2011.03.014.
Plessas, S., D. E. Kiousi, M. Rathosi, A. Alexopoulos, Y. Kourkoutas, I. Mantzourani, A. Galanis, and E. Bezirtzoglou. 2020. Isolation of a Lactobacillus paracasei Strain with probiotic attributes from kefir grains. *Biomedicines* 8(12):594. doi: 10.3390/biomedicines8120594.
Pogány Simonová, M., Ľ. Chrastinová, and A. Lauková. 2020. Autochtonous strain *Enterococcus faecium* EF2019(CCM7420), its bacteriocin and their beneficial effects in broiler rabbits—A review. *Animals (Basel)* 10(7):1188. doi: 10.3390/ani10071188.
Pupa, P., P. Apiwatsiri, W. Sirichokchatchawan, N. Pirarat, N. Muangsin, A. A. Shah, and N. Prapasarakul. 2021. The efficacy of three double-microencapsulation methods for preservation of probiotic bacteria. *Sci Rep* 11(1):13753. doi: 10.1038/s41598-021-93263-z.
Raveschot, C., F. Coutte, M. Fremont, M. Vaeremans, J. Dugersuren, S. Demberel, D. Drider, P. Dhulster, C. Flahaut, and B. Cudennec. 2020. Probiotic Lactobacillus strains from Mongolia improve calcium transport and uptake by intestinal cells in vitro. *Food Res Int* 133:109201. doi: 10.1016/j.foodres.2020.109201.
Ray, B. 1992. Cells of lactic acid bacteria as food biopreservatives. In Ray, B., and Daeschel, M. A. (Eds.), *Food Biopreservatives of Microbial Origin*. Boca Raton, FL: CRC Press, 81–101.
Ray, B., and W. E. Sandine. 1992. Acetic, propionic and lactic acids of starter culture bacteria as biopreservatives. In: Ray, B., and Daeschel, M. A. (Eds.), *Food Biopreservatives of Microbial Origin*. Boca Raton, FL: CRC Press, pp. 103–136.
Rehm, B. H. A., R. V. Antonio, P. Spiekermann, A. A. Amara, and A. Steinbüchel. 2002. Molecular characterization of the poly (3-hydroxybutyrate) (PHB) synthase from Ralstonia eutropha: *In vitro* evolution, site-specific mutagenesis and development of a PHB synthase protein model. *Biochim Biophys Acta* 1594(1):178–190.
Rezac, S., C. R. Kok, M. Heermann, and R. Hutkins. 2018. Fermented foods as a dietary source of live organisms. *Front Microbiol* 9:1785. doi: 10.3389/fmicb.2018.01785.
Ricci, A., A. Levante, M. Cirlini, L. Calani, V. Bernini, D. Del Rio, G. Galaverna, E. Neviani, and C. Lazzi. 2018. The influence of viable cells and cell-free extracts of *Lactobacillus casei* on volatile compounds and polyphenolic profile of elderberry juice. *Front Microbiol* 9:2784. doi: 10.3389/fmicb.2018.02784.
Ricci, A., G. Bertani, A. Maoloni, V. Bernini, A. Levante, E. Neviani, and C. Lazzi. 2021. Antimicrobial activity of fermented vegetable byproduct extracts for food applications. *Foods (Basel, Switzerland)* 10(5):1092. doi: 10.3390/foods10051092.
Rodríguez-Sánchez, S., P. Fernandez-Pacheco, S. Sesena, C. Pintado, and M. L. I. Palop. 2021. Selection of probiotic *Lactobacillus* strains with antimicrobial activity to be used as biocontrol agents in food industry. *LWT* 143:111142.
Romero-Luna, H. E., A. Peredo-Lovillo, A. Hernandez-Mendoza, H. Hernandez-Sanchez, P. I. Cauich-Sanchez, R. M. Ribas-Aparicio, and G. Davila-Ortiz. 2020. Probiotic potential of Lactobacillus paracasei CT12 isolated from water kefir grains (tibicos). *Curr Microbiol* 77(10):2584–2592. doi: 10.1007/s00284-020-02016-0.
Salazar, J. K., L. J. Gonsalves, M. Fay, P. Ramachandran, K. M. Schill, and M. L. Tortorello. 2021. Metataxonomic profiling of native and starter microbiota during ripening of Gouda cheese made with Listeria monocytogenes-contaminated unpasteurized milk. *Front Microbiol* 12:642789. doi: 10.3389/fmicb.2021.642789.
Sarkar, D., M. Yabusaki, Y. Hasebe, P. Y. Ho, S. Kohmoto, T. Kaga, and K. Shimizu. 2010. Fermentation and metabolic characteristics of *Gluconacetobacter oboediens* for different carbon sources. *Appl Microbiol Biotechnol* 87(1):127–136. doi: 10.1007/s00253-010-2474-x.
Saviano, A., M. Brigida, A. Migneco, G. Gunawardena, C. Zanza, M. Candelli, F. Franceschi, and V. Ojetti. 2021. *Lactobacillus reuteri* DSM 17938 (*Limosilactobacillus reuteri*) in diarrhea and constipation: Two sides of the same coin? *Med (Kaunas Lith)* 57(7):643. doi: 10.3390/medicina57070643.
Selhub, E. M., A. C. Logan, and A. C. Bested. 2014. Fermented foods, microbiota, and mental health: Ancient practice meets nutritional psychiatry. *J Physiol Anthropol* 33(1):2. doi: 10.1186/1880-6805-33-2.
Sengun, Y. I., and S. Karabiyikli. 2011. Importance of acetic acid bacteria in food industry. *Food Control* 22(5):647–656.

Shah, N., and J. B. Prajapati. 2014. Effect of carbon dioxide on sensory attributes, physico-chemical parameters and viability of probiotic *L. helveticus* MTCC 5463 in fermented milk. *J Food Sci Technol* 51(12):3886–3893. doi: 10.1007/s13197-013-0943-9.

Shan, L. S., P. Hou, Z. J. Wang, F. R. Liu, N. Chen, L. H. Shu, H. Zhang, X. H. Han, X. X. Han, X. X. Cai, Y. X. Shang, and Y. Vandenplas. 2013. Prevention and treatment of diarrhoea with *Saccharomyces boulardii* in children with acute lower respiratory tract infections. *Benef Microbes* 4(4):329–334. doi: 10.3920/bm2013.0008.

Shehata, H. R., and S. G. Newmaster. 2020. A validated real-time PCR method for the specific identification of probiotic strain *Lactobacillus rhamnosus* GG (ATCC 53103). *J AOAC Int* 103(6):1604–1609. doi: 10.1093/jaoacint/qsaa063.

Sheoran, P., and S. K. Tiwari. 2019. Anti-staphylococcal activity of bacteriocins of food isolates Enterococcus hirae LD3 and *Lactobacillus plantarum* LD4 in pasteurized milk. *3 Biotech* 9(1):8. doi: 10.1007/s13205-018-1546-y.

Shyntum, D. Y., N. P. Nkomo, N. L. Shingange, A. R. Gricia, D. Bellieny-Rabelo, and L. N. Moleleki. 2019. The impact of type VI secretion system, bacteriocins and antibiotics on bacterial competition of Pectobacterium carotovorum subsp. brasiliense and the regulation of carbapenem biosynthesis by iron and the ferric-uptake regulator. *Front Microbiol* 10:2379. doi: 10.3389/fmicb.2019.02379.

Silva, D. R., J. C. O. Sardi, N. de Souza Pitangui, S. M. Roque, A. C. B. da Silva, and P. L. Rosalen. 2020. Probiotics as an alternative antimicrobial therapy: Current reality and future directions. *J Funct Foods* 73:104080.

Simoes da Silva, T. M., A. C. M. Piazentin, C. M. N. Mendonca, A. Converti, C. S. B. Bogsan, D. Mora, and R. P. de Souza Oliveira. 2020. Buffalo milk increases viability and resistance of probiotic bacteria in dairy beverages under in vitro simulated gastrointestinal conditions. *J Dairy Sci* 103(9):7890–7897. doi: 10.3168/jds.2019-18078.

Siragusa, S., M. De Angelis, M. Calasso, D. Campanella, F. Minervini, R. Di Cagno, and M. Gobbetti. 2014. Fermentation and proteome profiles of *Lactobacillus plantarum* strains during growth under food-like conditions. *J Proteomics* 96:366–380. doi: 10.1016/j.jprot.2013.11.003.

Sivamaruthi, B., P. Kesika, and C. Chaiyasut. 2019. Toxins in fermented foods: Prevalence and preventions-A mini review. *Toxins* 11(1):4. doi: 10.3390/toxins11010004.

Solheim, M., S. L. La Rosa, T. Mathisen, L. G. Snipen, I. F. Nes, and D. A. Brede. 2014. Transcriptomic and functional analysis of NaCl-induced stress in *Enterococcus faecalis*. *PLOS ONE* 9(4):e94571. doi: 10.1371/journal.pone.0094571.

Song, W., H. Chung, D. Kang, and J. Ha. 2019. Microbial quality of reduced-sodium napa cabbage kimchi and its processing. *Food Sci Nutr* 7(2):628–635. doi: 10.1002/fsn3.898.

Spinnler, H. E., and J. C. Gripon. 2004. Surface mould-ripened cheeses. In *Cheese Chemistry, Physics, and Microbiology.*. Elsevier, pp. 157–174.

Sridhar, A., M. Ponnuchamy, P. S. Kumar, and A. Kapoor. 2021. Food preservation techniques and nanotechnology for increased shelf life of fruits, vegetables, beverages and spices: A review. *Environ Chem Lett* 19(2):1735. doi: 10.1007/s10311-020-01126-2.

Sturino, J. M., M. Rajendran, and E. Altermann. 2013. Draft genome sequence of the pediocin-encoding biopreservative and biocontrol strain *Pediococcus acidilactici* D3. *Genome Announc* 1(3):e00208-13. doi: 10.1128/genomeA.00208-13.

Suesse, A. R., G. A. Norton, and J. van Leeuwen. 2016. Pilot-scale continuous-flow hydrothermal liquefaction of filamentous fungi. *Energy Fuels* 30(9):7379–7386. doi: 10.1021/acs.energyfuels.6b01229.

Supriyanto, Y. F., and I. Hayakawa. 1994. Statistical characterization of tempeh starter from the aroma components of soybean tempeh. In *Developments in Food Engineering: Proceedings of the 6th International Congress on Engineering and Food*. Boston, MA: Springer US, pp. 522–524.

Surono, I. S. 2016. Ethnic fermented foods and beverages of Indonesia. In Tamang, J. (Ed.), *Ethnic Fermented Foods and Alcoholic Beverages of Asia*. New Delhi: Springer. https://doi.org/10.1007/978-81-322-2800-4_14

Suskovich, J., B. Kos, J. Beganovich, A. Pavunc, K. Habjanic, and S. Matosich. 2010. Antimicrobial activity – The most important property of probiotic and starter lactic acid bacteria. *Food Technol Biotechnol* 48(3):296–307.

Sybesma, W., J. Hugenholtz, W. M. De Vos, and E. J. Smid. 2006. Safe use of genetically modified lactic acid bacteria in food: Bridging the gap between consumers, green groups, and industry. *Electron J Biotechnol* 9(4). doi: 10.2225/vol9-issue4-fulltext-12.

Szajewska, H., and J. Mrukowicz. 2005. Meta-analysis: Non-pathogenic yeast *Saccharomyces boulardii* in the prevention of antibiotic-associated diarrhoea. *Aliment Pharmacol Ther* 22(5):365–372. doi: 10.1111/j.1365-2036.2005.02624.x.

Tamang, J., and D. Samuel. 2010. Dietary cultures and antiquity of fermented foods and beverages. In *Fermented Foods and Beverages of the World*. Boca Raton: CRC Press, pp. 1–40. https://doi.org/10.1201/EBK1420094954-c1

Tamang, J., N. Thapa, B. Tamang, A. Rai, and R. Chettri. 2015. Microorganisms in fermented foods and beverages , Chap. 1. In *Health Benefits of Fermented Foods and Beverages*. CRC Press.

Tamang, J. P., W. H. Holzapfel, and K. Watanabe. 2016. Review: Diversity of microorganisms in global fermented foods and beverages. *Front Microbiol* 7:377.

Tamang, J. P., D.-H. Shin, S.-J. Jung, and S.-W. Chae. 2016. Functional properties of microorganisms in fermented foods. *Front Microbiol* 7:578doi. doi: 10.3389/fmicb.2016.00578.

Tanasupawat, S., J. Thongsanit, S. Okada, and K. Komagata. 2002. Lactic acid bacteria isolated from soy sauce mash in Thailand. *J Gen Appl Microbiol* 48(4):201–209. doi: 10.2323/jgam.48.201.

Taverniti, V., C. Scabiosi, S. Arioli, D. Mora, and S. Guglielmetti. 2014. Short-term daily intake of 6 billion live probiotic cells can be insufficient in healthy adults to modulate the intestinal bifidobacteria and lactobacilli. *J Funct Foods* 6:482–491. doi: 10.1016/j.jff.2013.11.014.

Tenea, G. N., and J. Suárez. 2020. Probiotic potential and technological properties of bacteriocinogenic *Lactococcus lactis* Subsp. Lactis UTNGt28 from a native Amazonian fruit as a yogurt starter culture. *Microorganisms* 8(5):733. doi: 10.3390/microorganisms8050733.

Terefe, N. S., and M. A. Augustin. 2020. Fermentation for tailoring the technological and health related functionality of food products. *Crit Rev Food Sci Nutr* 60(17):2887–2913.

Thapa, S., and J. P. Tamang. 2004. Product characterization of kodo ko jaanr: Fermented finger millet beverage of the Himalayas. *Food Microbiol* 21(5):617–622. doi: 10.1016/j.fm.2004.01.004.

Tharmaraj, N., and N. P. Shah. 2009. Antimicrobial effects of probiotics against selected pathogenic and spoilage bacteria in cheese-based dips. *Int Food Res J* 16:261–276.

Thierry, A., F. Valence, S.-M. Deutsch, S. Even, H. Falentin, Y. Le Loir, G. Jan, and V. Gagnaire. 2015. Strain-to-strain differences within lactic and propionic acid bacteria species strongly impact the properties of cheese–A review. *Dairy Sci Technol* 95(6):895–918. doi: 10.1007/s13594-015-0267-9.

Thygesen, J. B., H. Glerup, and B. Tarp. 2012. *Saccharomyces boulardii* fungemia caused by treatment with a probioticum. *BMJ Case Rep* 2012. doi: 10.1136/bcr.06.2011.4412.

Toğay, S. Ö., M. Ay, O. Güneşer, and Y. K. Yüceer. 2016. Investigation of antimicrobial activity and entA and entB genes in Enterococcus faecium and *Enterococcus faecalis* strains isolated from naturally fermented Turkish white cheeses. *Food Sci Biotechnol* 25(6):1633–1637. doi: 10.1007/s10068-016-0251-z.

Toshima, H., A. Yoshimura, K. Arikawa, A. Hidaka, J. Ogasawara, A. Hase, H. Masaki, and Y. Nishikawa. 2007. Enhancement of Shiga toxin production in enterohemorrhagic Escherichia coli Serotype O157:H7 by DNase colicins. *Appl Environ Microbiol* 73(23):7582–7588. doi: 10.1128/aem.01326-07.

Toussaint-Samat, M. 2009. *A History of Food*. Oxford, UK: Wiley-Blackwell.

Trapecar, M., T. Leouffre, M. Faure, H. E. Jensen, P. E. Granum, A. Cencic, and S. P. Hardy. 2011. The use of a porcine intestinal cell model system for evaluating the food safety risk of *Bacillus cereus* probiotics and the implications for assessing enterotoxigenicity. *APMIS* 119(12):877–884. doi: 10.1111/j.1600-0463.2011.02797.x.

Trivedi, D., P. K. Jena, and S. Seshadri. 2014. Colicin E2 expression in *Lactobacillus brevis* DT24, A vaginal probiotic isolate, against uropathogenic *Escherichia coli*. *ISRN Urol* 2014:869610. doi: 10.1155/2014/869610.

Uraipan, S., and T. Hongpattarakere. 2015. Antagonistic characteristics against food-borne pathogenic bacteria of lactic acid bacteria and bifidobacteria isolated from feces of healthy Thai infants. *Jundishapur J Microbiol* 8(6):e18264. doi: 10.5812/jjm.8(5)2015.18264.

Valledor, S. J. D., J. E. V. Bucheli, W. H. Holzapfel, and S. D. Todorov. 2020. Exploring beneficial properties of the bacteriocinogenic *Enterococcus faecium* ST10Bz strain isolated from Boza, a Bulgarian cereal-based beverage. *Microorganisms* 8(10):1474s. doi: 10.3390/microorganisms8101474.

Vasquez, E. C., R. Aires, A. M. M. Ton, and F. G. Amorim. 2020. New insights on the beneficial effects of the probiotic kefir on vascular dysfunction in cardiovascular and neurodegenerative diseases. *Curr Pharm Des* 26(30):3700–3710. doi: 10.2174/1381612826666200304145224.

Velázquez-Suárez, C., R. Cebrián, C. Gasca-Capote, A. Sorlózano-Puerto, J. Gutiérrez-Fernández, M. Martínez-Bueno, M. Maqueda, and E. Valdivia. 2021. Antimicrobial activity of the circular bacteriocin AS-48 against clinical multidrug-resistant *Staphylococcus aureus*. *Antibiotics (Basel)* 10(8):925. doi: 10.3390/antibiotics10080925.

Venkatachalam, K. V., W. Huang, M. LaRocco, and T. Palzkil. 1994. Characterization of TEM-1 beta-lactamase mutants from positions 238 to 241 with increased catalytic efficiency for ceftazidim. *J Boil Chem* 269(38):23444–43450.

VidyaLaxme, B., A. Rovetto, R. Grau, and R. Agrawal. 2014. Synergistic effects of probiotic Leuconostoc mesenteroides and Bacillus subtilis in malted ragi (*Eleucine corocana*) food for antagonistic activity against *V. cholerae* and other beneficial properties. *J Food Sci Technol* 51(11):3072–3082. doi: 10.1007/s13197-012-0834-5.

Vieira, A. T., M. M. Teixeira, and F. S. Martins. 2012. The role of probiotics and prebiotics in inducing gut immunity. *Front Immunol* 4. doi: 10.3389/fimmu.2013.00445.

Vijayakumar, P. P., and P. M. Muriana. 2017. Inhibition of Listeria monocytogenes on ready-to-eat meats using bacteriocin mixtures based on mode-of-action. *Foods* 6(3):22. doi: 10.3390/foods6030022.

Vijayendra, S. V. N., K. Rajashree, and P. M. Halami. 2010. Characterization of a heat stable anti-listerial bacteriocin produced by vancomycin sensitive *Enterococcus faecium* isolated from idli batter. *Indian J Microbiol* 50(2):243–246. doi: 10.1007/s12088-010-0030-0.

Vitetta, L., S. Hall, and A. W. Linnane. 2014. Live probiotic cultures and the gastrointestinal tract: Symbiotic preservation of tolerance whilst attenuating pathogenicity. *Front Cell Infect Microbiol* 4:143. doi: 10.3389/fcimb.2014.00143.

Voidarou, C., M. Antoniadou, G. Rozos, A. Tzora, I. Skoufos, T. Varzakas, A. Lagiou, and E. Bezirtzoglou. 2021a. Fermentative foods: Microbiology, biochemistry, potential human health benefits and public health issues. *Foods* 10(1). doi: 10.3390/foods10010069.

Voidarou, C., Î. Antoniadou, G. Rozos, A. Tzora, I. Skoufos, T. Varzakas, A. Lagiou, and E. Bezirtzoglou. 2021b. Fermentative foods: Microbiology, biochemistry, potential human health benefits and public health issues. *Foods* 10(1):69. doi: 10.3390/foods10010069.

Walter, J. 2008. Ecological role of lactobacilli in the gastrointestinal tract: Implications for fundamental and biomedical research. *Appl Environ Microbiol* 74(16):4985–4996. doi: 10.1128/aem.00753-08.

Wang, M. C., A. I. Zaydi, W. H. Lin, J. S. Lin, M. T. Liong, and J. J. Wu. 2020. Putative probiotic strains isolated from kefir improve gastrointestinal health parameters in adults: A randomized, single-blind, placebo-controlled study. *Probiotics Antimicrob Proteins* 12(3):840–850. doi: 10.1007/s12602-019-09615-9.

Wang, S.-Y., K.-N. Chen, Y.-M. Lo, M.-L. Chiang, H.-C. Chen, J.-R. Liu, and M.-J. Chen. 2012. Investigation of microorganisms involved in biosynthesis of the kefir grain. *Food Microbiol* 32(2):274–285. doi: 10.1016/j.fm.2012.07.001.

Witthuhn, R. C., T. Schoeman, and T. J. Britz. 2005. Characterisation of the microbial population at different stages of Kefir production and Kefir grain mass cultivation. *Int Dairy J* 15(4):383–389. doi: 10.1016/j.idairyj.2004.07.016.

Woskow, S. A., and B. A. Glatz. 1991. Propionic acid production by a propionic acid-tolerant strain of *Propionibacterium acidipropionici* in batch and semicontinuous fermentation†. *Appl Environ Microbiol* 57(10):2821–2828.

Wyder, M.-T., L. Meile, and M. Teuber. 1999. Description of *Saccharomyces turicensis* sp. nov., a new Species from Kefyr. *Syst Appl Microbiol* 22(3):420–425. doi: 10.1016/s0723-2020(99)80051-4.

Xia, P., J. Zhu, and G. Zhu. 2013. *Escherichia coli* Nissle 1917 as safe vehicles for intestinal immune targeted therapy—A review. *Acta Microbiol Sin* 53(6):538–544.

Xiao, J.-Z., S. Takahashi, M. Nishimoto, T. Odamaki, T. Yaeshima, K. Iwatsuki, and M. Kitaoka. 2010. Distribution of *in vitro* fermentation ability of lacto- N -Biose I, a major building block of human milk oligosaccharides, in bifidobacterial strains. *Appl Environ Microbiol* 76(1):54–59. doi: 10.1128/aem.01683-09.

Yerlikaya, O., D. Saygili, and A. Akpinar. 2021. Evaluation of antimicrobial activity and antibiotic susceptibility profiles of Lactobacillus delbrueckii subsp. Bulgaricus and Streptococcus thermophilus strains isolated from commercial yoghurt starter cultures. *Food Sci Technol, (Campinas)* 41(2):418–425.

Yu, H. H., Y. W. Chin, and H. D. Paik. 2021. Application of natural preservatives for meat and meat products against food-borne pathogens and spoilage bacteria: A review. *Foods* 10(10):2418. doi: 10.3390/foods10102418.

Yusuf, M. A., and H. Thata. 2013. Lactic acid bacteria: Bacteriocin producer: A mini review. *IOSR J Pharm* 3(4):44–50.

Zelaya, H., K. Tsukida, E. Chiba, G. Marranzino, S. Alvarez, H. Kitazawa, G. Agüero, and J. Villena. 2014. Immunobiotic lactobacilli reduce viral-associated pulmonary damage through the modulation of inflammation–coagulation interactions. *Int Immunopharmacol* 19(1):161–173. doi: 10.1016/j.intimp.2013.12.020.

Zhai, Z., F. Zhang, R. Cao, X. Ni, Z. Xin, J. Deng, G. Wu, W. Ren, Y. Yin, and B. Deng. 2019. Cecropin A alleviates inflammation through modulating the gut microbiota of C57BL/6 mice with DSS-induced IBD. *Front Microbiol* 10:1595. doi: 10.3389/fmicb.2019.01595.

Zhang, D., R. Y. Gan, A. K. Farha, G. Kim, Q. Q. Yang, X. M. Shi, C. L. Shi, Q. X. Luo, X. B. Xu, H. B. Li, and H. Corke. 2019. Discovery of antibacterial dietary spices that target antibiotic-resistant bacteria. *Microorganisms* 7(6):157. doi: 10.3390/microorganisms7060157.

Zhang, M., Y. Shan, H. Gao, B. Wang, X. Liu, Y. Dong, X. Liu, N. Yao, Y. Zhou, X. Li, and H. Li. 2018. Expression of a recombinant hybrid antimicrobial peptide magainin II-cecropin B in the mycelium of the medicinal fungus *Cordyceps militaris* and its validation in mice. *Microb Cell Fact* 17(1):18. doi: 10.1186/s12934-018-0865-3.

Zhang, Y., L. Zhu, P. Dong, R. Liang, Y. Mao, S. Qiu, and X. Luo. 2018. Bio-protective potential of lactic acid bacteria: Effect of Lactobacillus sakei and *Lactobacillus curvatus* on changes of the microbial community in vacuum-packaged chilled beef. *Asian-Australas J Anim Sci* 31(4):585–594. doi: 10.5713/ajas.17.0540.

Zhao, C. M., T. Du, P. Li, X. J. Du, and S. Wang. 2021. Production and characterization of a novel low-sugar beverage from red jujube fruits and bamboo shoots fermented with selected Lactiplantibacillus plantarum. *Foods* 10(7):1439. doi: 10.3390/foods10071439.

Zhu, R., X. Liu, X. Li, K. Zeng, and L. Yi. 2021. Transformation of inferior tomato into preservative: Fermentation by multi-bacteriocin producing *Lactobacillus paracasei* WX322. *Foods (Basel, Switzerland)* 10(6):1278. doi: 10.3390/foods10061278.

Zimina, M., O. Babich, A. Prosekov, S. Sukhikh, S. Ivanova, M. Shevchenko, and S. Noskova. 2020. Overview of global trends in classification, methods of preparation and application of bacteriocins. *Antibiotics (Basel)* 9(9):553. doi: 10.3390/antibiotics9090553.

Zommiti, M., E. Bouffartigues, O. Maillot, M. Barreau, S. Szunerits, K. Sebei, M. Feuilloley, N. Connil, and M. Ferchichi. 2018. In vitro assessment of the probiotic properties and bacteriocinogenic potential of *Pediococcus pentosaceus* MZF16 isolated from artisanal Tunisian meat “dried Ossban”. *Front Microbiol* 9:2607. doi: 10.3389/fmicb.2018.02607.

Zugic Petrovic, T. D., P. D. Ilic, M. Z. Grujovic, K. G. Mladenovic, S. D. Kocic-Tanackov, and L. R. Comic. 2020. Assessment of safety aspect and probiotic potential of autochthonous *Enterococcus faecium* strains isolated from spontaneous fermented sausage. *Biotechnol Lett* 42(8):1513–1525. doi: 10.1007/s10529-020-02874-510.

3 Applications of antimicrobial agents in food science and technology

V. Manju Meena, B.S. Dhanya, K.R. Preethy, and M. Chamundeeswari

LIST OF ABBREVIATIONS

LAB: Lactic Acid Bacteria
FDA: Food and Drug Administration
DNA: Deoxyribonucleic Acid
RNA: Ribonucleic Acid
ATP: Adenosine Triphosphate
EPA: Eicosapentaenoic Acid
DHA: Docosahexaenoic Acid
pH: Potential of Hydrogen

3.1 INTRODUCTION

3.1.1 Current Global Trends

The demands from customers for a higher quality of life have grown in the current global market. A few elements influencing this requirement are the need for items that meet high-quality standards and minimal chemical use. The process of food contamination that leads to the loss of colour, texture, and nutritional content, as well as the growth of harmful bacteria, degrades the product's quality and renders it unfit for human consumption, is known as food spoilage. Contamination occurs when food is exposed to the environment during processing, packaging, and slaughter. Traditional food preservation methods, including drying, freezing, heating, fermenting, and salting, can increase the shelf life of food goods. Still, they also run the risk of recontamination, which could make the food tasteless to consumers (Malhotra et al., 2015). This has prompted businesses and researchers to investigate various strategies for increasing productivity using sustainable materials in food packaging. Recently, there has been an increased focus on creating materials with antimicrobial qualities that increase health safety and shelf life. Antimicrobial packaging is one of the best methods for preventing the growth of pathogenic microorganisms that contaminate food. Food safety is greatly influenced by food packaging. One of the critical elements affecting societal well-being and public health is food safety. Good antimicrobial packaging could be a potential remedy for foodborne infections (Ahmed et al., 2022).

3.1.2 Antimicrobials

Antimicrobials are substances that are used in the prevention and treatment of infections not only in humans but also in animals and plants. These are the agents used against many microorganisms.

DOI: 10.1201/9781003268949-3

They include antifungals, antibiotics, antiparasitic, and antiviral compounds. Their mode of action includes cidal/static. Cidal means it functions by killing the microorganisms, and static means preventing or retarding the growth of microorganisms in the log phase. They assault metabolic processes, such as protein synthesis, target DNA and its associated activities, or interfere with cell wall construction and function (Mohamad Asif, 2017).

3.1.2.1 Types of antimicrobials

The antimicrobials cover antibiotics/antibacterial, antiparasitic, antifungal, and antiviral agents.

3.1.2.1.1 Antibiotic agents

An antibiotic is described as "a chemical produced by a microorganism and in a diluted solution having the capacity to inhibit the growth of or kill another microorganism" in Webster's Third International Dictionary from 1981 (Julian). Examples include penicillin, cephalosporins, tetracyclines, aminoglycosides, etc. The use of antibiotics as preservatives is still controversial. An approved food preservative should not endanger the public's health or mask a product's inadequacies. It should considerably improve quality retention during harvesting, processing, storage, and distribution while remaining inexpensive enough to be used commercially. Antibiotic preservatives meet some of these requirements, but other issues might arise. Antibiotics are a persistent risk to human health, mainly when used with substances also used to treat human sickness (Vaughn et al., 1960).

3.1.2.1.2 Antiparasitic agents

Antiparasitics is recommended for treating parasitic disorders, including those brought on by helminths, amoebas, ectoparasites, parasitic fungi, and protozoa, among others. Antiparasitics kill or stop the growth of the parasitic infection-causing agents, and they are typically only effective against a small subset of parasites within a given class. Malaria, leishmaniasis, sleeping sickness, and filariasis are among the parasitic diseases that annually affect millions of people (Pink et al., 2005).

3.1.2.1.3 Antifungal agents

Food products' naturally nutrient-rich composition makes them subject to fungal deterioration. However, this susceptibility is dependent on several variables, including the nature of the food matrix, its composition, biological, physical, and chemical parameters, management during fruit and vegetable harvesting, and raw material storage. The antifungal activity of several different microorganisms against food spoilers has been studied. Microorganisms have been effectively used in food products as bioprotective instruments. Many strains from various microbial species that possess antifungal activities have been discovered in previous years. They have been isolated from various foods, including meat, milk, cereals, fruits, vegetables, and other food-related goods. In order to find microorganisms that could create new antifungal compounds, the isolation of new bioprotective cultures has recently been expanded to different habitats, such as deep-sea, hot springs, and cold region Antarctic soil samples. The two primary food industries investigated for biopreservation against fungi use lactic acid bacteria (LAB) as the primary microorganisms examined for application in dairy and bakery manufacturing. Additionally, these microorganisms have been investigated for use in fermented vegetables, brewing (the malting process), fruit, grain, and seed preservation (Salus et al., 2017).

3.1.2.1.4 Antiviral agents

The Food and Drug Administration (FDA) in the USA has approved drugs known as antiviral agents for treating or managing viral infections. Most antiviral treatments on the market only work against actively reproducing viruses. An ideal antiviral treatment would also work against latent viruses. These antiviral agents can be naturally/artificially produced. Recent studies show almonds, apples, bael, blackberry, blackcurrants, crane berries, citrus, grapes, mango, mulberry, pomegranate, strawberries, and other edible fruits have antiviral qualities. Many of these ingestible fruit extracts have antiviral properties that may work by preventing the generation of viral DNA/RNA or viral replication inside the host cell (Santhi et al., 2021).

3.1.3 Pathogens in Food

A wide variety of microorganisms can contaminate human food and water supplies, resulting in diseases when consumed or even when their poisons are ingested. Numerous enteric bacteria, aerobes, anaerobes, viral pathogens, parasites, marine dinoflagellates, and bacteria that generate biotoxins in fish and shellfish are all included in the spectrum of foodborne pathogens. The various causative organisms include Norwalk-like virus, *Campylobacter, Salmonella, Clostridium perfringens, Giardia lamblia, Staphylococcus, Escherichia coli* OH157:H7, *Shigella, Yersinia pestis*, and so on (Tauxe et al., 2002). These pathogens cause various impacts, such as those in Table 3.1 (Fuglsang et al., 1995).

3.1.4 Use of Antimicrobials in the Food Industry

Farmers and primary producers can use antimicrobials to maintain the health of their livestock. As a result, some foods we eat may contain insufficient quantities of antimicrobial residues. The prolonged usage of antimicrobials leads to antimicrobial resistance, which poses a threat to the environment. The activity of antimicrobial compounds depends on the physical and chemical characteristics, including volatility, polarity, hydrophilicity, and hydrophobicity. The chosen antimicrobial agent should be effective against the target organism, have the least interaction with the food components, and be stable during the processing stages. Antimicrobials are used in beverages, bakery goods, seafood, dairy, meat, and poultry products. Antimicrobials are preferred because they have limited challenges when utilized (Keymanesh et al., 2009).

3.2 ANTIMICROBIAL AGENTS

The food industry and consumers are paying close attention to using natural antimicrobial agents in food. They are the compounds that kill or suppress the growth of microorganisms, even at lower concentrations. These antimicrobial agents are obtained naturally from plants, plant byproducts, animals, yeasts, moulds, and other microorganisms using chemical synthesis (Gyawali et al., 2014).

3.2.1 Plant Byproducts

Plants produce various aromatic compounds, most of which are phenol derivatives. These are generally secondary metabolites. These metabolites serve as defence mechanisms for various species.

TABLE 3.1
Impacts of pathogens in food

Plant/animal health	Socio-economic impact	Food safety and security	Consumer health
Mortality/morbidity	Maintenance costs	Foodborne diseases	Human morbidity/ mortality
Production loss	Compensation costs to producers	Reduction in protein intake	Spread of infectious diseases
Depopulation	Farmers' costs of risk to mitigation	Food insecurity	The psychological impact of loss on the family
Reproductive loss	Ongoing market impacts	Decreased supply	Malnutrition and deficiency diseases

The structural diversity of plant-derived substances is vast, and their structural configuration determines the impact of the antimicrobial action they produce against microbes. It includes essential oils, terpenoids, alkaloids, flavonoids, lactones, tannins, coumarins, anthraquinones, polyacetylene, polyphenols, lignans, resins, organic acids, and sulphoxides. All these compounds are effective against various organisms like Gram-positive bacteria, fungi, *E. coli*, *Staphylococcus aureus*, *Helicobacter pylori*, viruses, etc. (Cowan et al., 1999).

3.2.2 Plant-Derived

Most antimicrobials are derived from herbs, spices, and their related compounds. These have been used since ancient times for various purposes like texture, colour, flavour enhancers, valued medicinal products, and preservatives. They are reported to have various antioxidant, antimicrobial, and pharmaceutical properties. All these properties are due to their structures and their positions. The presence of hydroxyl groups and double bonds plays an essential role in the antioxidant nature of the compounds. Plant antibacterial activity may also vary based on the type of microorganisms, extraction process, culture medium, inoculum size, and method of determination. The hydroxyl groups in phenols enhance electron delocalization, allowing them to act as proton exchangers and lower the gradient across the cytoplasmic membrane of bacterial cells. This will result in the depletion of the ATP pool and the collapse of the proton motive force, ultimately leading to cell death. They can even bind with the active sites of the enzymes by alteration of the metabolism of the microbes (Cowan et al., 1999).

3.2.3 Animal-Based

There are few antimicrobials derived from animal sources used in food industries. Lactoferrin is obtained majorly from milk. These target organisms like *E. coli* O157:H7, *Salmonella*, and *Pseudomonas*. They are preferred because of the presence of an iron-binding glycoprotein. Chitosan is obtained from the exoskeleton of crustaceans and arthropods and used for targeting organisms like *E. coli*, *S. aureus*, *Listeria monocytogenes*, and *Pseudomonas* because of the polycationic biopolymer compound. Lipids obtained from fish and shellfish have Eicosapentaenoic acid (EPA) and Docosahexaenoic acid (DHA), which are used against *Bacillus subtilis*, *L. monocytogenes*, and *S. aureus* ATCC 6538 and KCTC 1916. Defensins from mammalian epithelial cells of chickens and turkeys with cationic peptides are used against Gram-positive and Gram-negative bacteria, fungi, and viruses. Lactoperoxidase from raw milk colostrum, saliva, and other biological secretions containing glycoprotein is used against *Salmonella*, *E. coli*, *L. monocytogenes*, and *Yersinia enterocolitis*. Ovatransferrin obtained from egg albumin is used to target *Micrococcus*, *Bacillus*, and *E. coli* species due to the presence of glycoprotein. Lysozyme from hen's eggs and mammalian milk contain glutamic acid and aspartic acids that are used against *C. tyrobutyricum*, *Bacillus*, *Micrococcus*, and *L. monocytogenes* (Cowan et al., 1999).

3.2.4 Bacterial Derivatives

Antimicrobials are derived even from beneficial microbes. Microbes, mainly lactic acid bacteria (LAB), create a broad spectrum of antibacterial compounds. Proteinaceous substances known as bacteriocins, such as nisin, have been found to hinder the growth and development of other microbial species. Various antimicrobial compounds extracted from bacterial origin include nisin, natamycin, lactocin, acidophilin, bulgaricin, pediocin, reuterin, plantaricin, and helveticin (Cowan et al., 1999).

3.2.5 Algae and Mushroom Sources

There have been numerous reports in recent years of algae and mushrooms being natural sources of bioactive compounds with a wide range of biological activity, including antibiotic, antiviral, antioxidant, antifouling, anti-inflammatory, cytotoxic, antimitotic activity, and other health-promoting benefits. However, there needs to be more investigation into the antibacterial action of algae and mushrooms in food preservation. The enormous diversity of diverse algae and mushroom species may thus be a source of novel antimicrobials. Terpenoids, phlorotannins, acrylic acid, phenolic compounds, steroids, halogenated ketones and alkanes, cyclic polysulphides, and fatty acids have been reported to be present in algae. The presence of these chemicals suggests a variety of antibacterial activity mechanisms. Different bioactive secondary metabolites, volatile chemicals, certain phenols, gallic acids, free fatty acids, and their derivatives may be responsible for mushrooms' antibacterial action. They have significant components like polyphenols, carotenoids, amino acids, hydrocarbons, phytochemicals, lipophilic compounds, chlorophyll derivatives, etc. These are obtained from red, green, marine brown, brown, macrofungi, edible mushroom, and non-gilled fungal group species, which target various species such as *V. cholera*, *E. coli*, *S. aureus*, *B. subtilis*, *Salmonella*, *L. monocytogenes*, etc. (Cowan et al., 1999).

3.3 APPLICATIONS OF ANTIMICROBIAL AGENTS IN THE FOOD INDUSTRY

3.3.1 Antimicrobial Agents in Food Packaging

A recent innovation in the food industry is antimicrobial packaging, which inserts an antibacterial chemical into a polymer film to inhibit the growth of specific microbes. Numerous studies have discovered that when the right number of antimicrobial agents is put into a polymer film, it can effectively suppress targeted bacteria in food packaging. Due to two essential factors, antimicrobial agents incorporated in packaging are more effective than adding preservatives directly to food. Firstly, the delayed release of antimicrobial agents functions over a long time due to their adhesion to the polymer film. Secondly, when administered directly to the meal, the antimicrobial actions of preservation agents may be inactive by food matrixes and components (Muriel-Galet et al., 2012; Chausali et al., 2022, 2023). Additionally, the direct addition of preservatives to food can lower the quality of the food by altering its organoleptic and textural features. Thus, antimicrobial packaging is essential in preventing the spread of specific microorganisms on foods, enhancing food safety, and extending shelf life without degrading food quality. Food is often deemed to be spoiled when the total microbial count exceeds 10^7 CFU/g. An unpleasant odour develops when the total count reaches 10^8 CFU/g and above. Most researchers who evaluated packaging used a standard bacterium count of 10^7 CFU/ml, g, or cm^2 to indicate the shelf life of products (Sung et al., 2013).

By decreasing dangerous bacteria activity in food, antimicrobial packaging serves various purposes. Packaging resists the growth of microorganisms and helps to improve food safety. Antimicrobial packaging extends the shelf life of food and decreases food waste. Antimicrobial packaging materials made from biomaterials increase consumer health safety.

3.3.1.1 Plant extracts and essential oils as antimicrobial agents in food packaging

The antimicrobial actions of essential oils and plant extracts have long been known, and many study findings on the antimicrobial activities of plant essential oils against foodborne pathogens have been reported. Essential oils have a significant potential to inhibit a variety of bacteria because they are abundant in volatile terpenoids and phenolic particles. The active ingredients in plant essential oils typically prevent the growth of microbes by altering the proton motive force, active transport, cytoplasmic membrane, electron flow, and protein synthesis. Thymol, clove oil, linalool, cinnamaldehyde, carvacrol, and basic essential oils are a few examples of plant extracts and essential oils frequently used in food packaging (Sung et al., 2013).

Numerous studies have been conducted on the antimicrobial efficacy of plant extracts used in food packaging. According to Ha et al. (2001), when compared to films without antimicrobial agents, linear low-density polyethylene co-extruded films with 0.5% and 1% w/w grapefruit seed extract increased the shelf-time of beef and have been able to be stored at 3°C, increasing the total aerobic bacteria count from 13 to 14 days, or 10^7 CFU/g. The shelf life of beef is extended from 9 to 14 days when the same amount of grapefruit seed extract coated on linear low-density polyethylene film is tested. In the meantime, there is no discernible difference between the results of 0.5% and 1% grapefruit seed extract in linear low-density polyethylene film. *L. monocytogenes* and other harmful bacteria can be efficiently stopped from growing using grapefruit seed extract (Sivarooban et al., 2008).

Muriel-Galet et al. (2012) researched polypropylene film coated in citral and oregano essential oil. After being stored under 4°C in a packaging environment of 12% CO_2 and 4% O_2 for two days, oregano essential oil-coated polypropylene film with 6.7% carvacrol active ingredient successfully reduced the number of *L. monocytogenes*, *S. enteric*, and *E. coli*, in a salad with 0.36 log, 0.5 log, and 1.4 log, CFU/g, respectively, in comparison to the polypropylene film that is devoid of antimicrobial agent. Meanwhile, 0.36 log and 0.41 log CFU/g *E. coli* and *Salmonella enterica* are detected when 5% of citral-coated polypropylene film is used. However, compared to oregano, citral-coated polypropylene film performs less as an antimicrobial agent. Additionally, the polypropylene film with citral coating is ineffective against *L. monocytogenes*. Due to their strong affinity, carvacrol and citral are more efficient against Gram-negative bacteria. Gram-negative bacteria's cell wall lipids are particularly sensitive to carvacrol. This was in contrast to some researchers' claims that there is no difference between the efficiency of antimicrobial agents against Gram-positive and Gram-negative bacteria (Quattara et al., 1997).

Other extensively researched essential oil-coated films include rosemary (Han et al., 2007), garlic oil (Gamage et al., 2009), allyl isothiocyanate (Nadarajah et al., 2005), and cinnamaldehyde (Han et al., 2007). Oriented polypropylene/polyethylene film with 1.2% w/w of allyl isothiocyanate dramatically reduced the microbial count on sprouts kept at 10°C (Gamage et al., 2009). In addition, Nadarajah et al. (2005) noted that allyl isothiocyanate reduced the amount of *E. coli* on ground meat patties by 3 log CFU/g within ten days of storage. Only when a high concentration of garlic oil is utilized does film with garlic oil shows high effectiveness. According to Gamage et al. (2009), 1.2% weight-per-weight of garlic oil prevented the growth of microorganisms on sprouts, but 0.6, 0.8, and 1.0% w/w of garlic oil had no discernible effects. However, when used in modest quantities in edible film, garlic oil has been shown to lower the number of Gram-negative and Gram-positive bacteria successfully. In particular, 0.2% weight/weight of garlic oil-incorporated alginate film is shown to prevent the development of *S. aureus* and *B. cereus* (Pranoto et al., 2005). To assess the antimicrobial efficacy of protein-based film, essential oils are also included. Seydim and Sarikus (2006) developed whey protein isolate films containing ratios of rosemary, oregano, and garlic essential oils ranging from 1.0 to 4.0% (w/v). It is used to test against *E. coli* O157:H7 (ATCC 35218), *Lactiplantibacillus plantarum* (DSM 20174), *S. aureus* (ATCC 43300), *L. monocytogenes* (NCTC 2167), and *Salmonella enteritidis* (ATCC 13076). The scientists discovered that all tested strains are significantly inhibited when essential garlic oil in whey protein isolate films is raised to 3% w/v. Compared to rosemary essential oil-incorporated films, oregano, and garlic essential oil-added films showed increased inhibitory zones on *S. aureus*, *S. enteritidis*, *L. monocytogenes*, *E. coli*, and *L. plantarum*. Additionally, beef's shelf life is increased by a factor of 2 using whey protein isolate films that include 1.5% w/w oregano oil. When tested with fresh ground beef patties, Emiroglu et al. (2010) found that the antimicrobial actions of edible soy films infused with thyme and oregano essential oils revealed a lack of significant inhibition on *S. aureus*. When thyme (TH), oregano (OR) or thyme essential oil (ORT) were added to the film, *Pseudomonas aeruginosa* showed greater results in terms of colony reduction. On the other hand, Oussalah et al. (2004) investigated how milk protein-based film affected *E. coli* and *Pseudomonas* spp. They claimed that during seven days at 4°C of storage, films containing essential oils dramatically reduced the number of microorganisms in meat. The essential oil blend that inhibits the growth of the relevant bacteria is oregano oil. While a pimento oil-coated film is equally effective as an antimicrobial against the development of *Pseudomonas* spp., oregano oil and pimento oil in a 1:1 (w/w) ratio are more effective against the growth of *E. coli*.

3.3.1.2 Enzymes as antimicrobial agents in food packaging

Lysozyme, a known natural antibacterial agent, is derived from hen egg white and is frequently used in food packaging. Due to a protective lipopolysaccharide layer on the Gram-negative bacteria cell wall, lysozyme is less efficient against Gram-negative bacteria than other antimicrobial drugs. Gram-positive bacteria are particularly vulnerable to lysozyme because this enzyme hydrolyses the glycosidic bond between N-acetyl muramic acid and N-acetyl glucosamine in their membrane. Other antimicrobial agents, such as detergents and chelators, are frequently added to enzymes to increase their antimicrobial efficacy. Ellison and Gehl (1991) suggested using lactoferrin to increase lysozyme's activity against Gram-negative bacteria. Large cationic patches found in lactoferrin, a naturally occurring component of milk, interfere with the permeability of the outer membrane and cause the release of lipopolysaccharide. A decade ago, it was utilized in an antibacterial spray for cattle carcasses. Barbiroli et al. (2012) created a paper containing lysozyme and lactoferrin and tested it on *E. coli* (DSMZ 50902) and *Listeria innocua* (DSMZ 20649). When tested on the Gram-positive bacterium *L. innocua*, both antimicrobial drugs significantly increased the length of the lag phase, causing it to extend from 1.86 to 6.5 hours. Still, lysozyme alone only caused it to extend from 1.86 to 5.81 hours. Lysozyme alone only marginally lengthens the lag phase for *E. coli*, increasing it from 1.08 to 1.79 h; lactoferrin-containing papers demonstrated an increase from 1.8 to 3.25 h (Mecitoglu et al., 2006) suggested combining lysozyme and disodium-Ethylene-diamine-Tetra-Acetic acid (Na_2EDTA) as an antimicrobial agent. The chelating chemical EDTA can destabilize the lipopolysaccharide layer of Gram-negative bacteria. Zein films that included lysozyme demonstrated the impact of antimicrobials on *B. subtilis* and *L. plantarum*. Na_2EDTA is added to the films, which makes them effective against *E. coli*, a Gram-negative bacterium. Similarly, Gemili et al. (2009) found that utilizing cellulose acetate film containing both antimicrobial agents significantly formed an inhibitory zone during an agar disk diffusion test for *E. coli*. No zones emerged when lysozyme and Na_2EDTA were used alone. Through an agar disk diffusion test, Unalan et al. (2011) discovered that zein films containing lysosomes (700 gcm^{-1}) and Na_2EDTA (300 gcm^{-1}) are effective. The zein films containing lysozyme and Na_2EDTA alone and the combination are efficient against *Salmonella typhimurium* and *E. coli* O157:H7. After five days of storage, the films containing both antimicrobial agents reduced total viable counts and coliform counts on beef patties compared to control patties using the challenge test. In another work, Gucbilmez et al. (2007) used functional protein extracts – chicken albumin extract – to improve zein film's antioxidant activity while controlling the lysozyme's distribution and release rate. Zein films containing 277 g/cm^2 Na_2EDTA and 1400 $\mu g/cm^2$ lysozyme were tested on *E. coli* and revealed an uneven distribution of these antimicrobial agents. The antimicrobial efficacy did not alter considerably when the films contained 530 g/cm^2 protein extracts – chicken albumin extract.

3.3.1.3 Chitosan as antimicrobial agents in food packaging

Researchers are particularly drawn to chitosan because of its intriguing qualities, which include being a natural polymer, non-toxicity, capable of producing films, and biodegradability (Torlak & Nizamlioglu, 2011). Additionally, chitosan contains antimicrobial characteristics that can stop a variety of food pathogens from growing. However, the application of chitosan is only efficient in acidic media due to the limited solubility of chitosan at high pH values. In a study by Liu et al. (2004), chitosan solutions inhibited the populations of *E. coli* and *S. aureus* by 1 log in five minutes. In 120 minutes, almost all *E. coli* was destroyed, although *S. aureus* numbers remained unaffected. Its only flaw is the inferior mechanical qualities of chitosan film relative to the synthetic polymer. Thus, to alter mechanical characteristics and to improve antimicrobial activities, chitosan can be coated on a plastic sheet. The efficiency of 2% w/v chitosan coated on polypropylene films against a bacteria-inoculated Kashar cheese slice is investigated by Torlak & Nizamlioglu, (2011). For 14 days, the cheese slices are kept at 4°C in vacuum packaging. As a control, polypropylene films without an antimicrobial agent are employed. After 14 days, the control cheese had 5.11 log CFU/g of *L. monocytogenes*, 4.92 log CFU/g of *S. aureus*, and 4.88 log CFU/g of *E. coli* O157:H7. Compared to polypropylene films without an antimicrobial agent, chitosan-coated polypropylene films showed a reduction in the number of *L. monocytogenes*, *S. aureus*, and *E. coli* 157:H7 at 0.7, 0.61, and 0.49 log CFU/g, respectively. Similarly, Duan et al. (2007) also noted a notable decline in *L. monocytogenes* in camembert and mozzarella cheese. Within 14 days of storage in the refrigerator, packaging coated with 2% and 2.5% v/v chitosan under vacuum conditions reduced the population of the total aerobic count in pork from 6 log (control) to 3.75 and 3.61 log CFU/g, respectively. On the other hand, Joerger et al. (2009) tested a chitosan-coated film that is made utilizing a high-pressure treatment on turkey breast. After one week of storage, compared to plastic film without chitosan and high-pressure treatment, they reported a 3 log lower count of *L. monocytogenes.* According to most studies, chitosan is quite effective against Gram-negative bacteria. This is so that the negatively charged teichoic acid backbone in the cell wall of Gram-negative bacteria and the positively charged chitosan molecules can interact more effectively. Other antimicrobial agents are also included in chitosan-coated films to improve chitosan's antimicrobial characteristics further. Since chitosan can act as a carrier for additional chemicals, these synergistic combinations can kill a larger range of microorganisms. Zivanovic et al. (2005) researched chitosan-wrapped bologna and kept at 10°C for five days. They found that adding 1% oregano oil to chitosan films reduced the number of *L. monocytogenes* by 3.6 log, compared to chitosan films alone, which reduced the number of *L. monocytogenes* by 2 log. The inhibitory effects against bacteria are notable when 1% oregano and 1% clove essential oils are applied to chitosan-incorporated polypropylene film, particularly when essential oregano oil is introduced. After two weeks of storage, vacuum-sealed cheese kept at 4°C had a 1.31, 1.4, and 1.18 log CFU/g reduction in *L. monocytogenes*, *S. aureus*, and *E. coli* O157:H7 count compared with control film (Torlak & Nizamlioglu, 2011).

Due to its biodegradability property, chitosan has been extensively researched as an antimicrobial and edible packaging material. The most recent developments in edible chitosan films involve adding antimicrobial agents like nisin and essential oils and combining chitosan with other substances to develop very effective antimicrobial films (Du et al., 2008). According to the reports of Mehdizadeh et al. (2012), *Thymus kotschyanus* essential oil inhibits varying levels of *L. monocytogenes, E. coli, S. aureus*, and *S. typhimurium* growth in vitro when used with edible chitosan-starch films. Films devoid of essential oils failed to inhibit *S. typhimurium* when a definite zone of inhibition was absent. The films with *T. kotschyanus* essential oil inhibited *L. monocytogenes* the most, followed by *E. coli*, *S. aureus*, and *S. typhimurium.*

3.3.1.4 Bacteriocin as antimicrobial agents in food packaging

In addition to naturally occurring antimicrobial agents derived from plants, animals, and essential oils, bacteria-produced antimicrobial agents known as bacteriocin have been steadily gaining favour due to their resistance to high temperatures and acidic environments. Bacteriocin is a metabolic by-product synthesized by all forms of bacteria's defence mechanisms. Because of this naturally occurring activity, one strain of bacteria can stop the growth of another strain in close proximity. Due to lactic acid bacteria's long history as essential for food fermentation, the public has a high acceptability for the bacteriocin. The Food and Drug Administration has categorized nisin, a product of *Lactococcus lactis* bacteria frequently found in milk, as a "Generally Recognized as Safe" antimicrobial agent, and it is used commercially for food preservation. It is the first isolated bacteriocin to stop *Clostridium botulinum* growth in cheese.

It has been demonstrated that bacteriocins like sakacin, pediocin AcH, enterocins A and B, and enterocin K1 can stop *L. monocytogenes* from multiplying in intentionally contaminated meals. When used for vacuum packaging of meat inoculated with *Brochothrix thermosphacta*, Siragusa et al. (1999) observed that bacteriocin-coated polyethylene film made by the blown film extrusion process significantly suppressed the growth of this bacterium. An et al. (2000) reported polyethylene film coated with bacteriocin-inhibited *Micococcus flavus* and *L. monocytogenes.* They researched LDPE film coated with bacteriocin and revealed that using nisin and lacticin NK24-coated films increased the shelf life of beef and oysters kept at 3°C from 9 to 13 days and 10 to 12 days, respectively. In the meantime, when stored at 10°C, the shelf life of beef and oysters increased from 5 to 12 days and 5 to 9 days, respectively.

Additionally, it has been shown that pediocin-plastic film has inhibitory effects against several pathogenic and Gram-positive bacteria in foods. Their effectiveness against *L. monocytogenes* is particularly notable. In Goff et al. (1996) research, a pediocin-containing preparation is added to sterilize the inoculated raw chicken. After 28 days of storage at 5°C, the results revealed a decrease in *L. monocytogenes* from about 3.0 log CFU/g to below detectable levels, whereas the control treatment of counting reached 1.9 log CFU/g. A cellulose film with pediocin was created by Santiago-silva et al. (2009), who discovered that the film with 50% pediocin promoted an inhibition against *L. innocua* counts on sliced ham. Still, films with 25% pediocin and control cannot suppress the microbe growth after storage at 12°C.

Compared to the control, the inhibition is reduced by 2 log cycles for the sliced ham in contact with the 50% pediocin film and by less than 1 log cycle for the 25% pediocin film at the end of storage. Ming et al. (1997) observed that cellulosic casings sprayed with pediocin on the meats for 12 storage weeks at 4°C had total inhibitory effects towards *L. monocytogenes*. According to Iseppi et al. (2008), infected hot dogs held at 4°C had a lower *L. monocytogenes* count using electrocin-coated low-density polyethylene. There was a noticeable decline at more than one log during the first day. Whereby, up until the 14th day, this amount of diminution is maintained. Lactocin-infused multilayer plastic films were developed by Blanco Massani et al. (2012) and tested on *L. plantarum*. An external polypropylene layer, an interior polyamide-polyethylene structure, an ethylene vinyl alcohol copolymer barrier layer, and a linear low-density polyethylene food contact layer constitute the films. The researchers found that temperature and time had an impact on the antimicrobial activity of the films. An agar disk diffusion test on *L. plantarum* revealed that the inhibitory zones reached 32, 66, and 100%, respectively, after seven days of storage at 30, 10, and 5°C. No antimicrobial activity is seen on the multilayer film kept at 30 and 10°C after 14 days.

3.3.1.5 Inorganic nanoparticles as antimicrobial agents in food packaging

Due to their excellent stability under challenging processing circumstances, such as high pressures or temperatures in the manufacture of plastics, inorganic nanoparticles have recently gained interest as antimicrobial agents. TiO_2 and ZnO are widely researched inorganic nanoparticles used as antimicrobial agents (Jassal et al., 2022; Kavitha et al., 2023). TiO_2 is non-toxic and FDA-approved for use in materials that come into contact with food, pharmaceuticals, and beverages. It is a photocatalyst that is frequently used to render a variety of bacteria inactive. TiO_2 produces reactive oxygen species and hydroxyl radicals through the light reaction, which can make bacteria inactive by oxidizing the phospholipids in the cell membrane. TiO_2-coat-oriented polypropylene films were developed by Chawengkijwanich and Hayata (2008) to study the effects of antimicrobial agents on *E. coli* in vitro and real-world situations. The results demonstrated that after 180 minutes of two 20W black-light illuminations, the film coated with TiO_2 powder had stronger antimicrobial effects, resulting in a 3 log CFU/ml reduction of *E. coli* as opposed to the uncoated film, which had a 1 log CFU/ml reduction. In the actual test, a more significant result was found when cut lettuce was stored in a TiO_2-coated film bag under the irradiation of ultraviolet A light. A decrease in *E. coli* count from 6.4 on day 0 to 4.9 log CFU/g on day 1 is observed with the use of coated film, while the uncoated film only showed a slight decrease from 6.4 to 6.1 log CFU/g after one day of storage. By adopting the antibacterial drop-test method, TiO_2-incorporated polyethylene film developed by Xing et al. (2012) using a blown film extruder demonstrated that the blank film lacked antibacterial properties. In contrast, untreated TiO_2 showed inhibition ratios of 50.4% for *E. coli* and 58% for *S. aureus*. Contrarily, after being exposed to UV light for one hour, the antibacterial activity of TiO_2-incorporated polyethylene film dramatically increased, reaching 89.3% for *E. coli* and 95.2% for *S. aureus*. Bodaghi et al. (2013) developed a TiO_2-LDPE film using a blown film extruder and tested it against *Rhodotorula mucilaginosa* and several *Pseudomonas* species. After three hours of ultraviolet A irradiation on TiO_2 film and blank film, respectively, the quantity of *Pseudomonas* spp. considerably decreased by 4 log CFU/ml and

1.35 log CFU/ml using the modified in vitro test method by Chawengkijwanich and Hayata (2008). On TiO_2 film and blank film, the amount of *R. mucilaginosa* reduced by 2 log CFU/ml and 0.64 log CFU/ml. The number of mesophilic bacteria decreased from 3.14 to < 2 log CFU/g in an in vivo test on fresh pears packaged in TiO_2 nanocomposite film and stored under illumination by a fluorescent light lamp at 5°C for 17 days, while the bacteria count increased from 3.19 to 4.02 log CFU/g for pears packaged in blank films. On the other hand, it was seen that fruits packaged in TiO_2 films reduced the yeast count from 2.45 to < 2 log CFU/g, but the yeast count increased for the blank film from 2.1 to 3.37 log CFU/g. These findings demonstrated the viability of using plastic films made by extrusion with nanoparticles for fruit packing. Due to its antimicrobial activity for both Gram-positive and Gram-negative bacteria, ZnO nanoparticle coating technologies have received much attention recently. In another study, when compared to blank film, ZnO-coated PVC films significantly reduced the growth of *S. aureus* and *E. coli*. When ZnO nanoparticles are added to films exposed to ultraviolet light for three hours, the inhibition of *E. coli* and *S. aureus* increased by 29.8% and 26.0%, respectively. According to this finding, ZnO breakdown products are the main antimicrobial particles that stop bacterial development (Li et al., 2010).

3.3.1.6 Bio-nanocomposites as antimicrobial agents in food packaging

Bio-nanocomposites are bio-based polymers with dimensions between 1 and 100 nm. One of the main components that serve as a matrix is a biopolymer (continuous phase). The second component is a reinforcement agent (dispersed phase). They have common properties, including flexibility, biocompatibility, biodegradability, eco-friendliness, and cost-effectiveness, which the reinforcing components can further improve (Chausali et al., 2022; 2023). Nanofillers like montmorillonite, AgO, TiO_2, SiO_2, and ZnO and biodegradable polymers like polyhydroxy butyrate, polybutylene succinate, polylactic acid, and poly-caprolactone can be combined with natural biopolymers like starch and chitosan to develop modified bio-nanocomposites. Antimicrobial compounds are incorporated into the food packaging materials to prevent the growth of germs on food surfaces, extending the shelf life of the food. Polylactic acid/halloysite, polylactic acid/Ag nanoparticles, hydroxyapatite/titania, layered silicate, and cellulose nanofibres are some of the antimicrobial nanocomposites that have been experimentally developed. These materials can inhibit the growth of various pathogens, including *Listeria monocytogenes*, *Escherichia coli*, and *B. subtilis*. These favourable characteristics of antimicrobial bio-nanocomposites favour the industry's potential application in food packaging (Sharma et al., 2020).

The first polymer nanocomposite to hit the market was clay, a revolutionary material for food packaging, with montmorillonite being the most popular type. According to reports, the nano clay contained in food packaging sheets lowers the gas transmission rate, preserving the freshness and extending the shelf life of goods susceptible to oxygen. When nano clay is added to beer bottles, the storage time of beer rises from 11 to 30 weeks. Hence this technique has been used in many breweries (Silvestre et al., 2011). They also effectively enhanced the retention of antibacterial compounds by polymer matrices and increased the tortuosity effect of distributed nano clays on stabilizing additives (Huang et al., 2015). The broad-spectrum inhibitory properties of silver nanoparticles prevent the growth of bacteria. In addition to having antibacterial properties to extend shelf life, nano-Ag can catalyse the absorption and breakdown of ethylene, an ethylene blocker, released during fruit metabolism. ZnO nanoparticles are added to the polymer of a material and used to package food, thereby allowing interactions between the packaging and the food and forming a significant impact on food preservation (Espitia et al., 2012).

Since TiO_2 nanoparticles can filter UV light, they can be used to develop packaging films to protect the food components from UV-induced oxidation while maintaining good transparency (Duncan, 2011). Cerradaetal (2008) investigated the biocidal effects of TiO_2 nanocomposites with polymer materials, such as ethylene vinyl alcohol and chitosan, in the food processing industry. TiO_2 has a higher antibacterial effect on Gram-positive than Gram-negative germs, according to Xing et al. (2012). These nanoparticles can also create oxygen scavenger films (Xiao-e et al., 2004).

Due to their ability to pierce microbial cells and cause permanent harm, carbon nanotubes have strong antibacterial effects. They are employed in food packaging because they can improve the mechanical characteristics of polymer matrices (Kang et al., 2007).

3.3.2 Antimicrobial Agents in Food Preservation

3.3.2.1 Combination of drug and antimicrobial peptide as antimicrobial agents in food preservation

Combining a drug with one or more antimicrobial agents is one of the best ways to boost its antibacterial activity. The antibacterial effects of buforin I and nisin and its combined form are assessed to combat the most significant food-rotting pathogens such as *Staphylococcus epidermidis, Escherichia coli, Aspergillus oryzae, Geotrichum candidum, Bacillus subtilis, Listeria innocua, Salmonella* serovar Enteritidis, and *Rhodotorula glutinis.* Buforin I had stronger antimicrobial activity than nisin on all microbial strains, confirmed by the results of MIC and MBC or minimum fungicidal concentration tests. Both antimicrobial drugs are most effective against *E. coli*, while nisin and buforin I were most effective against *L. innocua* and *S. epidermidis*, respectively. The results of the synergistic interaction between buforin I and nisin suggested that, while buforin I and nisin did not affect *Salmonella* serovar Enteriditis and *G. candidum*, they had a synergistic effect on *B. subtilis, S. epidermis*, and *A. oryzae. L. innocua, E. coli*, and *R. glutinis* responded partially synergistically to the combination of buforin I and nisin. Utilizing these antimicrobial agents in combination increases antimicrobial activity at lower concentrations. Buforin I's antimicrobial activities against food spoilage microbes revealed that its use alone or in combination with nisin might open up new possibilities for using antimicrobial peptides as natural preservatives. Consequently, preventing contamination, deterioration, and infestation of food and beverage goods may benefit from combining antimicrobial peptides and conventional food preservatives (Roshanak et al., 2020).

3.3.2.2 Thyme essential oil-based nanoemulsions as antimicrobial agents in food preservation

Disc diffusion is used to assess the antibacterial effects of nanoemulsions based on thyme essential oil and its purified form on fish spoilage bacteria and foodborne pathogens. GC-MS is used to identify the flavouring elements in thyme oil that had been extracted. The physical characteristics of nanoemulsions are studied, and the outcomes demonstrated that thyme-based nanoemulsions not subjected to purification are more effective against foodborne microorganisms. Following nanoemulsion treatments, damage to bacterial cell membranes is noticed. The antibacterial activity of thyme oil is enhanced with its conversion into a nanoemulsion. Its nano-form can be employed as a substitute antimicrobial agent in processed or packed fish (Ozogul et al., 2020).

3.3.2.3 Extracts of *Melisa officinalis* as antimicrobial agents in food preservation

The Lamiaceae plant *Melissa officinalis* is a source of biologically active chemicals used extensively in traditional medicine, cosmetics, and food. Due to increased demand for more natural and risk-free food preservation agents, *Melissa officinalis* is utilized as a food preservative. The antibacterial properties of Melissa *officinalis* are studied, and the results indicated that it has antimicrobial effects on various bacteria, including significant foodborne pathogens. The plant's potential to be employed as a food preservative in the future has been established by the antibacterial impact in various food matrices (Carvalho et al., 2021).

3.3.2.4 Combination of antimicrobial agents and other preservation techniques in food preservation

To increase the shelf life of foods, hurdle technology combines various preservation techniques. Here, two or more preservation methods are applied, with one method specifically designed to counteract the effects of various microorganisms (Leistner, 2000). It has been suggested that natural antibacterial agents could be used with various thermal and non-thermal preservation techniques, such as pulsed electric fields and high hydrostatic pressure. This approach looks efficient, allowing for the appropriate level of preservation (Smid & Gorris, 1999). Using an electric pulse field, mild heating, and natural antimicrobials such as lysozyme and nisin against naturally occurring bacteria in milk showed synergistic activity with higher microbial inactivation than the individual food preservation techniques (Smith et al., 2002). It has been documented that combining high hydrostatic pressure and bacteriocin has synergistic benefits in decreasing microorganisms. For instance, the use of bacteriocin resulted in a 2.2-log decrease of *S. aureus* in reconstituted skim milk at a pressure of 250 MPa. In contrast, using 10,000 AU/ml lacticin alone produced a 1-log reduction. But when both methods are used together, the microbial population is reduced by more than 6 log cycles, which is more than either treatment did on its own (Morgan et al., 2000). As a result of the overall effectiveness of several antimicrobial compounds combined with various technologies, there has been a significant decrease in the population of microorganisms with little impact on food quality. As a result, these technologies can be further recommended for use in new and emerging food industries that call for little to no microbial presence in food systems.

3.3.2.5 Nanoparticles as antimicrobial agents in food preservation

In the production of food, nanotechnology is a young sector that is becoming more important. In recent years, numerous types of research on the application of nanotechnology in the food business have been conducted. Recent studies have preferred this technique to improve the effectiveness of natural antibacterial agents and the functionality and quality of food. This has been accomplished by developing delivery systems that use nanoencapsulation and nano-laminate materials (Aloui & Khwaldia, 2016). In addition to improving stability and solubility and reducing direct interactions of antimicrobials with food components during food processing and storage, nanoencapsulated delivery systems for antimicrobial compounds offer a sound barrier against some environmental influences, and chemical interactions may affect the quality (Quinto et al., 2019). Nanoencapsulating technology in food preservation is currently gaining popularity in the food business. One example is nano-based emulsions. This is so that they can be used in natural food ingredients that require rigorous mixing and less complicated production methods (Donsi et al., 2011). They aid in maintaining food's sensory and organoleptic qualities and are efficient antimicrobial agents. Their use as inhibitory compounds against pathogenic and food spoilage microorganisms and as preservatives for some quality attributes of different foods has some restrictions, and their use in solid foods, such as fruits and meat products, is typically associated with challenges in immobilizing nano-droplets on food surfaces (Mate et al., 2016). They are also appropriate for inclusion in foods that lack specific functional qualities, like visual qualities. The development of in situ nanoemulsions from biopolymer-based edible coatings has recently advanced. These materials have been examined and proven to be an effective way to place antimicrobial agents on the surface of solid items. These studies' findings appear to hold promise for enhancing the quality and safety of meat-based products, fresh fruits, and vegetables (Severino et al., 2015) and the meat itself (Wu et al., 2016).

3.4 DEVELOPING NEW ANTIBIOTICS

The classic screening methods are discovering newer types of antibiotics; each has a distinct class or mechanism of action against the target organisms. Novolactamycin falls under the category of glycosylated macrolactams. Lycostaphin is a kind of Staphylococcal lytic enzyme. Fusidic acid performs by inhibiting the release of elongation factor G. Fidaxomicin is a macrocytic antibacterial inhibitor of RNA polymerase. VIC 200107 is a lantibiotic active on the cell membrane. All these are newer antibiotics developed by different pharmaceutical developers (Mollering et al., 2011).

3.5 CHALLENGES

Despite these beneficial qualities, certain obstacles still prevent their widespread use, including high production costs, potential human toxicity, and sensitivity to harsh environmental conditions (proteases and extreme pH). Lack of selectivity against particular strains and folding concerns with antimicrobials may cause them to lose some antimicrobial capabilities when employed as surface coatings. Because of this, many antimicrobials, including natural ones, are only effective at very high dosages in animal models of infection, despite being active in vitro. Additionally, it has been noted that some antimicrobials are resistant to some microorganisms (Zhu et al., 2017).

Carbonated beverages are homogeneous and acidic. They are mainly made of water and have deficient protein and fat content. These aid in even dispersion of antimicrobials throughout the product – generally, the lower the pH higher the antimicrobial activity. There are a few disadvantages, such as carbohydrates reducing antimicrobial activity. Antimicrobials in bakery products such as bread, muffins, and cakes have lesser performance because they are non-homogeneous and have a low value of proteins and neutral pH. The efficacy of antimicrobials is reduced in fruits and vegetables because they have low levels of fats and proteins. Natural antimicrobials are used in fresh fruits and vegetables as a spray. This application is utilized due to higher chances of surface contamination due to the rough and harder surfaces, which provide room for various microbial contaminants. Dairy products' higher pH and water activity aid in the growth of several microorganisms.

Moreover, they also contain protein and fat, which reduces antimicrobial activity. Non-homogeneous, higher pH, protein, and fat nature of meat, seafood, and poultry render antimicrobials challenging. Further studies or specialized packaging have been required to enhance the added compound's antimicrobial nature (Davidson et al., 2015).

Multiple antimicrobials in food can act as synergists or even create adverse product effects. All the advancements and studies benefit the food processing industry. Growing demands for cleaner and healthy foods increase the utilization of antimicrobials in food. Future research should focus on the sources (from nature or synthesis) and mechanisms of action of antimicrobials. New techniques that can decode the structure-function interaction should be developed to achieve practical synthesis and manipulation of antimicrobials.

3.6 CONCLUSION

The application of antimicrobial agents in food science and technology offers various advantages. It aids in the prevention of the growth of dangerous microbes such as bacteria, yeast, and mould, which can cause food spoiling and foodborne illnesses. Furthermore, antimicrobial compounds can improve the sensory quality of food products by inhibiting the growth of microorganisms that create off-flavours and odours. However, applying antimicrobial compounds in food science and technology has several limitations. Overuse or inappropriate use of antimicrobial drugs can develop antibiotic resistance in bacteria, making it more difficult to treat illnesses in humans and animals. Furthermore, if antimicrobial compounds are present in hazardous concentrations in food products, they might negatively affect the environment and consumer health. To summarize, the use of antimicrobial compounds in food science and technology is a two-edged sword with both advantages and disadvantages. It is critical to use these agents responsibly, minimizing their possible negative effects while optimizing their advantages.

REFERENCES

Ahmed, S., Sameen, D.E., Lu, R., Li, R., Dai, J.W., Qin, W. and Liu, Y.W., 2022. Research progress on antimicrobial materials for food packaging. *Critical Reviews in Food Science and Nutrition*, 62(11), pp. 3088–3102. doi:10.1080/10408398.2020.1863327.

Aloui, H. and Khwaldia, K., 2016. Natural antimicrobial edible coatings for microbial safety and food quality enhancement. *Comprehensive Reviews in Food Science and Food Safety*, 15(6), pp. 1080–1103.

An, D.S., Kim, Y.M., Lee, S.B., Paik, H.D. and Lee, D.S., 2000. Antimicrobial low-density polyethylene film coated with bacteriocins in binder medium. *Food Science and Biotechnology*, 9(1), pp. 14–20.

Asif, M., 2017. Review on Antimicrobial Agents, *Organic & Medicinal Chemistry International Journal, Juniper Publishers Inc.*, 1(5), pp. 179–185.

Barbiroli, A., Bonomi, F., Capretti, G., Iametti, S., Manzoni, M., Piergiovanni, L. and Rollini, M., 2012. Antimicrobial activity of lysozyme and lactoferrin incorporated in cellulose-based food packaging. *Food Control*, 26(2), pp. 387–392.

Blanco Massani, M., Morando, P.J., Vignolo, G.M. and Eisenberg, P., 2012. Characterization of a multilayer film activated with Lactobacillus curvatus CRL705 bacteriocins. *Journal of the Science of Food and Agriculture*, 92(6), pp. 1318–1323.

Bodaghi, H., Mostofi, Y., Oromiehie, A., Zamani, Z., Ghanbarzadeh, B., Costa, C., Conte, A. and Del Nobile, M.A., 2013. Evaluation of the photocatalytic antimicrobial effects of a TiO_2 nanocomposite food packaging film by in vitro and in vivo tests. *LWT – Food Science and Technology*, 50(2), pp. 702–706.

Carvalho, F., Duarte, A.P. and Ferreira, S., 2021. Antimicrobial activity of Melissa officinalis and its potential use in food preservation. *Food Bioscience*, 44, p. 101437.

Cerrada, M.L., Serrano, C., Sánchez-Chaves, M., Fernández-García, M., Fernández-Martín, F., de Andres, A., Rioboo, R.J.J., Kubacka, A., Ferrer, M. and Fernández-García, M., 2008. Self-sterilized EVOH-TiO_2 nanocomposites: Interface effects on biocidal properties. *Advanced Functional Materials*, 18(13), pp. 1949–1960.

Chausali, N., Saxena, J. and Prasad, R., 2022. Recent trends in nanotechnology applications of bio-based packaging. *Journal of Agriculture and Food Research*, https://doi.org/10.1016/j.jafr.2021.100257.

Chausali, N., Saxena, J. and Prasad, R., 2023. Nanotechnology as a sustainable approach for combating the environmental effects of climate change. *Journal of Agriculture and Food Research*, https://doi.org/10.1016/j.jafr.2023.100541.

Chawengkijwanich, C. and Hayata, Y., 2008. Development of TiO_2 powder-coated food packaging film and its ability to inactivate Escherichia coli in vitro and in actual tests. *International Journal of Food Microbiology*, 123(3), pp. 288–292.

Cowan, M.M., 1999. Plant products as antimicrobial agents. *Clinical Microbiology Reviews*, 12(4), p. 564. doi:10.1128/cmr.12.4.564.

Davidson, P., Cekmer, H.B., Monu, E. andTechathuvanan, C., 2015. The use of natural antimicrobials in food: An overview. Edited by M Taylor In *Handbook of Natural Antimicrobials for Food Safety and Quality*, pp. 1–27.

Dònsì, F., Annunziata, M., Sessa, M. and Ferrari, G., 2011. Nanoencapsulation of essential oils to enhance their antimicrobial activity in foods. *LWT – Food Science and Technology*, 44(9), pp. 1908–1914.

Du, W.X., Olsen, C.W., Avena-Bustillos, R.J., Mchugh, T.H., Levin, C.E. and Friedman, M., 2008. Storage stability and antibacterial activity against Escherichia coli O157: H7 of carvacrol in edible apple films made by two different casting methods. *Journal of Agricultural and Food Chemistry*, 56(9), pp. 3082–3088.

Duan, J., Park, S.I., Daeschel, M.A. and Zhao, Y., 2007. Antimicrobial chitosan-lysozyme (CL) films and coatings for enhancing microbial safety of mozzarella cheese. *Journal of Food Science*, 72(9), pp. M355–M362.

Duncan, T.V., 2011. Applications of nanotechnology in food packaging and food safety: Barrier materials, antimicrobials and sensors. *Journal of Colloid and Interface Science*, 363(1), pp. 1–24.

Ellison, R. and Giehl, T.J., 1991. Killing of gram-negative bacteria by lactoferrin and lysozyme. *The Journal of Clinical Investigation*, 88(4), pp. 1080–1091.

Emiroğlu, Z.K., Yemiş, G.P., Coşkun, B.K. and Candoğan, K., 2010. Antimicrobial activity of soy edible films incorporated with thyme and oregano essential oils on fresh ground beef patties. *Meat Science*, 86(2), pp. 283–288.

Espitia, P.J.P., Soares, N.D.F.F., Coimbra, J.S.D.R., de Andrade, N.J., Cruz, R.S. and Medeiros, E.A.A., 2012. Zinc oxide nanoparticles: Synthesis, antimicrobial activity and food packaging applications. *Food and Bioprocess Technology*, 5(5), pp. 1447–1464.

Fuglsang, C.C., Johansen, C., Christgau, S. and AdlerNissen, J., 1995. Antimicrobial enzymes: Applications and future potential in the food industry. *Trends in Food Science and Technology*, 6(12), pp. 390–396. doi:10.1016/s0924-2244(00)89217-1.

Gamage, G.R., Park, H.J. and Kim, K.M., 2009. Effectiveness of antimicrobial coated oriented polypropylene/polyethylene films in sprout packaging. *Food Research International*, 42(7), pp. 832–839.

Gemili, S., Yemenicioğlu, A. and Altınkaya, S.A., 2009. Development of cellulose acetate based antimicrobial food packaging materials for controlled release of lysozyme. *Journal of Food Engineering*, 90(4), pp. 453–462.

Goff, J.H., Bhunia, A.K. and Johnson, M.G., 1996. Complete inhibition of low levels of Listeria monocytogenes on refrigerated chicken meat with pediocinAcH bound to heat-killed Pediococcusacidilactici cells. *Journal of Food Protection*, 59(11), pp. 1187–1192.

Güçbilmez, Ç.M., Yemenicioğlu, A. and Arslanoğlu, A., 2007. Antimicrobial and antioxidant activity of edible zein films incorporated with lysozyme, albumin proteins and disodium EDTA. *Food Research International*, 40(1), pp. 80–91.

Gyawali, R. and Ibrahim, S.A., 2014. Natural products as antimicrobial agents. *Food Control*, 46, pp. 412–429. doi:10.1016/j.foodcont.2014.05.047.

Ha, J.U., Kim, Y.M. and Lee, D.S., 2001. Multilayered antimicrobial polyethylene films applied to the packaging of ground beef. *Packaging Technology and Science: An International Journal*, 14(2), pp. 55–62.

Han, J., Castell-Perez, M.E. and Moreira, R.G., 2007. The influence of electron beam irradiation of antimicrobial-coated LDPE/polyamide films on antimicrobial activity and film properties. *LWT – Food Science and Technology*, 40(9), pp. 1545–1554.

Huang, J.Y., Li, X. and Zhou, W., 2015. Safety assessment of nanocomposite for food packaging application. *Trends in Food Science and Technology*, 45(2), pp. 187–199.

Iseppi, R., Pilati, F., Marini, M., Toselli, M., de Niederhäusern, S., Guerrieri, E., Messi, P., Sabia, C., Manicardi, G., Anacarso, I. and Bondi, M., 2008. Anti-listerial activity of a polymeric film coated with hybrid coatings doped with enterocin 416K1 for use as bioactive food packaging. *International Journal of Food Microbiology*, 123(3), pp. 281–287.

Jassal, P.S., Kaur, D., Prasad, R. and Singh, J., 2022. Green synthesis of titanium dioxide nanoparticles: Development and applications. *Journal of Agriculture and Food Research*, 10: 100361. https://doi.org/10.1016/j.jafr.2022.100361.

Joerger, R.D., Sabesan, S., Visioli, D., Urian, D. and Joerger, M.C., 2009. Antimicrobial activity of chitosan attached to ethylene copolymer films. *Packaging Technology and Science: An International Journal*, 22(3), pp. 125–138.

Kang, S., Pinault, M., Pfefferle, L.D. and Elimelech, M., 2007. Single-walled carbon nanotubes exhibit strong antimicrobial activity. *Langmuir: The ACS Journal of Surfaces and Colloids*, 23(17), pp. 8670–8673.

Kavitha, A., Doss, A., Pole, R.P.P., Pushpa Rani, T.P.K., Prasad, R. and Satheesh, S., 2023. A mini review on plant-mediated zinc oxide nanoparticles and their antibacterial potency. *Biocatalysis and Agricultural Biotechnology*. https://doi.org/10.1016/j.bcab.2023.102654.

Keymanesh, K., Soltani, S. and Sardari, S., 2009. Application of antimicrobial peptides in agriculture and food industry. *World Journal of Microbiology and Biotechnology*, 25(6), pp. 933–944. doi:10.1007/s11274-009-9984-7.

Leistner, L., 2000. Basic aspects of food preservation by hurdle technology. *International Journal of Food Microbiology*, 55(1–3), pp. 181–186.

Li, X.H., Xing, Y.G., Li, W.L., Jiang, Y.H. and Ding, Y.L., 2010. Antibacterial and physical properties of poly (vinyl chloride)-based film coated with ZnO nanoparticles. *Food Science and Technology International*, 16(3), pp. 225–232.

Liu, H., Du, Y., Wang, X. and Sun, L., 2004. Chitosan kills bacteria through cell membrane damage. *International Journal of Food Microbiology*, 95(2), pp. 147–155.

Malhotra, B., Keshwani, A. and Kharkwal, H., 2015. Antimicrobial food packaging: Potential and pitfalls. *Frontiers in Microbiology*, 6. doi:10.3389/fmicb.2015.00611.

Maté, J., Periago, P.M. and Palop, A., 2016. Combined effect of a nanoemulsion of d-limonene and nisin on Listeria monocytogenes growth and viability in culture media and foods. *Food Science and Technology International*, 22(2), pp. 146–152.

Mecitoğlu, Ç., Yemenicioğlu, A., Arslanoğlu, A., Elmacı, Z.S., Korel, F. and Çetin, A.E., 2006. Incorporation of partially purified hen egg white lysozyme into zein films for antimicrobial food packaging. *Food Research International*, 39(1), pp. 12–21.

Mehdizadeh, T., Tajik, H., Rohani, S.M.R. and Oromiehie, A.R., 2012. Antibacterial, antioxidant and optical properties of edible starch-chitosan composite film containing Thymus kotschyanus essential oil. In *Veterinary Research Forum* Summer (Vol. 3, No. 3, pp. 167–173). PMID: 25610564; PMCID: PMC4299978.

Ming, X., Weber, G.H., Ayres, J.W. and Sandine, W.E., 1997. Bacteriocins applied to food packaging materials to inhibit Listeria monocytogenes on meats. *Journal of Food Science*, 62(2), pp. 413–415.

Moellering, R.C., 2011. Discovering new antimicrobial agents. *International Journal of Antimicrobial Agents*, 37(1), pp. 2–9. doi:10.1016/j.ijantimicag.2010.08.018.

Morgan, S.M., Ross, R.P., Beresford, T. and Hill, C., 2000. Combination of hydrostatic pressure and lacticin 3147 causes increased killing of Staphylococcus and Listeria. *Journal of Applied Microbiology*, 88(3), pp. 414–420.

Muriel-Galet, V., Cerisuelo, J.P., López-Carballo, G., Lara, M., Gavara, R. and Hernández-Muñoz, P., 2012. Development of antimicrobial films for microbiological control of packaged salad. *International Journal of Food Microbiology*, 157(2), pp. 195–201.

Nadarajah, D., Han, J.H. and Holley, R.A., 2005. Inactivation of Escherichia coli O157: H7 in packaged ground beef by allyl isothiocyanate. *International Journal of Food Microbiology*, 99(3), pp. 269–279.

Oussalah, M., Caillet, S., Salmieri, S., Saucier, L. and Loicroix, M., 2004. Antimicrobial and antimicrobial effect of milk based protein-based film containing essential oils for the preservatives of whole beef muscle. *Journal of Agricultural and Food Chemistry*, 52(18), pp. 5598–5605.

Ozogul, Y., Boğa, E.K., Akyol, I., Durmus, M., Ucar, Y., Regenstein, J.M. and Köşker, A.R., 2020. Antimicrobial activity of thyme essential oil nanoemulsions on spoilage bacteria of fish and food-borne pathogens. *Food Bioscience*, 36, p. 100635.

Pink, R., Hudson, A., Mouries, M.A. andBendig, M., 2005. Opportunities and challenges in antiparasitic drug discovery. *Nature Reviews Drug Discovery*, 4(9), pp. 727–740. doi:10.1038/nrd1824.

Pranoto, Y., Rakshit, S.K. and Salokhe, V.M., 2005. Enhancing antimicrobial activity of chitosan films by incorporating garlic oil, potassium sorbate and nisin. *LWT – Food Science and Technology*, 38(8), pp. 859–865.

Quattara, B., Simard, R.E., Holley, R.A., Piette, G.J.P. and Bégin, A., 1997. Antibacterial activity of selected fatty acids and essential oils against six meat spoilage organisms. *International Journal of Food Microbiology*, 37(2–3), pp. 155–162.

Quinto, E.J., Caro, I., Villalobos-Delgado, L.H., Mateo, J., De-Mateo-Silleras, B. and Redondo-Del-Río, M.P., 2019. Food safety through natural antimicrobials. *Antibiotics*, 8(4), p. 208.

Roshanak, S., Shahidi, F., Yazdi, F.T., Javadmanesh, A. and Movaffagh, J., 2020. Evaluation of antimicrobial activity of Buforin I and Nisin and the synergistic effect of their combination as a novel antimicrobial preservative. *Journal of Food Protection*, 83(11), pp. 2018–2025.

Salas, M.L., Mounier, J., Valence, F., Coton, M., Thierry, A. and Coton, E., 2017. Antifungal microbial agents for food biopreservation-A review. *Microorganisms*, 5(3). doi:10.3390/microorganisms5030037.

Santhi, V.P., Sriramavaratharajan, V., Murugan, R., Masilamani, P., Gurav, S.S., Sarasu, V.P., Parthiban, S. and Ayyanar, M., 2021. A review of edible fruit extracts and fruit juices as a potential source of antiviral agents. *Journal of Food Measurement and Characterization*, 15(6), pp. 5181–5190. doi:10.1007/s11694-021-01090-7.

Santiago-Silva, P., Soares, N.F., Nóbrega, J.E., Júnior, M.A., Barbosa, K.B., Volp, A.C.P., Zerdas, E.R. and Würlitzer, N.J., 2009. Antimicrobial efficiency of film incorporated with pediocin (ALTA® 2351) on preservation of sliced ham. *Food Control*, 20(1), pp. 85–89.

Severino, R., Ferrari, G., Vu, K.D., Donsì, F., Salmieri, S. and Lacroix, M., 2015. Antimicrobial effects of modified chitosan based coating containing nanoemulsion of essential oils, modified atmosphere packaging and gamma irradiation against Escherichia coli O157: H7 and Salmonella Typhimurium on green beans. *Food Control*, 50, pp. 215–222.

Seydim, A.C. and Sarikus, G., 2006. Antimicrobial activity of whey protein based edible films incorporated with oregano, rosemary and garlic essential oils. *Food Research International*, 39(5), pp. 639–644.

Sharma, R., Jafari, S.M. and Sharma, S., 2020. Antimicrobial bio-nanocomposites and their potential applications in food packaging. *Food Control*, 112, p. 107086.

Silvestre, C., Duraccio, D. and Cimmino, S., 2011. Food packaging based on polymer nanomaterials. *Progress in Polymer Science*, 36(12), pp. 1766–1782.

Siragusa, G.R., Cutter, C.N. and Willett, J.L., 1999. Incorporation of bacteriocin in plastic retains activity and inhibits surface growth of bacteria on meat. *Food Microbiology*, 16(3), pp. 229–235.

Sivarooban, T., Hettiarachchy, N.S. and Johnson, M.G., 2008. Physical and antimicrobial properties of grape seed extract, nisin, and EDTA incorporated soy protein edible films. *Food Research International*, 41(8), pp. 781–785.

Smid, E.J. and Gorris, L.G.M., (1999). *Natural Antimicrobials for Food Preservation. Handbook of Food Preservation*; Rahman, MS, Ed. New York: Marcel Dekker, pp. 285–308.

Smith, K., Mittal, G.S. and Griffiths, M.W., 2002. Pasteurization of milk using pulsed electrical field and antimicrobials. *Journal of Food Science*, 67(6), pp. 2304–2308.

Sung, S.Y., Sin, L.T., Tee, T.T., Bee, S.T., Rahmat, A.R., Rahman, W.A.W.A., Tan, A.C. and Vikhraman, M., 2013. Antimicrobial agents for food packaging applications. *Trends in Food Science and Technology*, 33(2), pp. 110–123.

Tauxe, R.V., 2002. Emerging foodborne pathogens. *International Journal of Food Microbiology*, 78(1–2), pp. 31–41. doi:10.1016/s0168-1605(02)00232-5.

Torlak, E. and Nizamlioğlu, M., 2011. Antimicrobial effectiveness of chitosan-essential oil coated plastic films against foodborne pathogens. *Journal of Plastic Film and Sheeting*, 27(3), pp. 235–248.

Ünalan, İ.U., Korel, F. and Yemenicioğlu, A., 2011. Active packaging of ground beef patties by edible zein films incorporated with partially purified lysozyme and Na2EDTA. *International Journal of Food Science and Technology*, 46(6), pp. 1289–1295.

Vaughn, R.H. and Stewart, G.F., 1960. Antibiotics as food preservatives. *JAMA*, 174(10), pp. 1308–1310. doi:10.1001/jama.1960.63030100001016.

Wu, C., Wang, L., Hu, Y., Chen, S., Liu, D. and Ye, X., 2016. Edible coating from citrus essential oil-loaded nanoemulsions: Physicochemical characterization and preservation performance. *RSC Advances*, 6(25), pp. 20892–20900.

Xiao-e, L., Green, A.N., Haque, S.A., Mills, A. and Durrant, J.R., 2004. Light-driven oxygen scavenging by titania/polymer nanocomposite films. *Journal of Photochemistry and Photobiology A: Chemistry*, 162(2–3), pp. 253–259.

Xing, Y., Li, X., Zhang, L., Xu, Q., Che, Z., Li, W., Bai, Y. and Li, K., 2012. Effect of TiO_2 nanoparticles on the antibacterial and physical properties of polyethylene-based film. *Progress in Organic Coatings*, 73(2–3), pp. 219–224.

Zhu, M., Liu, P. and Niu, Z.W., 2017. A perspective on general direction and challenges facing antimicrobial peptides. *Chinese Chemical Letters*, 28(4), pp. 703–708. doi:10.1016/j.cclet.2016.10.001.

Zivanovic, S., Chi, S. and Draughon, A.F., 2005. Antimicrobial activity of chitosan films enriched with essential oils. *Journal of Food Science*, 70(1), pp. M45–M51.

4 Antimicrobials
An imperative food industry agent

Monika Chauhan, Lovlish Gupta,
Diwakar Chauhan, and Ajay Kumar

4.1 INTRODUCTION

Food industries play a significant part in the maintenance of the nutritional demand for the continuously growing population. As a result, food industries are using various means for food production and manufacturing. However, the production of food is not sufficient; there also needs to be a focus on the maintenance of food for its longevity and procurement needs various special processes, machinery, and chemicals, such as antimicrobials and preservatives. Because of this, the food industries are one of the most economically growing industries. Antimicrobials are one of the most important food industry agents; these antimicrobials help in the processing of food to preserve it from various microbial growth, to conserve food (Sadiku et al., 2019)

Food items must be preserved from deterioration throughout the preparation, storage, and distribution to prevent monetary loss and also to increase their shelf life throughout their use and to prevent its loss. The interest in effortlessly ready and prepared-to-eat new food items has increased the demand for foods that have a longer shelf life. Food items can be exposed to pollution by microorganisms and their growths. A large number of these microorganisms can cause bothersome responses that disintegrate the flavour, smell, variety, and tangible and textural properties of food varieties (Sadiku et al., 2019; Huang et al., 2021).

Research has shown that the US Department of Agriculture, Food Safety, and Inspection Service depicted the ability of antimicrobials to utilize them to clean and sterilize food products. The department has supported and shown interest in the use of antimicrobial substances such as added preservatives, additives, and conservatives. These various approaches towards the use of antimicrobials in food are not only helping the food industry in increasing the longevity of food but also with the increasing demands of the population. They are one of the largest sources of commercialization for economic growth. The food antimicrobial additive market has become a vast sector, which has been shown in various statistical analyses done by DuPont de Nemours Inc., Corbion NV, Koninklijke DSM NV, and Galactic, the major players in the market (Lucera et al., 2012).

The use of antimicrobials of any class relies upon a few variables such as their side effects and lawful constraints. Antimicrobials are categorized into three groups:

I. Processing acids
II. Secondary food additives
III. Direct food additives

Other than this, antimicrobials are of various types. They are as follows:

1. Acid-based antimicrobials: they consist of various organic acids, prominently lactic acid, acetic acid, and citric acid.

DOI: 10.1201/9781003268949-4

2. Chemicalized antimicrobials: they include trisodium phosphate (TSP), chlorine dioxide, peracetic acid (peroxyacetic acid), sodium nitrite, and nitrate.
3. Ovo – antimicrobials: lysozymes
4. Lacto – antimicrobials: lactoferrin (a glycoprotein found in milk)
5. Bacto – antimicrobials: they contain nisin (a polypeptide of antimicrobial obtained from certain strains of *Lactococcus lactis* subsp., prominently a lactic acid producing bacterium), ediocin, sakacin, reuterin, lacticin, macedocin, and colicin.
6. Phyto – antimicrobials: spices, essential oils, herbs

Various researchers have obtained a rough figure for the medically suggested daily milligrams intake per kilogram of body mass, for the sake of concerning human health as a factor. A large amount of antimicrobial intake might become toxic and its overdose can be lethal. This is shown in Table 4.1.

There are various methods and mechanisms for the preservation of food and other edible items. The methods involve various ancient and advanced methodologies as shown in Figure 4.1. Various ancient methods employed are drying, refrigeration, and fermentation (can be done by alcohol or by acid). Drying involves sun drying, mechanical drying, and freeze drying; freeze drying is one such methodology that involves refreshing and vacuuming the food items from the environmental entity. The older/ancient methods are time-consuming, and the restraint from such methodologies may not preserve the food for a longer time. However, the advanced methods work on the principle of preservation of food materials for a longer period. For example, the use of chemicals such as sodium benzoate helps in the preservation of pickles, and the addition of white vinegar to various food items helps preserve them for a longer time (Pathare et al., 2013). Apart from this, vacuum packing, pasteurization, irradiation, and canning are additional methodologies beneficial to modern/advanced methodology. All these methods help in achieving the conservation of food resources, which is not only time narrative but beneficial from both quantitative and economical perspectives. The major mechanism of this food preservation includes minimizing pathogenic growth, maintaining the qualitative and nutritional content of specific food items, and reducing foodborne pathogens. This methodology also helps in ensuring the better taste, texture, and appearance of food items and is proven to be economically beneficial (Worsfold et al., 2006).

4.2 ANTIMICROBIALS

Antimicrobials as the name proposes assist with halting the development and spread of microorganisms that could carry damage to us. The revelation of new foodborne microorganism specialties and possibly dangerous food sources, alongside a basic need to lessen food decay squander, has expanded the requirement for reasonable antimicrobial mixtures. Traditional food protection

TABLE 4.1
Medically accepted daily intake (MADI) per kilogram of body weight

Preservatives/antimicrobial	MADI (mg/kg body weight)
Sodium benzoate (C_6H_5COONa)	5
Sulphur dioxide (SO_2)	0.7–0.8
Acetic acid (CH_3COOH)	Mentioned acids can be taken in excess amounts, hence their intake has no such limit
Lactic acid ($C_3H_6O_3$)	
Citric acid ($C_6H_8O_7$)	
Potassium sorbate ($C_6H_7KO_2$)	25–27
Sodium propionate ($C_3H_5NaO_2$)	10
Sodium nitrite ($NaNO_2$)	0.2–0.4

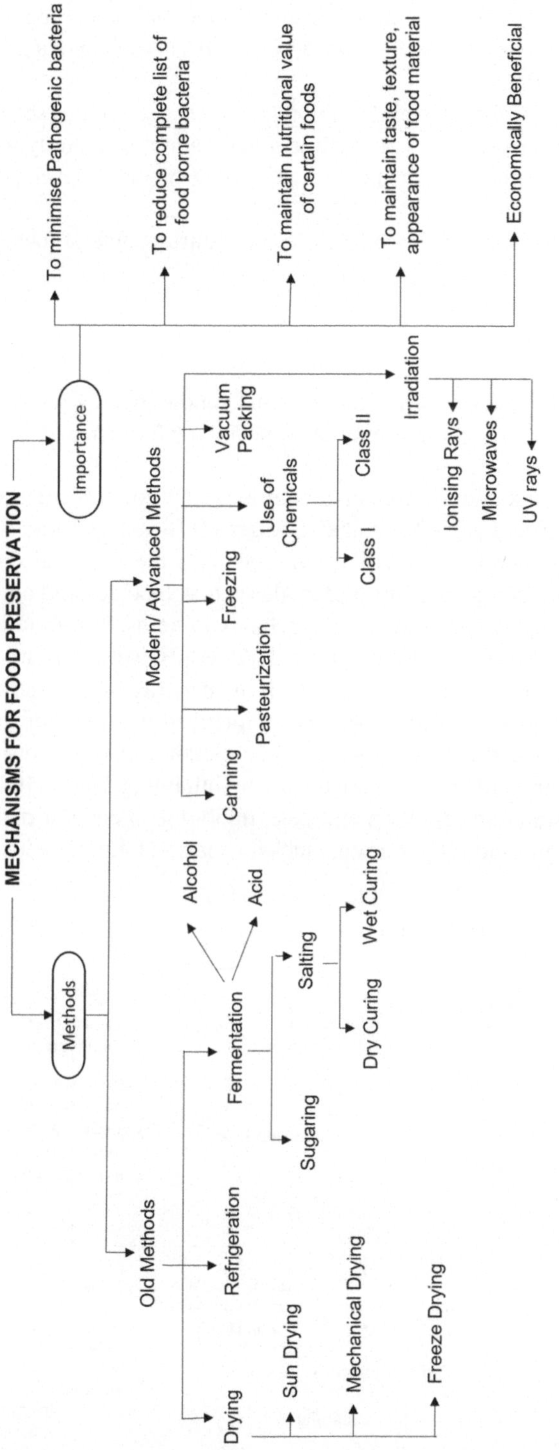

FIGURE 4.1 Classification of food preservation methodologies

strategies are less successful in forestalling the development of these food microbes in new food. New strategies are required by the food business to defeat these difficulties. Utilization of antimicrobials is an arising innovation that could be utilized by the business to expand the capacity of life of food and defeat these food handling issues. Novel antimicrobials are being found, and exploration has done on customary mixtures (Sultanbawa et al., 2011; Arshad et al., 2017).

Antimicrobial mechanisms

An antimicrobial agent performs explicit inhibitory functions, and has set up various systems to target particular microorganism. This approach is greatly and significantly used today by various scientists, antimicrobial specialists, and microbiologists in order to develop an idea of objective microbiologists.

Microorganisms are characterized based on several features, some of them are as follows:

- Aerobic or anaerobic microbes
- Cell wall composition
- Development stages
- Ideal development temperature and other optimal conditions
- Their presence as corrosive or assimilation opposition (Gumienna et al., 2021)

Microbial growth is the commonest source of contamination of food, as microorganisms grow very easily in moist and warm places; added to it, they also get a sufficient quantity of nutrition in the food material. Hence, fighting microbial growth antimicrobials is the only major approach left behind for biologists and food technologists. Antimicrobials such as biocides and antiseptics are meant to treat injuries and infections; hence, they are not reliable to be used in food industries because of their various associated side effects (Gumienna et al., 2021). However, antimicrobials such as antibacterial, antiviral, antiparasitic, and antifungal help in encouraging the anti-infective mechanism of antimicrobials, as mentioned in Figure 4.2. This biological treatment of the food enhances the longevity, climatic resistance, thermostability, and pH tolerance of food. Antimicrobials kill foodborne pathogens and other spoilage microorganisms by inhibiting their cellular processes such as the synthesis of the cell wall and cellular membrane, inhibition of cellular metabolism, and inhibition of synthesis of proteins and other various nucleic acids and amino acids. Anti-infectives like

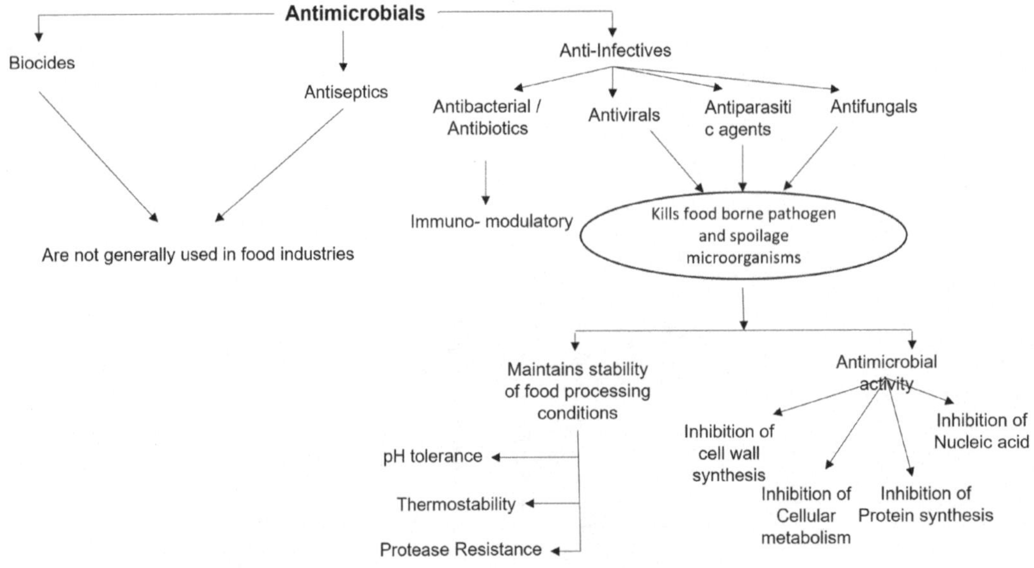

FIGURE 4.2 Action mechanism of antimicrobials for food preservation

antibacterial, antiviral, antiparasitic, and antifungal are useful antimicrobial agents that help not only in the prevention of food resources but also boost the immune system and act as immunomodulatory signals. They act on the physicochemical processes of microbes by inhibiting their cell wall synthesis, cellular metabolism, and protein synthesis, inhibiting nucleic acid formation, etc. in turn stopping the biological growth of the pathogen (Ayseli et al., 2020).

4.3 SOURCES/TYPES OF ANTIMICROBIALS

It is well known that the utilization of antimicrobials is another innovation by food business experts to expand the timeframe of the realistic utility of food without compromising the security, quality, and quantitative measures of food. However, antimicrobial agents can be either natural or man-made, i.e., synthetic type. Natural antimicrobials are acquiring greater significance rather than artificial or man-made ones. Even though synthetic additives are endorsed by government organizations for human use, a considerable lot of these additives compromise human health in a certain manner (Arshad et al., 2017).

4.3.1 Natural Antimicrobials

Natural antimicrobials are those which are produced by living organisms that have some properties to resist the growth of microbes. They contain the properties of antimicrobials as well as antioxidants and hence are considered a superior choice for food preservation when contrasted with synthetic additives. Plant extracts and edible oils are the main examples of natural antimicrobials; also extracts from fruit peels have also shown some antimicrobial properties in recent studies (Ibarra et al., 2023). In fruits and freshly cut byproducts, tartaric, succinic, malic, and acidic acids are frequently present. They have typically been used as additives in the eatery, with the pH alterations of the treated media being credited with their antimicrobial efficacy (Crozier-Dodson et al., 2004).

Antimicrobial properties are commonly seen in herbs, spices, fruits, veggies, seeds, and leaves.

4.3.1.1 Plant-based antimicrobials

1. Herbs and Spices: The ability of herbs and spices to generate antimicrobial is basically dependent upon the oil, their inbuilt content present inside them. Herbs and spices have often been used as flavouring agents and preservatives; now, as times have changed, they are being used as antimicrobial agents for keeping food for a long time without spoilage (Pellerito et al., 2018)

   For example, clove, oregano, rosemary, thyme, sage, and vanillin contain phenolic group which has high antimicrobial properties.

   The antibacterial properties of oregano essential oils prevent the growth of *Saccharomyces cerevisiae*, *B. subtilis*, *Escherichia coli*, and *S. aureus*.
2. Fruits and Vegetables: Because of the presence of phenolic and organic acids fruits and vegetables tend to have antimicrobial properties within them.

   The addition of capsicum annum extracts inhibits the growth of *S. typhimurium*; similarly, pomegranate extracts interfere with the growth of *Escherichia coli*. Garlic, peels of lime, lemon grass, etc. are some of the other examples (Huang, E et al., 2021; Arshad et al., 2017).

4.3.2 Synthetic/Artificial Antimicrobials

Synthetic antimicrobials are substances that are manufactured synthetically rather than using natural extracts and are designed to have specific antimicrobial effects. These are the chemical preservatives that have been introduced to slow the growth of bacteria (Taylor et al., 2019). The several types of synthetic antimicrobials include benzoates, nitrites, sorbates, propionates, lactates, and acetates (Souza et al., 2022).

- Benzoates: used to keep fruit juices, drinks, pickles, etc. fresh. They interfere with the operation of cell membranes and stop growth.
- Nitrites are added to processed meats like bacon, ham, and sausages to preserve them.
- Sorbates: moulds and yeast are inhibitors.
- Propionates: these are used in cheese, frostings, gelatins, and baked goods.
- Lactates: produced by LAB (Lactic Acid Bacteria), they prevent food spoiling by slowing the growth of infections.
- Acetates: found in salad dressings, mayonnaise, and pickled foods (Sung et al., 2013).

4.4 ANTIMICROBIAL CLASSES

Antimicrobials are divided into two types in addition to natural and synthetic ones.

4.4.1 Class 1 Antimicrobials

These include antimicrobials such as honey, sugar, table salts, glucose, fructose, alcohol, spices, vinegar or acetic acid, and other sugars and carbohydrates. These fall under the category where there are no legal restrictions on their addition to food ingredients.

This class of additives includes various antimicrobials such as oil, various spices, and sugars:

a) Sugars: by preventing mould, yeast, and bacteria, sugar preserves jellies, jams, fruit preserves, candied fruit, and chutneys.
b) Salts: microorganisms' growth is inhibited by high salt concentrations. Ham and ox tongue are pickled meat. Pickling and curing are two classic uses for salt.
c) Smoke: by combining an intricate blend of acids, phenols, alcohols, and toxic chemicals, smoke inhibits the growth of microorganisms. The right temperature is crucial when smoking.
d) Vinegar: to pickle vegetables, 4% acetic acid and water are mixed with shallots, red cabbages, and onions.

4.4.2 Class 2 Antimicrobials

These are primarily synthetic additives that are only added in the allotted amounts and are further restricted by legislation. They consist of fumaric acid, sulphuric acid, sodium and potassium nitrates and nitrites, sorbic acid, propionic acid, acetic acid, and nisin.

This class includes salts of various acids. These salts are of utmost help in performing the following functions:

a) Tomato sauce, squash, and syrups contain benzoic acid and its salts.
b) Sulphurous acids and their salts: fruit juices with potassium metabisulphite, bulk storage of fruit pulp.
c) Meat preservation with potassium and sodium nitrates and nitrates.
d) Propionic acid and its salts: mould-inhibiting agents adsorbed over the cheese surfaces.
e) Compounds such as propionic acid, sorbic acid, sodium diacetate, etc., help in the fermentation and prevention of yeast and moulds over various food items such as bread.. The best example is sorbic acid and its various salts help in the preservation of fruit juices, sauces, margarine, etc.
f) Sorbic acid and its salts: in fruit juices, sauces, in margarine to prevent yeast and mould growth.

4.5 ANTIMICROBIALS VS. SYNTHETIC/CONVENTIONAL METHODS

Microorganisms (microbes, parasites, and moulds) are responsible for delivering compounds known as antimicrobials, which can either prevent the growth of microorganisms or even eradicate them. Some of the mixes were initially discovered as the outcomes of microorganisms have been blended thanks to the advancement of designed procedures (Mahmud et al., 2018).

Antimicrobial have been so significantly traditionally used and for their use following criteria has to be fulfil.

(i) It is in use for a longer time.
(ii) Their use in food industries has been promoted and encouraged by various countries and administrative agencies, or
(iii) Is provided by designed procedures rather than being regular concentrations (e.g., Advance methods for the maturation of nisin from the bacterium *Lactococcus lactis* and its subspecies).

Innumerable manufactured and natural edible antimicrobials have been discovered by various scientists and are obtained from nature only; examples are acidic corrosiveness in vinegar and lactoperoxidase in fluid milk.

Traditional ways to keep food fresh longer don't involve adding antimicrobials, but the way the food is packaged protects it and keeps it from going bad. Food supplies can be preserved via traditional methods such as freezing, refrigeration, drying, ageing, salting, canning, and a controlled environment (Steinka et al., 2006).

4.5.1 Drying, Ageing, and Canning

Before processing, food sources that will be dried, aged, or canned are briefly soaked in hot, boiling water. In addition to having a longer shelf life, these preservation techniques also require less preparation and do not need the addition of any chemicals (Malhotra et al., 2015; Lueck et al., 1997; Abdalla et al., 1993).

(i) **Drying:** Microorganisms require moisture to grow. Food sources won't become spoiled by bacteria, yeast, or moulds if they are dried, or in the unlikely event that the moisture is removed from the food. Sun drying, mechanical dryers, and freeze drying can all be used to remove moisture.
   - **Sun drying:** This method is used for certain fruits that are placed on trays, such as grapes, figs, and apricots. During drying, the fruits could be rotated. Sulphur is added to light-coloured fruits to prevent enzymatic browning.
   - **Mechanical dryers:** The food to be dried is passed through hot air that is adjusted to be somewhat damp, or hot air is passed through the food, on transport lines. Milk, for instance, is dried either by being sprayed into a current of dry, heated air or by being passed over warmed rollers. Dried food needs to be placed right away in air- and water-tight containers.
   - **Freeze drying:** This method of dehydration involves freezing the food to be dried in a cabinet first. By sucking the air out of the cabinet, a vacuum is created, and as ice sublimates, it turns into vapour. Freeze-dried foods don't need additional refrigeration or preservation. It is little, retains its shape and size, and when wet, returns to its original size and flavour. Fruits, vegetables, meat, poultry, and shellfish are all dried using this method.

(ii) **Ageing:** In terms of food or beverages, ageing refers to letting anything out for a long time (sometimes months or years) to enhance its flavour and preserve it for a long time. Ageing should be achievable in various situations and for various reasons, such as more savoury and delicate flavours that are more grounded. Foods are aged to protect them against food-borne germs. There are several ways to age food, including salting, sugar/crystallization curing, and dehydration.

(iii) **Canning:** Canning involves preparing food, sealing it in sterile pots or canisters, and then heating the containers to sterilize the food containers and kill or weaken any lingering bacteria. Since different foods have varying levels of preservation, it could be necessary to cook them in a pressure cooker.

While limited natural products like tomatoes demand lengthier boiling times and the potential inclusion of additional acidic chemicals, high-corrosive organic items like strawberries only need a brief duration of boiling and no additives.

Vegetables and meats that have a low acidity level must be handled with canning pressure. Food that has been canned or packaged is immediately at risk of becoming bad after being unsealed.

The two types of safe canning methods are:

i) The water bath method
ii) The pressure canning method
- **The water bath method:** This technique works well for foods that are highly corrosive, such as fruits, jams, tomatoes, salsas, pickles, chutneys, pie fillings, vinegar, and sauces. It entails immersing the food containers in fully covered boiling water for a set amount of time to cook.
- **The pressure canning method:** This is the only risk-free way to preserve low-acid food sources like vegetables, meats, fish, and poultry. The containers are cooked in a special pressure cooker while submerged in water. Pressure cooking is the greatest method for eliminating the *Clostridium botulinum* bacteria, which can persist in the food types previously described (Katz et al., 2016; Brown et al., 2018)

4.5.2 Salting

Since ancient times, salt has been used as an addition. Salt was one of the main methods used to protect food before the period of refrigeration and cutting-edge heat treatments that are used to extend the timeframe of realistic usability of food supplies. Active salt granules are added directly to the food or through brine during this process. Salted pickles and fish are two examples (Abdalla et al., 1993).

Food is preserved by salt in two ways.

1. **Salt dries food:** Salt causes food to lose water and dries it out. All living creatures, including the germs that might contaminate food, need water to survive and flourish. Salts are used as an antimicrobial in beef jerky to keep it dry and also in margarine to prevent it from its spoilage by dehydrating it and isolating the fact molecules in it.
2. **Salt kills the microbes:** This is well known that osmolarity has adverse effects on a variety of microorganisms present in the environment; however, not all microorganisms are affected by such water pressure as they can preferably process the osmosis through their cell membranes. The metabolism of various bacteria supports this osmolarity, whereas in certain bacteria the excessively concentrated solutions create pressure differences between the exterior and interior of the cell or prokaryotic organism which results in lyses of the bacterium. Increased concentration of cell salts also affects other chemical compositions and produces a negative impact on the microbial life cycle. Therefore, various food products such as jams and jellies are incorporated with the salt to cause microbial lysis, which increases the shelf life of the food (Abee et al., 1995).

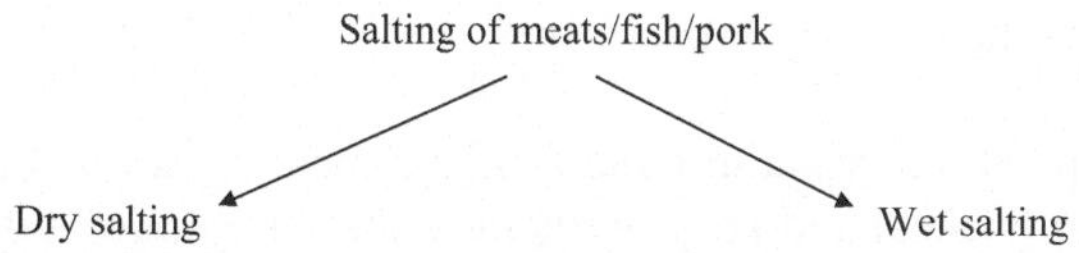

Dry salting: With dry salting meat is stuffed in dry salt or salt is scoured with a covering; however, this strategy doesn't safeguard the meat insofar as wet salting.

Wet salting: In wet salting meat is first scoured with endless salt and is likewise positioned between the layers of meat; then, at that point, brine is poured over the stuffed salted meat and kept lowered in a brine solution.

4.5.3 Use of Low Temperature (Freezing, Refrigeration)

Low temperatures safeguard food by hindering compounds, enzymatic activity, and the development and movement of microorganisms. The food will be better preserved at lower temperatures. Food varieties contain microbial concentration, low temperatures do not kill microbes, and once the temperature is favourable, microorganisms will increase (Figure 4.3).

Pros of cooking and freezing:

1. Maximum use of man-hours.
2. Maximum utilization of all equipment.
3. Minute spoilage and portion control.
4. Fewer deliveries from the commissary.
5. When in season and prices are low, foods can be prepared.
6. In advance, menus can be planned and prepared.
7. More variety menu with less number of staffs.
8. Overall, there are saving in staff, equipment, space, fuel, and costs (Lueck et al., 1997).

4.5.4 Modified Atmosphere

Fruits and vegetables are still alive after they have been harvested, and they respire by taking in oxygen and releasing carbon dioxide. When the respiration rate increases, the products ripen quickly.

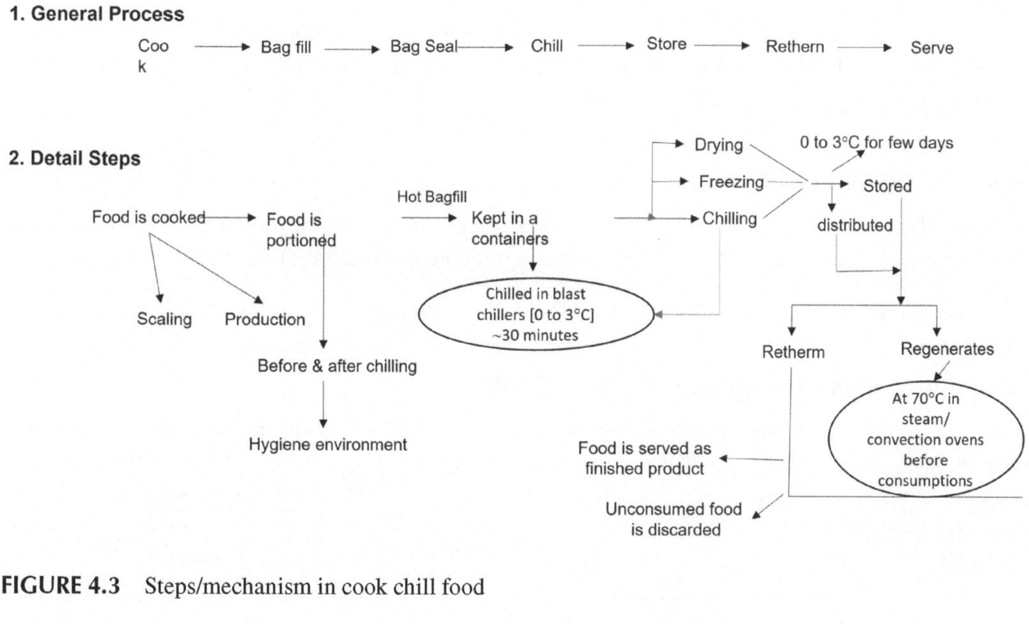

FIGURE 4.3 Steps/mechanism in cook chill food

The methods to reduce the ripening process are:

1. Reducing the temperature by natural and artificial means, such as keeping the fruits and vegetables in water buckets and keeping them in the refrigerator, respectively as cooling procedures.
2. Preventing oxidation is one of the supporting methods.
3. Introducing the carbon dioxide effect may delay the ripening process.

All these methods are utilized in modified atmosphere storage.

In this atmosphere, food packages are kept in specific mixtures of gases to inhibit the growth of microorganisms (Lueck et al., 1997) (Table 4.2).

Figure **4.3** illustrates food processing through various methods. The basic and mostly encountered methodology is putting the food sample in a bag fill, and then sealing it, after that, the bag has to be stored in a cool and damp place to prevent microbial growth. The food material can then be returned and served. In general, food is chilled in containers with 0–3°C temperature. For the regeneration, a temperature of around 70°C should be given, or it should be steamed. The process sometimes seems to be even more complicated from cooking to processing. A hygienic environment is one of the most important conditions to be maintained as the food items can be prevented from additional contamination via entities present in the environment. The portioned food items are to be kept in different containers for drying, freezing, and chilling and are then distributed or commercialized as finished, packaged food. Various natural and artificial antimicrobials such as nanostructured antimicrobials are used for the packaging of the food, whereas unconsumed foods are supposed to be discarded.

4.6 ADVANTAGES OF ANTIMICROBIALS

4.6.1 Cost-Effectiveness

Due to the abundance of antimicrobial sources and their simplicity of handling, these are incredibly inexpensive, which enables the food sector to make them on a large scale. A food antibacterial should also be affordable to be used in the food industry. For everyone to have access to antimicrobials for their personal needs, they are therefore produced at extremely low costs and sold on the market (Quinto et al., 2019).

Measures to prevent illnesses with more sophisticated administration, such as microbe inhibitors in food, have been considered more effective in preserving economic productivity than using antimicrobials as growth boosters in food animals.

4.6.2 Eco-Friendliness

Antimicrobials that are often taken from natural sources are more environmentally and biodegradably friendly. For instance, the adoption of biodegradable antimicrobial food packaging materials

TABLE 4.2
Composition of the atmosphere (Lueck et al., 1997)

Gases present	General composition of atmosphere in normal conditions (in percentage)	Modified atmospheric composition (in percentage)
Oxygen (O_2)	20.9	2.5–5
Carbon dioxide (CO_2)	0.3–0.4	2–5
Nitrogen (N_2)	79	90–95

has raised the bar for antimicrobials beyond their standing as simple food additives. The development of antimicrobial packaging has been driven by the growing demand for healthier, less processed foods with reduced added substance contents (Taylor et al., 2014; Motelica et al., 2020; Haque et al., 2023).

4.6.3 Maintenance of Optimal and Sustainable Environment and Good Health

Since the beginning of medicine and human health, around 40 years ago, the discovery of the first particular antibacterial specialist was a major accomplishment. The emergence and dissemination of antimicrobial blockage seriously jeopardize the efficacy of antibiotics, which are crucial treatments (Marshall et al., 2011; National Research Council et al., 1980).

Antimicrobial have been so significantly traditionally used and for their use following criteria has to be fulfil. The United Nations defined the eradication of hunger as the second goal for sustainable development, and it poses considerable challenges for improving global food production and pecking order. Antimicrobials have so been used to ensure the safe handling of food and to increase output (National Research Council et al., 1980; Rai et al., 2016).

4.6.4 Quality Maintenance

Food spoiling is the process of contaminating different food types that results in the loss of variety, surface, and nutritional value as well as the growth of pathogenic microbes, which undermines the item's quality and renders it unusable. Industrial and professional uses of antimicrobial have proven to be beneficial for preserving food (Rydlo et al., 2006; Thakali et al., 2022). Due to the greater understanding of their carcinogenic effect, synthetic preservatives like nitrites and nitrates have been banned by many industries. This has improved the quality of antimicrobials used in the food sector.

4.6.5 Safety

Food structure, handling methods, and capacity conditions are only a few of the many factors that can affect an antimicrobial mixture's ability to be used in food applications. Because they can reduce microbial blockage and meet consumers' demands for improved food options, antimicrobials are protected (Teshome et al., 2022).

Even though there have been many improvements in the use of antimicrobials in and on food products, further study in the area is crucial. Where possible, including the food safety of an item enhances the manufacturer's efforts to simultaneously regulate and reduce microbiological risks. To ensure the safety of food, natural antimicrobials are therefore overrunning the market (Huang et al., 2021, Kim et al., 2023).

4.6.6 Enhancing Shelf Life

For the most part, the food sector depends on synthetics to protect food products and extend their shelf lives. Food items can also be preserved by using the microbes as bio-additives which not only increases the food shelf life but also enhances the mode of its handling and dispatch.

The contamination of food by microbes, such as *Salmonella* species, *Shigella* species, *Micrococcus* species, *Enterococcus* species, *Bacillus licheniformis* species, *Escherichia coli* species, *Listeria monocytogenes* species, *Staphylococcus aureus* species, *Campylobacter jejuni* species, *Yersinia enterocolitica* species, *Vibrio parahemolyticus* species, and others, is a serious problem, and antimicrobial drugs have been shown to operate against foodborne microorganisms and, as a result, can aid in food preservation, according to scientists (Rai et al., 2016).

4.7 APPLICATIONS OF ANTIMICROBIALS IN THE FOOD INDUSTRY

4.7.1 Industrial Units

The use of antimicrobial surfaces in food industrial units is generating significant interest in all parts of food handling tasks. Antimicrobials can also be referred to as active substances that are introduced during manufacturing processes. These additives lessen microorganisms on safeguarded surfaces in various ways:

- Harm is incurred for microorganisms' protein parts, which causes the failure of fundamental capabilities like energy creation.
- By disrupting the microorganism's membrane layer, its structural integrity is compromised, which can make fundamental supplements spill out.
- Antimicrobials can cause expanded degrees of responsive oxygen species, which bring about harm to the inward frameworks of the organism.
- The hereditary material of the microorganism is upset, at last preventing the organism from having the option to reproduce by hindering the replication of its hereditary material (Gumienna et al., 2021).

4.7.2 Packaged Foods

The application of antimicrobial agents holds a significant role in packaged foods like pickles, frozen foods, instant noodles, etc.

Volatile and non-volatile antimicrobials are supposed to be used on polymers, which are generally used as a coat or an adsorbent material. This process is however very peculiar and has to be done by an antimicrobial specialist only. For antimicrobial mixtures or cell reinforcements to keep up with large concentrations of additives on food surfaces, such a coating can be used as a transport (Gumienna et al., 2021; Han et al., 2003).

There are two approaches to adding antimicrobials to packaged foods.

- The first method takes care of the packing materials, whose antimicrobial surfaces don't fall into contact with the food that has been saved and whose dynamic specialists can move into the food. Vacuum-sealed or foil-wrapping I traditionally used packaging substitutes.
- Another way involves including antibacterial chemicals inside the wrapping, but not within the food (Han et al., 2003).

Natural agents such as herb extracts, spices, enzymes, and bacteriocins are used in packaged foods. Spice and herb extracts contain various regular mixtures and are known as wide-range antimicrobial specialists. These agents hold an edge over different specialists utilized in antimicrobial bundling as these are palatable parts (Crozier-Dodson et al., 2004).

4.7.3 Composition of Antimicrobial Packaging

4.7.3.1 Antimicrobial packing material

There are two types of antimicrobial packing materials:

i) Biodegradable
ii) Non-biodegradable

Biodegradable packaging is in which the degradation process brings about lower sub-atomic weight sections created by the activity of naturally happening microorganisms like bacteria, fungi, and algae. Biodegradable packing incorporates edible coatings and films from proteins, lipids, starch, chitosan, polylactic acid, polyhydroxy butyrate, and polyhydroxyalkanoate. Non-biodegradable packaging doesn't go through corruption into parts by microorganisms like polymeric packaging materials from nylon, Low Thickness Polyethylene (LDPE), High Thickness Polyethylene (HDPE), and polystyrene, and so forth. Most engineered polymers are non-biodegradable and are predominant for use as food packing material with the upsides of minimal expense, low thickness, inactive, magnificent boundary properties, great mechanical strength, high straightforwardness, capacity to be heat-fixed, and easily printable (Moses et al., 2014; Chausali et al., 2022).

4.7.4 Antimicrobial Agents

Compounds, biotechnological products, antioxidants, antimicrobial polymers, natural antimicrobials, anti-antioxidants, and other gases are some of the antimicrobial agents that are incorporated into the packaging system. Antimicrobial agents work on the quality and keep up the safety of food packaging by decreasing surface contamination of processed food.

4.7.5 Processed Foods

Any food item which has undergone certain technical minor processing is referred as processed foods, such as sliced vegetables and fruits packed salads and leafy greens, roasted almonds, and so on.. Acid antimicrobials are the mainly used agents in processed foods. For example, sodium diacetate, a sub-atomic compound of acetic acid, is utilized to accomplish an antimicrobial impact in prepared merchandise, fats and oils, sauces and gravies, nibble food sources, meat items, and soups and soup blends, as well as to season these food sources (Gumienna et al., 2021; Han et al., 2003).

Due to its high oxidizing potential and low pH of 2.8, peracetic corrosive (peroxyacetic corrosive) has been considered a successful antimicrobial against vegetative cells, endospores, yeast, and form spores (Han et al., 2003).

Apart from using these acid antimicrobials, pasteurization, fermentation, freezing, vacuum-packing, refrigeration, etc. are some of the traditional methods utilized on increasing the timeframe of processed food and for inhibiting the growth of disease-causing microbes (Katz et al., 2016).

Food industries and biotechnology have all together assisted various food processing technologies which include the entire steps from production to packaging of food into suitable finished products till their commercialization. Technologists have assisted both natural and novel food processing techniques. Both techniques include thermal and non-thermal processes.

As Figure 4.4 (Toldra et al., 2017) indicates, thermal conventional methods include moist heating and dry heating, as general drying and dehydration increase the shelf life of various food items, and non-thermal conventional methods include fermentation, freezing, acidification, chemical additives, ultrafiltration, etc. These methods are part of various biological and chemical processes. Novel thermal techniques include microwaves, radio frequency dried stream treatments, restoring, etc. However, novel non-thermal techniques include high-pressure processing, ozonization, and ultrasonic and ultraviolet treatments, although these methods are costlier as compared to conventional techniques but can be effectively used in large-scale industries, as they have proven as savage techniques, preventing loss of food industries and economical loss.

4.8 ANTIMICROBIALS FROM DIFFERENT SOURCES

4.8.1 Natural Antimicrobial Compounds from Plant Source

Antimicrobial compounds derived from plant extracts have been steadily increasing as a result of the need to reduce the usage of artificially produced food additives. The majority of these substances come from spices, herbs, and essential oils that have been extracted from various herbs and

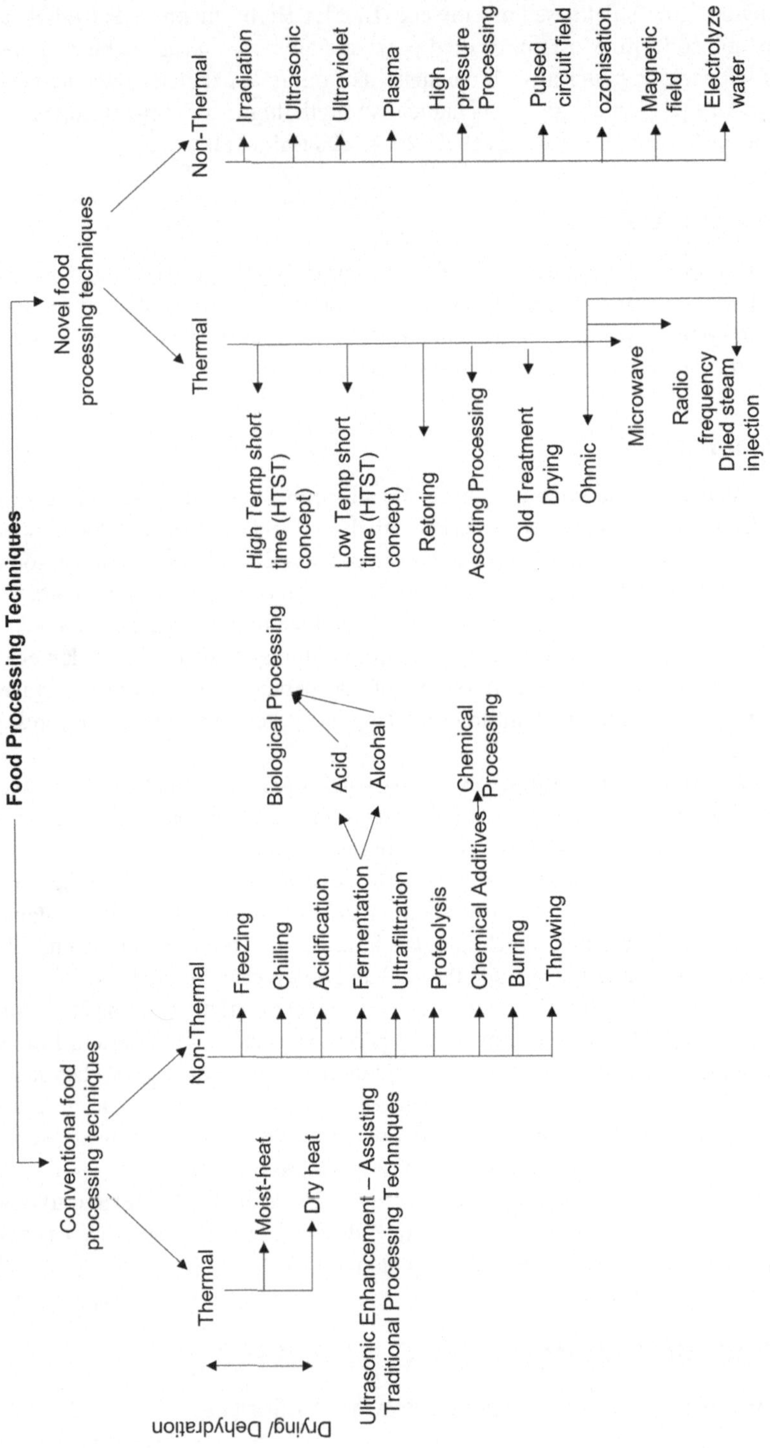

FIGURE 4.4 Different methods of processing food materials

spices, including oregano, basil, cinnamon, clove, thyme, rosemary, garlic, and many more (Nychas et al., 1995).

4.8.1.1 Spices

In previous research, the effectiveness of spices, herbs, and their extracts as natural antibacterial agents, most often utilized in whole, crushed, and encapsulated forms, was reported. In addition to having antimicrobial properties, these plant extracts also have high antioxidant capacities that help prevent oxidative spoilage. This is due to the presence of secondary metabolites like phenolic acids, saponins, tannins, alkaloids, alkenyl phenols, glycoalkaloids, flavonoids, sesquiterpenes, and terpenoids, as well as other secondary metabolites like alkenyl phenols. Aromatic and volatile chemicals found in spices and herbs protect food from microbial deterioration (Abdallah et al., 2011).

4.8.1.2 Essential oils

These are secondary aromatic metabolites that are obtained from plants and certain parts such as inner bark, pith, roots, and flowers and are extracted via various technical approaches, some of which involve solvent extraction and solvent-free extraction, distillation which includes both hydro distillations as well as diffusion processes, combined extraction using solvent and steam, etc. These essential oils, as the name suggests, are very essential in compromising the growth of microorganisms coming from nature as well as food which are hazardous to human health as well. These oils are supposed to have various antioxidant, antimicrobial, and food-preservative actions. It primarily oxidizes the lipid concentration of the pathogenic bacterium or other organism and lowers its population. This step is essential in every type of microorganism to prevent its growth order to prevent food deterioration and subsequent degradation. Physiologically active substances such as aldehydes, terpenoids, phenolic acids, and ketones are also essential to hinder the growth of infection-causing microbes (Pisoschi et al., 2018).

4.8.1.3 Antioxidants

Natural substances called antioxidants prevent oxidative food damage, which alters the nutritional and organoleptic qualities of food. They protect against several lifestyle diseases brought on by oxidative stress. Carotenoids, tocopherol (vitamin E), certain phenolic compounds, and ascorbic acid are a few examples of natural antioxidants found in the diet. Due to their potent ability to scavenge free radicals and potent anticancer and antidiabetic properties, phenolic compounds are the most well-known and significant class of naturally occurring antioxidant chemicals. In addition to regulating the immune system, antioxidants have a protective impact against cardiovascular and brain disorders. Common natural sources of antioxidants used as additions include blueberries, strawberries, blackberries, rosemary, turmeric, saffron, ginger, pepper, and green tea. Some essential oils are also capable of replacing the chemical antioxidants used as food preservatives, showing promise as antioxidants (Abeyrathne et al., 2022).

4.8.2 Natural Antimicrobial Compounds from Animal Sources

The most prevalent naturally occurring antibacterial substances in mammals are proteins and enzymes. These organic substances include things like lacto peroxidase, lactoferrin, and lysozyme. Due to their antibacterial capabilities, many of these compounds are primarily used by animals as a kind of defence against predators. They accomplish this by killing both Gram-positive and Gram-negative bacteria by rupturing the cell membranes of microorganisms, primarily bacteria (Gaber El-Saber Batiha et al., 2021).

4.8.2.1 Chitosan

Chitin, a naturally occurring polycationic linear polysaccharide primarily found in the shells of marine crustaceans, is converted into chitosan through partial deacetylation. Glucosamine and

N-acetyl glucosamine are polymers that make up chitosan. Through chemical and biological procedures, they are separated from crustaceans and shrimps; alternatively, they can be generated from fungi.

4.8.2.2 Lysozymes

Animals also have lysozymes, which are organic substances that work well against microbes that cause rotting. They are thought to be GRAS and may be added right away to a food matrix. Due to their stability over various pH and temperature ranges (4–95°C), lysozymes have drawn particular attention (2–10). Researchers have looked at the effectiveness of lysozymes towards a variety of pathogens and spoilage bacterium, including the Gram-positive *Listeria monocytogenes*, *Bacillus stearothermophilus*, *Micrococcus* spp., and *Clostridium tyrobutyricum*. The ability of lysozymes to damage the peptidoglycan layer of bacterial cell walls was attributed to their antimicrobial action. This was done by accelerating the hydrolysis of the bond between N-acetyl-muramic acid and N-acetyl-d-glucosamine (Board et al., 1995).

4.8.2.3 Lactoperoxidase

Another animal-based antibacterial is lactoperoxidase (LPS), an enzyme that is abundant in cow's milk and is released by the epithelium of the mammary gland. Salmonellae, Shigella, Pseudomonas, and Coliforms are just a few of the bacteria that LPS can effectively inhibit and eliminate (Saeed et al., 2019).

4.9 CONCLUSION

Without being handled or preserved, food will undoubtedly become stale. The composition, storage temperature, ideal duration, and prevention from contamination of food are essential factors that determine whether and how quickly food starts to deteriorate. Until it is consumed, food should be preserved from deterioration and contamination at all times. Thus, the addition of antimicrobials plays a vital role in food preservation.

The utilization of antimicrobial surfaces within the food sector is generating extensive interest in all parts of food handling tasks. In light of healthier food practices, antimicrobials help in preventing many foodborne diseases. Antimicrobials can be natural and synthetic (chemical additives), and they are added to different segments of food processing. Synthetic antimicrobials such as benzoates, nitrites, sorbates, propionates, lactates, and acetates can be effective in preventing pathogens, but as they are chemically manufactured, they can also harm our health. So, to maintain our health and the quality of food, food industries are promoting the use of natural antimicrobials in various stages. As it is naturally extracted, it also hikes up the economy of the industry.

Certain traditionally used food protection procedures like drying or dehydrating, freezing, warming, prevention from fermentation, and using salts have broadened the shelf life and the usability of food items; however, repeated pollution might happen that might deliver the food unpalatable for the buyers. A clever advancement called antimicrobial packaging framework incorporates antimicrobial agents into a polymer film to inhibit the activities of specific microorganisms which are the source of contamination of food. A slow assisted release of antimicrobials is a core part that makes the distinction between life and death because of the clear differences between lab scale and the relative food packaging without the use of antimicrobials. Antimicrobials have also been used for many decades and so far have various futuristic benefits. They will provide large sustainable preservation of food products and help food industries with various economic benefits. For meeting the demands of the upcoming population antimicrobials are one of the major sources that can even widely be used at nano levels, in nanotechnologies for encapsulated medicines, for example, liposome-based nanoparticles produce sustainable antibacterial stability. They are a boon to hurdle technologies which refer to the practices of combining various preservatives to generate a minimum inhibitory concentration of antimicrobials to degenerate microbes without causing harm to humans. Nanostructured antimicrobials promote a better quality of food products and proclaim their stability from various harsh conditions. Both inorganic and organic antimicrobials are playing a major role in accessing food industries and are potent future weapons against microbial agents.

ACKNOWLEDGEMENTS

The authors of this book chapter are grateful to all our all colleagues from Galgotias University and from outside of this university who have directly and indirectly helped and supported us in the completion of this book chapter.

REFERENCES

Abdalla, O.M., Davidson, P.M., Christen, G.L. 1993. Survival of selected pathogenic bacteria in white pickled cheese made with lactic acid bacteria or antimicrobials. *J. Food Prot.* 56(11): 972–976.

Abdallah, E.M. 2011. Plants: An alternative source for antimicrobials. *J. Appl. Pharm. Sci.* 1(6): 16–20.

Abee, T., Krockel, L., Hill, C. 1995. Bacteriocins: Modes of action and potentials in food preservation and control of food poisoning. *Int. J. Food Microbiol.* 28(2): 169–185.

Abeyrathne, E.D.N.S., Nam, K., Huang, X., Ahn, D.U. 2022. Plant- and animal-based antioxidants' structure, efficacy, mechanisms, and applications: A review. *Antioxidants* 11(5): 1–18.

Arshad, M.S., Batool, S.A. 2017. Natural antimicrobials, their sources, and food safety. *Food Addit.* 87(1): 87–102. http://doi.org/10.5772/intechopen.70197.

Ayseli, Y.I., Aytekin, N., Buyukkayhan, D., Aslan, I., Ayseli, M.T. 2020. Food policy, nutrition and nutraceuticals in the prevention and management of COVID-19: Advice for healthcare professionals. *Trends Food Sci. Technol.* 105: 186–199.

Board, R.G. 1995. Natural antimicrobials from animals. In Gould, G.W. (Ed.), *New Methods Food Preserv.* Boston, MA: Springer, 40–57.

Brown, A.C. 2018. *Understanding food: Principles and preparation.* Sixth ed. Boston MA: Cengage Learning.

Chausali, N., Saxena, J., Prasad, R. 2022. Recent trends in nanotechnology applications of bio-based packaging. *Journal of Agriculture and Food Research,* https://doi.org/10.1016/j.jafr.2021.100257.

Crozier-Dodson, B.A., Carter, M., Zheng, Z. 2004. Formulating food safety: An overview of antimicrobial ingredients. *J. Food Saf. (Magazine)* 10(6): 24–27.

Gaber El-Saber, B.D.E., Hussein, A.M., Algammal, T.T., George, P.J., Ali Esmail, A.-S., Tiwari, A., Jorge, P.P., Clara, M.L., Nanasaheb, D., Thorat, M.Z., El-Esawi, M., Dey, A., Saad, A., Amdi, H.F., Hetta, N.C., Cruz-Martins, N. 2021. Application of natural antimicrobials in food preservation: Recent views. *Food Control* 126: 108066.

Gumienna, M., Górna, B. 2021. Antimicrobial food packaging with biodegradable polymers and bacteriocins. *Molecules* 26(12): 3735.

Han, J.H. 2003. Antimicrobial food packaging. *Non. Food PG Tech*2003: 50–70.

Haque, S., Singh, R., Harakeh, S., Teklemariam, A.D., Tayeb, H.H., Deen, P.R., Srivastava, U.C., Srivastava, M. 2023. Green synthesis of nanostructures from rice straw food waste to improve the antimicrobial efficiency: New Insight. *Int. J. Food Microbiol.* 386: 110016.

Huang, E., Hussein, W.E., Campbell, E.P., Yousef, A.E. 2021. Applications in food technology: Antimicrobial. In Fidel Toldrá, Jianping Wu (Ed.), *Biologically Active Peptides.* Academic Press, 745–770.

Ibarra, A.F.C., Castaneda, P.S., Arbelaez, L.M., Sanchez, L.T., Pinzon, M.I., Villa, C.C. 2023. Natural antimicrobials. *Nat. Add. in Foods.* Springer, 69–85.

Katz, S.E. 2016. *Wild fermentation: The flavor, nutrition, and craft of live-culture foods* (2nd ed.). White River Junction, Vermont: Chelsea Green Publishing, 1–320.

Kim, Y., Ma, L., Huang, K., Nitin, N. 2023. Bio-based antimicrobial compositions and sensing technologies to improve food safety. *Curr. Opin. Biotechnol.* 79: 102871.

Lucera, A., Conte, C., Conte, A., Nobile, M.A.D. 2012. Food applications of natural antimicrobial compounds. *Front. Microbiol.* 3(287): 1–13.

Lueck, E., Jager, M. 1997. *Antimicrobial food additives: Characteristics, uses, effects* (2nd ed., Rev. and enlarged ed.). Berlin: Springer.

Mahmud, J., Khan, R.A. 2018. Characterization of natural antimicrobials in food system. *Adv. Microbiol.* 8(11): 894–916.
Malhotra, B., Keshwani, A., Kharkwal, H. 2015. Antimicrobial food packaging: Potential and pitfalls. *Front. Microbiol.* 6: 1–9.
Marshall, B.M., Levy, S.B. 2011. Food animals and antimicrobials: Impacts on human health. *Clin. Microbiol. Rev.* 24(4): 718–733.
Motelica, L., Ficai, D., Ficai, A., Oprea, O.C., Kaya, D.A., Andromescu, E. 2020. Biodegradable antimicrobial food packaging: Trends and perspectives. *Foods* 9(10): 1438.
Moses, J.A., Norton, T., Alagusundaram, K., Tiwari, B.K. 2014. Novel drying techniques for the food industry. *Food Eng. Rev.* 6(3): 43–55.
National Research Council (US) Committee to Study the Human Health Effects of Subtherapeutic Antibiotic Use in Animal Feeds. 1980. *The effects on human health of subtherapeutic use of antimicrobials in animal feeds.* Washington (DC): National Academies Press (US).
Nychas, G.J.E. 1995. Natural antimicrobials from plants. In Gould, G.W. (Ed.), *New methods of food preservation.* Boston, MA: Springer.
Pathare, P.B., Opara, U.L., Al-Said, F.A.J. 2013. Colour measurement and analysis in fresh and processed foods: A review. *Food Bioprocess Technol.* 6(1): 36–60.
Pellerito, A., Ameen, S.M., Micali, M., Caruso, G. 2018. Antimicrobial substances for food packaging products: The current situation. *J. AOAC Int.* 101(4): 942–947.
Pisoschi, A.M., Pop, A., Georgescu, C., Turcus, V., Olah, N.K., Mathe, E. 2018. An overview of natural antimicrobials role in food. *Eur. J. Med. Chem.* 143: 922–935.
Quinto, E.J., Caro, I., Villalobos-Delgado, L.H., Mateo, J., De-Mateo-Silleras, B., Redondo-Del-Río, M.P. 2019. Food safety through natural antimicrobials. *Antibiotics (Basel)* 8(4): 208.
Rai, M., Pandit, R., Gaikwad, S., Kovics, G. 2016. Antimicrobial peptides as natural bio-preservative to enhance the shelf-life of food. *J. Food Sci. Technol.* 53(9): 3381–3394.
Rydlo, T., Miltz, J., Mor, A. 2006. Eukaryotic antimicrobial peptides: Promises and premises in food safety. *J. Food Sci.* 71(9): 125–135.
Sadiku, M.N.O., Musa, S.M., Ashaolu, T.J. 2019. Food industry: An introduction. *Int. J. Trend Sci. Res. Develop* 3(4): 128–130.
Saeed, F., Afzaal, M., Tufail, T., Ahmad, A. 2019. Use of natural antimicrobial agents: A safe preservation approach. *Act. Antimicrob. Food Packaging*, IntechOpen. 1–18.
Souza, P.F.N. 2022. Synthetic antimicrobial peptides for controlling fungi in foods. *Curr. Opin. Food Sci.* 45: 100819.
Sultanbawa, Y. 2011. Plant antimicrobials in food applications: Minireview. In A. Méndez-Vilas (Ed.), *Science against microbial pathogens: Communicating current research and technological advances.* Badajoz, Spain: Formatex Research Center, 2: 1084–1099.
Sung, S.Y., Sin, L.T., Tee, T.T., Bee, S.T., Rahmat, A.R., Rahman, W., Tan, A.C., Vikhraman, M. 2013. Antimicrobial agents for food packaging applications. *Trends Food Sci. Technol.* 33(2): 110–123.
Steinka, I., Morawska, M., Rutkowska, M., Kukułowicz, A. 2006. The influence of biological factors on properties of some traditional and new polymers used for fermented food packaging. *J. Food Eng.* 77(4): 771–775.
Taylor, M. 2014. *Handbook of natural antimicrobials for food safety and quality.* Elsevier, Woodhead Publishing Series in Food Science, Technology and Nutrition, USA.
Taylor, T.M., Ravishankar, S., Bhargava, K., Juneja, V.K. 2019. Chemical preservatives and natural food antimicrobials. *Food Microbiology: Fundamental and Frontier* (eds. M.P. Doyle, F. Diez-Gonzalez and C. Hill). 705–731, ASM Press.
Teshome, E., Forsido, S.F., Rupasinghe, H.P.V., Keyata, E.O. 2022. Potentials of natural preservatives to enhance food safety and shelf life: A review. *Sci. World J.* 23: 9901018.
Toldra, F., Nollet, L.M.L. 2017. *Advances in food diagnostics* (2nd ed.). Chichester: John Wiley and Sons Ltd, pp. 231–250.
Thakali, A., MacRae, J.D., Isenhour, C., Blackmer, T. 2022. Composition and contamination of source separated food waste from different sources and regulatory environments. *J. Environ. Manag.* 314: 115043.
Worsfold, D. 2006. Consumer information on hygiene inspections of food premises. *J. Foodserv.* 17(1): 23–31.

5 Food preservation by natural antimicrobials
Current trends and future perspective

Sanchali Kundu, Somanjana Khatua, and Krishnendu Acharya

5 INTRODUCTION

It is expected that by 2050, the world's population will reach 9.7 billion people and approximately 70% extra food manufacturing will be necessary to feed them (Kumar and Kalita 2017). However, increasing urbanization, changed climate conditions, soil infertility, water scarcity, uncultivable land, and high-cost energy are presently affecting our food supply (Wilson 2013). On the other hand, fresh foods like seafood, meat, and horticultural products are extremely perishable and associated with foodborne disease (FBD) outbreaks triggered by pathogenic microorganisms. It results in an annual loss of a staggering one-third of all food produced (that equates to about 1.3 billion tonnes amounting to USD 1 trillion) globally during postharvest procedures (Kumar and Kalita 2017). In India alone, about 30% of the total food production is wasted during postharvest storage, handling, and transportation. The consequences not only impose adverse impacts on environment but also affect the income of growers and the national economy (Kumar et al. 2022). Meanwhile, ingestion of microbial-contaminated food may welcome many FBDs, and administration of antibiotics may worsen the situation necessitating advancement in food preservation techniques (Sweet et al. 2022).

To date, numerous methods have been developed assuring microbial food safety and nutritious benefits. However, the practices sometimes compromise with the organoleptic aspects of food products (Quinto et al. 2019). Besides, several chemical preservatives namely benzoic acid, sorbic acid, methyl paraben, and propyl paraben are commonly used in the dietary item and beverage industries that can control the growth of spoilage-related microorganisms and improve food quality. However, long-term use of these artificial antimicrobial agents may welcome serious health hazards such as hypersensitivity, asthma, and even cancer (Hussain et al. 2021). As sustainable alternatives, biopreservatives derived from plants (e.g., essential oil, phenolics), animals (e.g., chitosan, lysozyme, lactoferrin), and microbes (e.g., bacteriocins) are gaining increasing attention for food applications. These elements can eradicate microbial contamination improving food function and quality as well as prolonging the shelf life of dietary items. The majority of these substances own strong antimicrobial and antioxidant activities; despite that they remain underutilized (Mishra et al. 2022). Against this backdrop, the present chapter aimed to review the foremost natural food preservatives elucidating their mechanisms of action and application in food sectors. The concomitant articles were searched using online libraries such as Google Scholar, Scopus, PubMed, and Web of Science.

5.1 Food Preservation

With changes in the socio-economic scenario at the global scale, people are now leaning towards ready-to-cook fast meals arranged by food catering sectors outside the home. The trend parallels

DOI: 10.1201/9781003268949-5

demands for safe, nourishing, and high-quality foodstuffs packed with functional attributes as well as health benefits (Bekhit and Roohinejad 2016). However fruits, vegetables, fish, meat, and dairy products are perishable items that could easily be spoiled due to the growth of unwanted pathogenic microorganisms such as *Bacillus cereus*, *Escherichia coli*, *Listeria monocytogenes*, and *Staphylococcus aureus*. Several fungal species including *Alternaria*, *Aspergillus*, *Fusarium*, *Penicillium*, *Paecilomyces*, *Rhizopus*, *Ceratocystis*, *Rhizoctonia*, and *Botrytis* are also recognized to induce food deterioration (Faheem et al. 2022; Yousefi et al. 2022). Consumption of such nutrients may cause FBD resulting in nausea, anorexia, vomiting, and diarrhoea, with or without fever. Indeed, it has been postulated that each year more than 600 million people suffer from FBD leading to roughly 420,000 deaths worldwide (Milagres de Almeida et al. 2023). Thus food preservation is essential as it not only allows anticipated chemical, biochemical, and microbiological alterations during food processing but also enhances the shelf life of food and ensures food safety (Adegoke and Olapade 2012).

5.2 Physical Means for Food Preservation and Limitations

Preserving food is an extremely old phenomenon, which dates back to prehistoric times. Conventional techniques include drying, refrigeration, and fermentation; while modern physical methods encompass canning, pasteurization, freezing, and irradiation (Amit et al. 2017). Consideration of only one single technology may inhibit microbial growth, but the organisms may recover later during the food storage step. In this respect, most of the current research studies are focused on hurdle technology, a combination of preservation methods. The most promising amalgamations include non-thermal methods, namely ultrasound, high hydrostatic pressure, and pulsed electric fields. However, this costly expertise causes greenhouse gas emissions, significant energy consumption, and a negative impact on the environment; hence, they are not considered sustainable solutions (Putnik et al. 2020). Besides, nowadays consumer choice has shifted towards minimally processed foods with fewer additives intriguing exploitation resulting in the exploration of alternate food processing and preservation practices (Martínez-Graciá et al. 2015).

5.3 Chemical Food Preservatives and Limitations

Several chemical preservatives have shown the potential to retard spoilage and prevent food deterioration. The main synthetic antimicrobials used are butylated hydroxyanisole, butylated hydroxytoluene, parabens, ethoxyquin, tert-butylhydroquinone, sodium diacetate, potassium and calcium sorbates, benzoic acid, sorbic acid, sodium benzoate, sodium or potassium nitrate, propionic acid, and its salts and propyl gallate. They have the advantages of possessing strong antimicrobial activity, potent antioxidant effect, high industrial accessibility, and low cost (Liao et al. 2021). In contrast, they are known for probable risks of carcinogenesis and teratogenesis to humans raising negative concerns among consumers (Maurya et al. 2021). In fact, sodium benzoate, sodium nitrite, propionic acid, calcium/potassium/sodium propionates, propyl gallate, and more have been found to be responsible for hypersensitivity, asthma, and cancer. Consumption of foods containing monosodium glutamate can cause headaches, sweating, redness of skin, nausea, and fatigue (Hussain et al. 2021). Thus, there is a growing appeal for "green consumerism" and a "green, safe and clean" environment that has provoked food manufacturing free of synthetic preservatives (Angane et al. 2022).

5.4 Natural Food Preservatives

Based on safety issues, natural antimicrobial agents are more acceptable to consumers as they are directly derived from biological systems without alteration or modification in a laboratory setting (Liao et al. 2021). The preservatives are subdivided into three functional sets, namely antimicrobial, antioxidative, and anti-browning agents. The anti-browning components are employed in food matrices to terminate browning which is a common colour change during handling treatment and storage. Antimicrobials on the other hand impart two effects: retardation of normal spoilage and microbial contamination. Finally, antioxidative additives prevent food from becoming rancid and inhibit the development of undesirable odour, colour, and texture allowing the dietary item to remain in good condition for a longer time (Carocho et al. 2018). Today, the natural antioxidants used in food sectors are ascorbic acid, carotenoids, phenolic compounds, and tocopherols (Santos-Sánchez et al. 2017). Among the natural antimicrobials, essential oil isolated from plants and chitosans have received considerable interest for commercial applications (Lucera et al. 2012). For ease of further discussion, bio-resource–derived antimicrobials have been categorized into three main kinds based on their origin: plant, animal, and microorganism sources.

5.4.1 Food preservatives of plant origin

5.4.1.1 Herbs and spices

Herbs and spices have historically been employed to prevent the deterioration of foods and beverages as well as to inhibit FBD. Plants produce a plethora of phytochemicals, such as polyphenolics, flavonoids (e.g., apigenin, catechin, hesperetin, kaempferol, naringenin, and quercetin), terpenoids, phthalides, sulphides, phenolic acids (e.g., rosmarinic acid, gallic acid, and caffeic acid), carotenoids, diterpenes (e.g., carnosol and carnosic acid), saponins, coumarins, plant sterols, lignans, curcumins, and volatile oil (e.g., carvacrol, eugenol, menthol, and thymol) (Beya et al. 2021; Embuscado 2015). They impart effective antioxidant properties by free radical quenching, metal chelating, scavenging singlet oxygen, donating electrons, and increasing antioxidant enzyme activities (Khatua et al. 2013). As a result, extracts of herbs and spices have also been found to execute antioxidative properties in different food matrices as oxidative rancidity plays a key role in declining food quality resulting formation of objectionable flavour and odour (Embuscado 2015). These lipophilic metabolites are also responsible for strong antimicrobial prowess owing to their ability to retard the growth of spoilage and pathogenic microbes (Beya et al. 2021). Clove, cinnamon, rosemary, and oregano are considered to be the most common spices and herbs with strong antimicrobial activity attributed to the presence of pinenes, eugenol, and cinnamaldehyde. These functional benefits along with easy availability and no or less adverse effects have made plant-based preservatives to be generally recognized as safe (GRAS) and ideal food additives (Bag et al. 2022). The European Food Safety Agency (EFSA) has already granted the use of dry grape extract, catechins, rosemary extract, and anthocyanins in different nutrients (Sweet et al. 2022). Indeed, extracts from spices and herbs are today in great demand, and to date, a variety of fractions have been investigated for potent application in dietary industries to extend the shelf life, hinder food spoilage, and protect against foodborne pathogens (Table 5.1) (Premanath et al. 2022). However, there remains noteworthy ambiguity about the exact primary cellular targets for the diverse plant-derived metabolites and potent bacterial adaptation mechanisms (Sweet et al. 2022).

5.4.1.2. Essential oils (EOs)

EOs are oil-like, colourless, volatile chemical components synthesized by aromatic and medicinal plants from almost all parts including flowers, buds, leaves, seeds, stems, and fruits, corresponding to roughly <5% of the vegetable dry matter (Maurya et al. 2021). Chemically, they are a rich and complex mixture of numerous bioactive low-molecular weight secondary metabolites such as terpenes (e.g., pinene, myrcene, limonene, terpinene, *p*-cymene), terpenoids (e.g., oxygen-containing hydrocarbons), phenylpropanoids, and phenolics (e.g., carvacrol, thymol, safrole, eugenol) (Pandey et al. 2017). According to several previous studies, EOs exhibit broad-spectrum antimicrobial, fungitoxic, and anti-mycotoxigenic properties in postharvest stored food items (Maurya et al. 2021). EOs enriched with phenolic compounds, for instance thymol, carvacrol, or eugenol, have been found to exhibit more specificity towards foodborne pathogens (Falleh et al. 2020). The antibacterial efficacy of EOs is mostly facilitated by their hydrophobic or lipophilic nature. As such, EOs can effectively pass through the lipid bilayer of bacterial cell membranes resulting in subsequent

TABLE 5.1
Overview of previous studies dealing with the antimicrobial activity of plant extracts in different food matrices.

Sources	Susceptible microorganism	Food item	Reference
Acacia nilotica seed extract	*Bacillus cereus*, *Escherichia coli*, *Klebsiella pneumonia*, *Salmonella typhimurium*, *Staphylococcus aureus*, *Yersinia enterocolitica*	Minced chicken	Beya et al. 2021
Callistemon citrinus leaf and flower extract	*Listeria monocytogenes*	Burger	Beya et al. 2021
Citrus flower extract	*Bacillus subtilis*, *Escherichia coli* O157:H7, *Pseudomonas aeruginosa*, *Staphylococcus aureus*	Yoghurt stew	Bouarab Chibane et al. 2019
Mango seed kernel extract	Coliforms, *Escherichia coli*	Pasteurized cow milk	Bouarab Chibane et al. 2019
Punica granatum peel extract	*Listeria monocytogenes*	Chicken liver patties	Bouarab Chibane et al. 2019
Vitis vinifera seed extract	*Escherichia coli* O157:H7, *Listeria monocytogenes*, *Salmonella typhimurium*	Turkey Frankfurters	Beya et al. 2021
Rosemary extract	*Staphylococcus aureus*	Zucchini broth	del Campo et al. 2000
Rhus coriaria fruit extract	*Bacillus subtilis*, *Escherichia coli*, *Pseudomonas aeruginosa*, *Staphylococcus aureus*, *Aspergillus niger*, *Penicillium* sp.	Tomato paste and pasteurized cow milk	Atwaa et al. 2022

leakage of cellular ions (Na+, H+, and K+), alternation in proton motive force-mediated electron flow, and ultimately leading to apoptosis. Further, EOs inhibit cell-to-cell communication and biofilm production triggering the efficient breakdown of bacterial sensing communication (Maurya et al. 2021). It is also noteworthy that microbial adaptation to EOs or any of their constituents is not reported yet signifying exploitation of the metabolites as natural preservatives, exclusively in the agro-food trades (Falleh et al. 2020). Furthermore, the majority of the EOs are short-lived in nature and exhibit low mammalian toxicity portraying that they are comparatively safer for human health and the environment (Liao et al. 2021).

Out of the 300 known plant-derived EOs, some have found their way into food products owing to their immense antimicrobial potential; in fact, some have received the GRAS tag in the US. Of these, carvacrol, cinnamon oil, clove oil, oregano oil, and thymol are some of the most significant. They have been applied in a vast number of foodstuffs, namely meat, fish, dairy products, vegetables, rice, and fruit (Table 5.2). Overall, clove, oregano, cinnamon, and coriander EOs have been proposed as potent food preservatives in terms of their antibacterial effect followed by thyme, mint, rosemary, and mustard (Falleh et al. 2020). Furthermore, carvacrol and thymol have been utilized in the packaging sector by impregnating the EOs with some films (Carocho et al. 2015). Indeed EOs, either in pure or formulation forms, have been developed to employ in different storage containers such as those made of cardboard, tin, glass, polyethylene, or natural fabrics to enhance the shelf life of food commodities (Pandey et al. 2017). The major problem with low molecular weight EOs is that they are more degradable when subjected to environmental or processing situations limiting their real-time application. Numerous experimentations in this context have demonstrated the benefits of encapsulation as a method to lessen the sensitivity and regulate release over time. The tool also aids in maintaining physico-chemical features and medicinal effects of EOs in comparison with the free oil (Milagres de Almeida et al. 2023).

TABLE 5.2
Overview of previous studies dealing with the antimicrobial activity of EOs in different food matrices.

Source of EO	Susceptible microorganism	Food item	Reference
Artemisia nilagirica	*Aspergillus flavus*	Millets	Kumar et al. 2020
Cinnamomum verum, Lavandula stoechas, Myrtus communis, Syzygium aromaticum	*Escherichia coli*	UHT milk	Falleh et al. 2020
Cinnamomum verum, Pimenta racemosa, Syzygium aromaticum, Thymus vulgaris	*Salmonella enteritidis, Listeria monocytogenes*	Full-fat and low-fat soft cheese	Falleh et al. 2020
Cinnamomum zeylanicum, Cymbopogon citrates, Cymbopogon martini, Eugenia caryophyllata	*Salmonella enteritidis, Escherichia coli, Listeria innocua*	Fruit juices	Falleh et al. 2020
Laurus nobilis	Coliforms	Fresh Tuscan sausage	Teshome et al. 2022
Mentha cardiaca	*Aspergillus flavus*	Dry fruits	Dwivedy et al. 2017
Mentha piperita	*Salmonella enteritidis, Listeria monocytogenes*	Cucumber and yoghurt salad, fish roe salad, and pâté	Falleh et al. 2020
Origanum vulgare	*Escherichia coli* O157:H7	Eggplant salad	Falleh et al. 2020
Satureja horvatii	*Listeria monocytogenes*	Pork meat	Teshome et al. 2022
Trachyspermum ammi	*Aspergillus flavus*	Wheat and chickpea	Falleh et al. 2020
Thymus capitatus	*Staphylococcus aureus, Bacillus licheniformis, Enterococcus hirae*	Semi-skimmed UHT milk	Falleh et al. 2020
Thymus vulgaris, Verbena officinalis	*Monilinia laxa, Monilinia fructigena, Monilinia Fructicola*	Peaches	Teshome et al. 2022
Citrus fruit + mild heat	*Escherichia coli* O157: H7	Apple juice	Teshome et al. 2022

UHT: Ultra-high temperature

5.4.2 Food preservatives of animal origin

5.4.2.1 Lysozyme

Lysozyme (E.C 3.2.1.17), also known as muramidase or N-acetylmuramic hydrolase, is prevalent in nature and can be found in vegetables, insects, plants, fungi, egg white of birds and poultry, human colostrums, mammalian tissues, saliva, mucus, blood, and tears. This small (11–22 kDa), monomeric and alkaline protein mainly splits β-1,4-glycosidic bonds between N-acetyl-d-glucosamine and N-acetylmuramic acid residues in peptidoglycan cover leading to perforation of the cell wall and bacterial lysis (Dai and Li 2021). Gram-positive bacteria are thus more prone to be attacked by this polysaccharide-degrading enzyme as their cell walls consist of 90% peptidoglycan layers without any outer membrane. As a result, species of *Bacillus*, *Clostridium*, and *Micrococcus* have shown strong susceptibility to the protein, while Gram-negative bacteria remain ineffective. The antibacterial protein can also inhibit the growth of other microorganisms such as fungi namely *Aspergillus*, *Candida*, *Fusarium*, *Paecilomyces*, *Penicillium*, *Saccharomyces*, and *Sporothrix* (Barberis et al. 2018).

In the past few decades, the high demand for natural food preservatives has made lysozyme a progressively essential entity in the food processing and preservation sectors. Indeed, lysozyme is the only official antimicrobial derivative of animals in recent times, utilized both in the US and the EU (E-1105) (Carocho et al. 2015). Research has illustrated that lysozyme could be applied in dairy industries to inhibit the growth of spoilage bacteria and prevent the texture deterioration, blowing, and unpleasant taste of various kinds of cheese such as Edam, Gouda, and some Italian cheese (Khorshidian et al. 2022). In many countries, it is also used in the preservation of fruits, vegetables, beans, tofu, meat, seafood, soy bean products, sausages, and salads (Al-Baarri et al. 2018). Further, lysozyme has been conveyed as a potent natural preservative in alcoholic drinks, for instance, unpasteurized beer at pH 4.7. The protein as an antimicrobial reagent for the management of processed ham and bologna has been described as well. In another study, a commercially available lysozyme preparation was found to inhibit the growth of *L. monocytogenes* increasing the storage life of raw minced tuna and salmon roe products. Likewise, another commercially accessible lysozyme product, i.e., Inovapure, prolonged the shelf duration of raw and processed meat and milk products. Recent investigations showed that different types of modifications of lysozyme are possible with an aim to increase bioactivity. For instance, a heat-denatured lysozyme has been reported to exhibit potent antimicrobial properties against a wide range of bacteria and fungi. Lysozyme activity can also be improved by adding certain substances, e.g., EDTA and butylparaben; however, such chemical modifications might not be appropriate for food applications (Juneja et al. 2012). Lysozyme is commercially produced from hen egg white which restricts its application due to allergic reactions, and a huge amount of egg white is needed to reach a satisfactory purity. In that scenario, human lysozyme could be utilized which possesses higher antibacterial activity and thermal stability; but its usage has been limited due to restricted available resources. Microbial fermentation has been proposed as an alternate source (Pilevar et al. 2022).

5.4.2.2 Chitosan

Chitosan is the second most prevalent nature-derived polymer after cellulose and is composed of N-acetylglucosamine and D-glucosamine units linked by β-(1–4)glycosidic bond (No et al. 2007). It can be obtained by the deacetylation of chitin which is a major component of the shells of crustaceans (e.g., shrimp, lobster, and crab) and also found in exoskeletons of insects, in cell walls of most fungi and some algae (Hu and Gänzle 2019). This poly-cationic, linear, and bio-degradable polymer is water-insoluble but soluble in weak organic acid solutions. The versatile component has received increased attention for its commercial applications in bio-medical and chemical industries. Chitosan has recently attracted attention as a potential food preservative of natural origin as well due to its non-toxicity, bio-compatibility, and strong antimicrobial activity against a wide range of foodborne filamentous fungi, yeast, and bacteria (No et al. 2007). The mechanism behind the bioactivity has not yet been fully illuminated, but several suggestions have been proposed. It is hypothesized that chitosan is able to interact with negatively charged microbial cell membranes resulting in disruption in cell permeability and leakage of proteinaceous and other intracellular components. Besides, chitosan can inhibit transcription and translation processes and chelate metals that in turn trigger cell death. As the Gram-positive bacteria do not have an outer membrane, they facilitate better adherence with the long-chain molecule of chitosan. Hence, they have been found to be more susceptible to chitosan than Gram-negative bacteria (Duan et al. 2019).

Since 1983 and 1995, chitosan has been permitted to be produced as a food additive on an industrial scale in Japan and Korea respectively. Later in 2001, it was renowned as a GRAS component by the US FDA (Tantala et al. 2022). The utilization of chitosan has widely been explored as an edible coating, which provides protective palatable covering on the product to improve the quality and shelf life of fresh foodstuffs ranging from meat, poultry, fish, and dairy products. As such, chitosan-coated tomatoes have been reported to prevent the growth of *Aspergillus* spp., *Botrytis cinerea*, *Penicillium* spp., and *Rhizopus stolonifer* (Duan et al. 2019). The coating treatment with chitosan also resulted in the well-maintenance of freshness of eggs increasing concomitant commercial value (Hu and Gänzle 2019). The fungal chitosan-based edible coatings could be applied on fresh-cut melons and found to enhance fruit firmness, inhibit off-flavours, and reduce microbial counts (Poverenov et al. 2018). Despite the many benefits of chitosan, low solubility at neutral pH is limiting its application. In this backdrop, numerous chemical and physical processes have been suggested to escalate its solubility (Büyükyörük 2021).

5.4.2.3 Lactoferrin

Lactoferrin, a GRAS, is an iron-binding glycoprotein (80 kDa) and is present in high quantities in saliva, tear, seminal fluid, white blood cells, and milk of mammals. It has a high affinity for free ferric ions, and thus lactoferrin owns strong antioxidant activity (Niaz et al. 2019). It also possesses a potent antimicrobial effect against a wide range of deleterious foodborne microorganisms (coliforms, *B. subtilis*, *Bacillus stearothermophilus*, *E. coli* O157:H7, *L. monocytogenes*, *Klebsiella* spp., *Salmonella enteritidis*, and *S. aureus*) and viruses. The proposed antimicrobial mechanism of lactoferrin is carried out through the destabilization of the outer membrane of Gram-negative bacteria, enhancing membrane permeability and iron chelation. However, microorganisms with low iron requirements, such as lactic acid bacteria (LAB), are not susceptible to lactoferrin (Barberis et al. 2018). The oldest reports of lactoferrin use are in baby formulas (Carocho et al. 2015). Lactoferrin has been recently permitted for utilization on beef in the United States and employed as an antibacterial agent in a diverse range of meat foodstuffs (Juneja et al. 2012). Japan and other countries are exploiting the protein as a functional element as well as for biopreservation of yoghurt and other food items (Niaz et al. 2019).

5.4.2.4 Lactoperoxidase

Lactoperoxidase belongs to the peroxidase-cyclooxygenase superfamily and is very abundant in bovine milk (Carocho et al. 2015). It is a haeme-containing glycoprotein of 608 amino acids with a molecular mass of 78 kDa and found in plants and secretions of mammalian species (raw milk, saliva, colostrum, etc.) (Al-Baarri et al. 2018). It can serve as a natural antimicrobial in amalgamation with thiocyanate and H_2O_2, more familiar as lactoperoxidase system. The system catalyses the oxidation of thiocyanate, iodide, or bromide by H_2O_2, resulting in the production of reactive molecules such as hypothiocyanite and hypothiocyanous acid (Barberis et al. 2018). Hypothiocyanite can destroy sulfhydryl groups present in the microbial cell membrane, damage the membrane, and inhibit the proliferation of bacteria, fungi, and viruses leading to cellular death (Buys and Seifu 2022).

In recent times, the enzymatic system has been employed as a natural preservative in some foods, such as milk, fruits, chicken, and vegetables. The lactoperoxidase system has been found to be effective in diminishing the growth of *E. coli*, *Pseudomonas* sp., and *S. typhimurium* on cottage cheese (Al-Baarri et al. 2018). Lactoperoxidase is mainly used to keep raw milk consumable, particularly in places where refrigerators are not readily accessible. Addition of thiocyanate to the milk results in the initiation of the lactoperoxidase system to work portraying its antimicrobial activity. Further studies have suggested the application of the enzyme for preservation of fruit juices and also as coating of food items (Carocho et al. 2015).

5.4.3 Food preservatives of microbial origin

5.4.3.1 Bacteriocins

Bacteriocins are low molecular weight (2–300 kDa), multi-functional ribosomal active peptides synthesized by many microbial species including LAB or other bacteria, such as *Bacillus* spp., *Staphylococcus* spp., and *E. coli* to remove microbial competition in prokaryotes (Baptista et al. 2020). Several bacteriocins are colourless, odourless, tasteless, and stable at low pH, high temperature, and across a broad range of salt concentrations (Verma et al. 2014). They are protein toxins and display immense antimicrobial activities against deleterious bacteria and certain types of fungal species as well as viruses. The mode of action is generally mediated through adsorption and insertion of bacteriocins to cell membranes resulting in perforations. The consequence results in increased membrane permeability and losses of essential cellular constituents such as K, Mg, ATP, and amino acids,

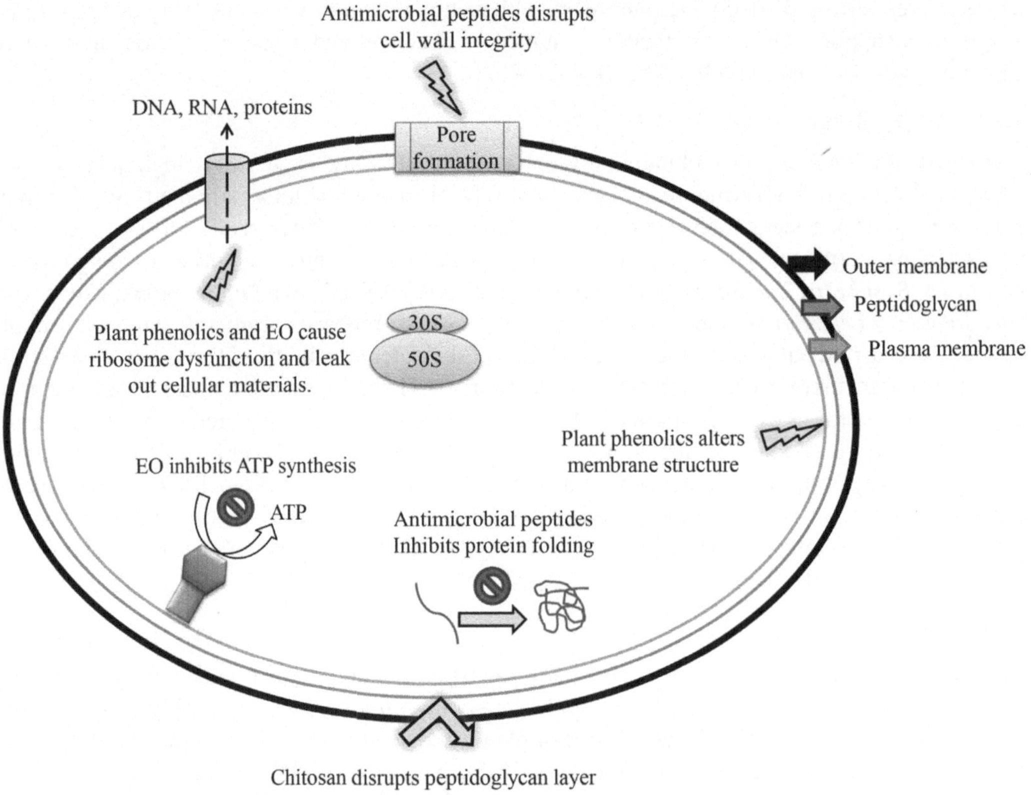

FIGURE 5.1 Mechanism of action behind antibacterial effect of different natural food preservatives

among others (Figure 5.1) (Baptista et al. 2020). Thus, bacteriocins offer several advantages as food preservatives such as extension of food shelf life, inhibition of transmission of foodborne pathogens through the food chain, reduction of economic losses, and preservation of nutrients, vitamin profiles, as well as sensory properties of food products (Verma et al. 2014). In addition, research has recommended the use of bacteriocins to boost food fermentation, hasten cheese ripening, and even increase its flavour. The recent trends are thus directed towards the incorporation of bacteriocins, as purified, semi-purified, or amalgamated form of bacteriocin-producing LAB into bioactive films and coatings for direct applications onto food surfaces and packaging (Silva et al. 2018).

5.4.3.2 Nisin

Nisin, a class I bacteriocin, is composed of 34 amino acid residues with a molecular weight of 3510 Daltons, an overall positive charge, and an amphipathic nature. This polycyclic antibacterial peptide is synthesized by some strains of *Lactococcus lactis* subsp. *lactis.* It has been used not only to prevent the development of >90% Gram-positive bacteria, for instance, *B. cereus*, *Clostridium botulinum*, *Clostridium perfringens*, *Enterococcus* sp. *L. monocytogenes*, *S. aureus*, and *Streptococcus* spp., but also against some Gram-negative pathogens such as *E. coli* and *Salmonella* sp. (Baptista et al. 2020). The mechanism of action is depicted to mediate through the formation of short-lived pores, interference with cell membrane integrity, and rapid efflux of amino acids and ATP leading to bacterial cell disruption. High tolerance to heat in addition to low pH, non-toxic nature, and quick digestibility has made nisin (E 234) an ideal food preservative (Gharsallaoui et al. 2016). In fact, it is the only bacteriocin approved for utilization as a preservative in many foods by the US FDA and is considered GRAS and licensed as a food additive in over 45 countries (Yang et al. 2014). It is exploited in various commercial formulations, namely Nisaplin® (Danisco, Copenhagen, Denmark),

Chrisin® (Chris Hansen, Horsholm, Denmark), and Delvo®Nis (DSM, Delft, the Netherlands). Nisin is the only bacteriocin possessing a prevalent utilization in related food industries including dairy products, fruits, meat and meat foodstuffs, juices and beverages, fish products and seafood, vegetables, and cereals. It is also extensively used in dairy industries to control *Clostridia* sp. and post-processing contamination from *Listeria* strains (Verma et al. 2014). Incorporation of nisin in liquid, especially beer, has been reported to kill around 90% of Gram-positive foodborne deleterious bacteria without influencing fermentation (Müller-Auffermann et al. 2015).

5.4.3.3 Natamycin

Natamycin ($C_{33}H_{47}NO_{13}$) (sometimes known as pimaracin or tennectin) is a polyene macrolide antimycotic agent produced by submerged fermentation of Actinomycetes strains such as *Streptomyces natalensis* and other closely related *Streptomyces* spp. This widely used natural preservative has been proven to possess beneficial effects, particularly against yeast, filamentous fungal pathogens, moulds, and several mycotoxins with no activity on bacteria, protozoa, and viruses (Meena et al. 2021). It inhibits microbial cell growth and spore germination and declines membrane-accompanying enzyme activity. As a result, natamycin preparations have long been employed as preservatives with an aim to protect foods and beverages from yeast and mould spoilage (Delves-Broughton and Weber 2011). At present, it is used in a range of foodstuffs such as cheese, sausages, yogurts, juices, and wines, as a free additive (E-235), encapsulated, and as a constituent of films. Indeed, research has shown that the exposure is very low, and absorption of this molecule by the human body is extremely poor ruling out any risk of its use (Carocho et al. 2015).

5.4.4 Synergistic effects of natural food preservatives

One main issue with the single use of natural food preservatives can be their inherent pungent flavour and odour when employed in high concentrations to impose their antimicrobial properties. Coupling of two or more compounds or combined technologies together with bio-resource-derived antimicrobials could be employed to improve bioactivity due to synergistic effects (Sweet et al. 2022). The concept has shown to be viable when oregano EO, thymol EO, lemon, and grapefruit seed fractions were amalgamated with the combined treatment of modified atmosphere packaging, and chilled storage was able to increase the storage duration of fresh chicken breast, fish burgers, and fresh sea bream fish significantly (Premanath et al. 2022). Addo et al. (2023) confirmed the synergistic effect of cineole and carvacrol against *E. coli* O157:H7 and also showed its antimicrobial effect as a sanitizer on fresh-cut cucumbers. On the other hand, lemongrass essential oil+citrus extract+lactic acid exhibited a growth inhibitory effect against a range of bacteria and exhibited potential for preservation of red pepper, cranberries, and pre-cut/pre-fried potatoes (Faheem et al. 2022). A combination of *Syzygium aromaticum*, *Cinnmomum cassia*, and *Origanum vulgare* extracts was reported to be very effective against microbial growth and lipid oxidation and has potential as a natural antioxidant in raw chicken meats (Radha Krishnan et al. 2014). A combination of chitosan and sodium lactate displayed a synergistic antibacterial property against fish spoilage flora in fish homogenate. The amalgamation of chitosan together with some common food preservatives such as sodium benzoate or potassium sorbate exhibited an extension of the mould-free shelf life of fresh rice noodles without negotiating with noodle quality (Tantala et al. 2022). Synergy has also been demonstrated between nisin, tea polyphenols, and chitosan at a definite combination prolonging the shelf life of chilled mutton. The antimicrobial blend also reduced viable counts of pathogenic and spoilage-related microorganisms in chilled pork samples maintaining colour, quality, water holding properties, and pH of the food (Sweet et al. 2022).

5.5 Conclusion and Future Perspectives

Food spoilage due to microbial growth is a critical issue worldwide where physical means of preservation are insufficient to meet growing needs. On the other hand, the use of chemical preservatives is revealing its own drawbacks. In this context, consumer demand for safe dietary items has enforced scientists to find better approaches where natural preservatives undoubtedly fit the concept. As a result, numerous herbal, animal, bacterial, and fungal additive ingredients have been discovered to date, although many more are still waiting to be revealed. Further research is thus needed to search for new bio-ingredients, determine appropriate concentration for application in food, identify a mode of action, and study protective effects on food with considerable market prospects.

REFERENCES

Addo, K. A., H. Li, Y. Yu, and X. Xiao. 2023. Unraveling the mechanism of the synergistic antimicrobial effect of cineole and carvacrol on *Escherichia coli* O157:H7 inhibition and its application on fresh-cut cucumbers. *Food Control* 144:109339. https://doi.org/10.1016/j.foodcont.2022.109339

Adegoke, G. O., and A. A. Olapade. 2012. Preservation of plant and animal foods: An overview. In *Progress in food preservation*, eds. R. Bhat, A. K. Alias, and G. Paliyath, 603–611. Wiley-Blackwell. https://doi.org/10.1002/9781119962045.ch28

Al-Baarri, A. N., A. M. Legowo, S. K. Arum, and S. Hayakawa. 2018. Extending shelf life of Indonesian soft milk cheese (dangke) by lactoperoxidase system and lysozyme. *International Journal of Food Science* 2018:4305395. https://doi.org/10.1155/2018/4305395

Amit, S. K., M. M. Uddin, R. Rahman, S. M. R. Islam, and M. S. Khan. 2017. A review on mechanisms and commercial aspects of food preservation and processing. *Agriculture and Food Security* 6(1):51. https://doi.org/10.1186/s40066-017-0130-8

Angane, M., S. Swift, K. Huang, C. A. Butts, and S. Y. Quek. 2022. Essential oils and their major components: An updated review on antimicrobial activities, mechanism of action and their potential application in the food industry. *Foods* 11(3):464. https://doi.org/10.3390/foods11030464

Atwaa, E. S. H., M. R. Shahein, H. A. Radwan et al. 2022. Antimicrobial activity of some plant extracts and their applications in homemade tomato paste and pasteurized cow milk as natural preservatives. *Fermentation* 8(9):428. https://doi.org/10.3390/fermentation8090428

Bag, B. B., C. Panigrahi, S. Gupta, and H. N. Mishra. 2022. Efficacy of plant-based natural preservatives in extending shelf life of sugarcane juice: Formulation optimization by MOGA and correlation study by principal component analysis. *Applied Food Research* 2(2):100164. https://doi.org/10.1016/j.afres.2022.100164

Baptista, R. C., C. N. Horita, and A. S. Sant'Ana. 2020. Natural products with preservative properties for enhancing the microbiological safety and extending the shelf-life of seafood: A review. *Food Research International* 127:108762. https://doi.org/10.1016/j.foodres.2019.108762

Barberis, S., H. G. Quiroga, C. Barcia, J. M. Talia, and N. Debattista. 2018. Natural food preservative against microorganisms. In *Food safety and preservation: Modern biological approaches to improving consumer health*, eds. A. M. Grumezescu, and A. M. Holban, 621–629. London: Elsevier Inc. https://doi.org/10.1016/B978-0-12-814956-0.00020-2

Bekhit, A. E., and S. Roohinejad. 2016. Cook-chilled and cook-frozen foods. In *Reference module in food science*, 1–10. Elsevier. https://doi.org/10.1016/B978-0-08-100596-5.03348-5

Beya, M. M., M. E. Netzel, Y. Sultanbawa, H. Smyth, and L. C. Hoffman. 2021. Plant-based phenolic molecules as natural preservatives in comminuted meats: A review. *Antioxidants (Basel)* 10(2):263. https://doi.org/10.3390/antiox10020263

Bouarab Chibane, L., P. Degraeve, H. Ferhout, J. Bouajila, and N. Oulahal. 2019. Plant antimicrobial polyphenols as potential natural food preservatives. *Journal of the Science of Food and Agriculture* 99(4):1457–1474. https://doi.org/10.1002/jsfa.9357

Buys, E. M., and E. Seifu. 2022. Enzymes indigenous to milk: Lactoperoxidase. In *Encyclopedia of dairy sciences*, 3rd ed., eds. P. L. H. McSweeney, and J. P. McNamara, 670–676. Oxford: Academic Press. https://doi.org/10.1016/B978-0-12-818766-1.00252-X

Büyükyörük, S. 2021. Chitosan for using food protection. In *Chitin and Chitosan-physicochemical properties and industrial applications*, ed. M. Berrada. IntechOpen. https://doi.org/10.5772/intechopen.99247

Carocho, M., P. Morales, and I. C. F. R. Ferreira. 2015. Natural food additives: Quo vadis? *Trends in Food Science and Technology* 45(2):284–295. https://doi.org/10.1016/j.tifs.2015.06.007

Carocho, M., P. Morales, and I. C. F. R. Ferreira. 2018. Antioxidants: Reviewing the chemistry, food applications, legislation and role as preservatives. *Trends in Food Science and Technology* 71:107–120. https://doi.org/10.1016/j.tifs.2017.11.008

Dai, P., and Q. Li. 2021. Research progress in the application of lysozyme in food and medicine field. *E3S Web of Conferences* 251:02048. https://doi.org/10.1051/e3sconf/202125102048

del Campo, J., M.-J. Amiot, and C. Nguyen-The. 2000. Antimicrobial effect of rosemary extracts. *Journal of Food Protection* 63(10):1359–68. https://doi.org/10.4315/0362-028x-63.10.1359

Delves-Broughton, J., and G. Weber. 2011. Nisin, natamycin and other commercial fermentates used in food biopreservation. In *Lacroix C, protective cultures, antimicrobial metabolites and bacteriophages for food and beverages biopreservation*, ed. C. Lacroix, 63–99. Woodhead Publishing. https://doi.org/10.1533/9780857090522.1.63

Duan, C., X. Meng, J. Meng et al. 2019. Chitosan as a preservative for fruits and vegetables: A review on chemistry and antimicrobial properties. *Journal of Bioresources and Bioproducts* 4(1):11–21. https://doi.org/10.21967/jbb.v4i1.189

Dwivedy, A. K., B. Prakash, C. S. Chanotiya, D. Bisht, and N. K. Dubey. 2017. Chemically characterized *Mentha cardiaca* L. essential oil as plant based preservative in view of efficacy against biodeteriorating fungi of dry fruits, aflatoxin secretion, lipid peroxidation and safety profile assessment. *Food and Chemical Toxicology* 106(Pt A):175–184. https://doi.org/10.1016/j.fct.2017.05.043

Embuscado, M. E. 2015. Spices and herbs: Natural sources of antioxidants—A mini review. *Journal of Functional Foods* 18:811–819. https://doi.org/10.1016/j.jff.2015.03.005

Faheem, F., Z. W. Liu, R. Rabail et al. 2022. Uncovering the industrial potentials of lemongrass essential oil as a food preservative: A review. *Antioxidants* 11(4):720. https://doi.org/10.3390/antiox11040720

Falleh, H., M. Ben Jemaa, M. Saada, and R. Ksouri. 2020. Essential oils: A promising eco-friendly food preservative. *Food Chemistry* 330:127268. https://doi.org/10.1016/j.foodchem.2020.127268

Gharsallaoui, A., N. Oulahal, C. Joly, and P. Degraeve. 2016. Nisin as a food preservative: Part 1: Physicochemical properties, antimicrobial activity, and main uses. *Critical Reviews in Food Science and Nutrition* 56(8):1262–1274. https://doi.org/10.1080/10408398.2013.763765

Hu, Z., and M. G. Gänzle. 2019. Challenges and opportunities related to the use of chitosan as a food preservative. *Journal of Applied Microbiology* 126(5):1318–1331. https://doi.org/10.1111/jam.14131

Hussain, M. A., T. A. Sumon, S. K. Mazumder et al. 2021. Essential oils and chitosan as alternatives to chemical preservatives for fish and fisheries products: A review. *Food Control* 129:108244. https://doi.org/10.1016/j.foodcont.2021.108244

Juneja, V. K., H. P. Dwivedi, and X. Yan. 2012. Novel natural food antimicrobials. *Annual Review of Food Science and Technology* 3:381–403. https://doi.org/10.1146/annurev-food-022811-101241

Khatua, S., S. Paul, and K. Acharya. 2013. Mushroom as the potential source of new generation of antioxidant: A review. *Research Journal of Pharmacy and Technology* 6(5):496–505.

Khorshidian, N., E. Khanniri, M. R. Koushki, S. Sohrabvandi, and M. Yousefi. 2022. An overview of antimicrobial activity of lysozyme and its functionality in cheese. *Frontiers in Nutrition* 9:833618. https://doi.org/10.3389/fnut.2022.833618

Kumar, D., and P. Kalita. 2017. Reducing postharvest losses during storage of grain crops to strengthen food security in developing countries. *Foods* 6(1):8. https://doi.org/10.3390/foods6010008

Kumar, M., A. K. Dwivedy, P. Sarma et al. 2020. Chemically characterised *Artemisia nilagirica* (Clarke) Pamp. essential oil as a safe plant-based preservative and shelf-life enhancer of milletsagainst fungal and aflatoxin contamination and lipid peroxidation. *Plant Bi Systems* 154(3):269–276. https://doi.org/10.1080/11263504.2019.1587539

Kumar, S., S. Kalita, A. Das et al. 2022. *Aloe vera*: A contemporary overview on scope and prospects in food preservation and packaging. *Progress in Organic Coatings* 166:106799. https://doi.org/10.1016/j.porgcoat.2022.106799

Liao, W., W. Badri, E. Dumas et al. 2021. Nanoencapsulation of essential oils as natural food antimicrobial agents: An overview. *Applied Sciences* 11(13):5778. https://doi.org/10.3390/app11135778

Lucera, A., C. Costa, A. Conte, and M. A. Del Nobile. 2012. Food applications of natural antimicrobial compounds. *Frontiers in Microbiology* 3:287. https://doi.org/10.3389/fmicb.2012.00287

Martínez-Graciá, C., C. A. González-Bermúdez, A. M. Cabellero-Valcárcel, M. Santaella-Pascual, and C. Frontela-Saseta. 2015. Use of herbs and spices for food preservation:advantages and limitations. *Current Opinion in Food Science* 6:38–43. https://doi.org/10.1016/j.cofs.2015.11.011

Maurya, A., J. Prasad, S. Das, and A. K. Dwivedy. 2021. Essential oils and their application in food safety. *Frontiers in Sustainable Food Systems* 5:653420. https://doi.org/10.3389/fsufs.2021.653420

Meena, M., P. Prajapati, C. Ravichandran, and R. Sehrawat. 2021. Natamycin: A natural preservative for food applications—A review. *Food Science and Biotechnology* 30(12):1481–1496. https://doi.org/10.1007/s10068-021-00981-1

Milagres de Almeida, J., B. L. Crippa, V. V. Martins Alencar de Souza et al. 2023. Antimicrobial action of Oregano, Thyme, Clove, Cinnamon and black pepper essential oils free and encapsulated against foodborne pathogens. *Food Control* 144:109356. https://doi.org/10.1016/j.foodcont.2022.109356

Mishra, B., A. K. Mishra, S. Kumar et al. 2022. Antifungal metabolites as food bio-preservative: Innovation, outlook, and challenges. *Metabolites* 12(1):12. https://doi.org/10.3390/metabo12010012

Müller-Auffermann, K., F. Grijalva, F. Jacob, and M. Hutzler. 2015. Nisin and its usage in breweries: A review and discussion. *Journal of the Institute of Brewing* 121(3):309–319. https://doi.org/10.1002/jib.233

Niaz, B., F. Saeed, A. Ahmed et al. 2019. Lactoferrin (LF): A natural antimicrobial protein. *International Journal of Food Properties* 22(1):1626–1641. https://doi.org/10.1080/10942912.2019.1666137

No, H. K., S. P. Meyers, W. Prinyawiwatkul, and Z. Xu. 2007. Applications of chitosan for improvement of quality and shelf life of foods: A review. *Journal of Food Science* 72(5):R87–R100. https://doi.org/10.1111/j.1750-3841.2007.00383.x

Pandey, A. K., P. Kumar, P. Singh, N. N. Tripathi, and V. K. Bajpai. 2017. Essential oils: Sources of antimicrobials and food preservatives. *Frontiers in Microbiology* 7:2161. https://doi.org/10.3389/fmicb.2016.02161

Pilevar, Z., K. Abhari, H. Tahmasebi et al. 2022. Antimicrobial properties of lysozyme in meat and meat products: Possibilities and challenges. *Acta Scientiarum: Animal Sciences* 44:e55262. https://doi.org/10.4025/actascianimsci.v44i1.55262

Poverenov, E., H. Arnon-Rips, Y. Zaitsev et al. 2018. Potential of chitosan from mushroom waste to enhance quality and storability of fresh-cut melons. *Food Chemistry* 268:233–241. https://doi.org/10.1016/j.foodchem.2018.06.045

Premanath, R., J. P. James, I. Karunasagar, E. Vaňková, and V. Scholtz. 2022. Tropical plant products as biopreservatives and their application in food safety. *Food Control* 141:109185. https://doi.org/10.1016/j.foodcont.2022.109185

Putnik, P., B. Pavlić, B. Šojić et al. 2020. Innovative hurdle technologies for the preservation of functional fruit juices. *Foods* 9(6):699. https://doi.org/10.3390/foods9060699

Quinto, E. J., I. Caro, L. H. Villalobos-Delgado, J. Mateo, B. De-Mateo-Silleras, and M. P. Redondo-Del-Río. 2019. Food safety through natural antimicrobials. *Antibiotics (Basel)* 8(4):208. https://doi.org/10.3390/antibiotics8040208

Radha Krishnan, K., S. Babuskin, P. Azhagu Saravana Babu et al. 2014. Antimicrobial and antioxidant effects of spice extracts on the shelf life extension of raw chicken meat. *International Journal of Food Microbiology* 171:32–40. https://doi.org/10.1016/j.ijfoodmicro.2013.11.011

Santos-Sánchez, N. F., R. Salas-Coronado, R. Valadez-Blanco, B. Hernández-Carlos, and P. C. Guadarrama-Mendoza. 2017. Natural antioxidant extracts as food preservatives. *Acta Scientiarum Polonorum, Technologia Alimentaria* 16(4):361–370. https://doi.org/10.17306/J.AFS.0530

Silva, C. C. G., S. P. M. Silva, and S. C. Ribeiro. 2018. Application of bacteriocins and protective cultures in dairy food preservation. *Frontiers in Microbiology* 9:594. https://doi.org/10.3389/fmicb.2018.00594

Sweet, R., P. A. Kroon, and M. A. Webber. 2022. Activity of antibacterial phytochemicals and their potential use as natural food preservatives. *Critical Reviews in Food Science and Nutrition*:1–12. https://doi.org/10.1080/10408398.2022.2121255

Tantala, J., S. Meethongchai, W. Suethong, S. Ratanasumawong, and C. Rachtanapun. 2022. Mold-free shelf-life extension of fresh rice noodles by synergistic effects of chitosan and common food preservatives. *Food Control* 133:108597. https://doi.org/10.1016/j.foodcont.2021.108597

Teshome, E., S. F. Forsido, H. P. V. Rupasinghe, and E. Olika Keyata. 2022. Potentials of natural preservatives to enhance food safety and shelf life: A review. *Scientific World Journal* 2022:9901018. https://doi.org/10.1155/2022/9901018

Verma, A., R. Banerjee, H. Dwivedi et al. 2014. Bacteriocins: Potential in food preservation. In *Encyclopedia of food microbiology*, eds. C. A. Batt, and M. L. Tortorello, 180–186. Academic Press, Elsivier Science. https://doi.org/10.1016/B978-0-12-384730-0.00029-X

Wilson, C. L. 2013. Establishment of a world food preservation center. *Agriculture and Food Security* 2(1):Article No 1. https://doi.org/10.1186/2048-7010-2-1

Yang, S. C., C. H. Lin, C. T. Sung, and J. Y. Fang. 2014. Antibacterial activities of bacteriocins: Application in foods and pharmaceuticals. *Frontiers in Microbiology* 5:241. https://doi.org/10.3389/fmicb.2014.00241

Yousefi, M., A. Nematollahi, M. Shadnoush, A. M. Mortazavian, and N. Khorshidian. 2022. Antimicrobial activity of films and coatings containing lactoperoxidase system: A review. *Frontiers in Nutrition* 9:828065. https://doi.org/10.3389/fnut.2022.828065

6 Efficacy of antimicrobial substances in food safety and quality

Recent advances and future trends

Sandhya Sharma, Sweta Sain, Shaily Mahur, Bhavna Choudhary, Pragati Saini, and Ajay Kumar

6.1 INTRODUCTION

Today, food safety is a general concern, and it is nearly impossible to find someone who has not experienced a foodborne disease within the past year. In 2005, diarrhoea (foodborne sickness) caused approximately 1.8 million deaths, and these infections were caused by contaminated food and drink. The primary cause of foodborne illnesses is the consumption of contaminated food containing microorganisms, poisons, or radioactive elements. When specific bacteria or pathogens infect food, they can lead to foodborne sickness, also known as "food poisoning". Foodborne infections are often mild, although they can occasionally be fatal. Due to their delicate nature, many foods need special care to prevent rotting throughout processing, storage, and transport to ensure they reach consumers in good condition. The demand for fresh food products that are easily prepared, little processed, and ready to eat and the globalization of the food trade and the distribution from centralized production all pose serious issues for food safety and quality. Heat treatment, salting, acidification, and drying are only some of the preservation processes used in the food business to keep perishable goods fresh and free from harmful bacteria and moulds (Quinto et al., 2019; Del Nobile Matteo Alessandro et al., 2012; Arshad, and Batool, 2017; Poonia, 2017). Natural methods of preventing the growth of bacteria and fungi in food have been the subject of extensive research and development. To stop the growth of pathogenic bacteria in food, antimicrobials can be integrated into the composition of the product, coated on the surface, or built into the packaging material. When added directly to foods of various origins, such as meat or vegetables, or to their packaging, natural antimicrobials provide food safety from a new angle while simultaneously increasing the shelf life of the product. Mustard and horseradish oil's main antibacterial component is allyl-isothiocyanate. Studies have shown that it is more efficient against Gram-negative bacteria and has little to no effect on lactic acid bacteria. Volatile molecules inhibit *E. coli* growth in particular, although its antibacterial efficacy varies greatly. Bacteriocins are tiny peptides produced by bacteria that exhibit potent antibacterial activity against closely related microorganisms. *Lactococcus lactis* spp. synthesize the polypeptide nisin. It has GRAS approval as a food additive in more than 50 countries globally. In addition, it is especially efficient against heat-resistant *Clostridium botulinum* spores and foodborne infections including *L. monocytogenes*, *S. aureus*, and *B. cereus*. For their intended purposes, secondary direct food additives are applied during production or processing but are ultimately eliminated before consumption. Inert additive residues of this sort are impossible (Quinto et al., 2019; Del Nobile Matteo Alessandro et al., 2012; Lucera et al., 2012). The Food and Drug Administration does not require labelling because this fits their definition of a processing aid. Food

DOI: 10.1201/9781003268949-6

additives that have technical impacts on the finished product must be disclosed on the label under their common names (Ibrahim et al., 2021; Ramos et al., 2014).

6.2 ANTIMICROBIAL AGENTS AND FOOD SAFETY

The widespread use of antimicrobials in the food industry is a result of the rising demand for chemical-free food and the inefficiency of conventional food preservation techniques. Antimicrobials are a relatively recent technology used by the food industry to address problems with food quality and safety, such as extending the duration of food's shelf life. Antimicrobials can be either natural or synthetic, with the former gaining prominence. There are various synthetic preservatives that pose risks to human health despite having been given the green light by regulatory organizations. Therefore, natural compounds with the potential for antibacterial properties are given higher priority by researchers (Arshad and Batool, 2017; Ramos et al., 2014; Quinto et al., 2019).

Natural products are substances made by living organisms that have pharmacological and biological effects. Metabolites, or chemical compounds, are created by all living things, both directly and indirectly. There are two types of metabolic products: primary metabolites and secondary metabolites. Primary metabolites provide a fundamental biological function in the organism, while secondary metabolites may be harmless by-products or serve an important biological role in their producers. Secondary metabolites can be utilized to treat disorders such as cancer and inflammation and also possess antibacterial activity. Secondary metabolites with antimicrobial activity that can be isolated from plants, animals, and microbes are known as natural antimicrobials. It has been demonstrated that secondary metabolites from plants act as antimicrobials and disease suppressants. The utilization of these secondary metabolites in food products has drawn a lot of attention in recent years due to their potential as antimicrobials against pathogenic and spoilage microbes (Ramos et al., 2014; Poonia, 2017). They're preferred over synthetic preservatives because they have the dual benefits of inhibiting microbial growth and protecting against oxidation in food. The flavourful and antibacterial properties of plant extracts and essential oils (EOs) have elevated their status in modern cooking. Extracts from a variety of fruit peels, including banana, apple, pomegranate, sweet lime, orange, mango, and papaya, have been studied for their antibacterial action and found to have a weak inhibitory effect on pathogenic bacteria. The secondary metabolites found in plants are loaded with antimicrobial compounds, making them highly inhibitive to both Gram-positive and Gram-negative bacteria. The effectiveness of an antimicrobial agent depends on its specific chemical makeup, concentration, and structure. Flavonoids, thiosulfinates, glucosinolates, phenolics, organic acids, flavonoids, and saponins are all examples of antimicrobial components derived from plants. Meat and meat products may benefit from the addition of essential oils because they are plant extracts that include antibacterial agents in addition to antioxidative and flavourful qualities (Ramos et al., 2014; Poonia, 2017).

6.3 ANTIMICROBIALS FROM DIFFERENT SOURCES

More research is being done to determine whether natural antimicrobials satisfy food safety regulations as a result of the harmful effects of some synthetic preservatives on consumer health; additionally, the emphasis on natural antimicrobials is further supported and justified by the misuse of antibiotics that leads to the emergence of multidrug-resistant pathogens. Natural antimicrobials assure food safety from a fresh angle while lengthening its shelf life. They also represent a workable substitute for the microbial resistance brought on by antibiotics (Quinto et al., 2019).

The different sources of antimicrobials are:

1. Plants
2. Animals
3. Microorganism

6.3.1 Plants

6.3.1.1 Herbs and spices

The majority of plant-based antimicrobials are found in spices and herbs. These substances have various structural arrangements and various antibacterial effects against foodborne infections. Since their contact with the bacterial cell membrane disturbs its structure and enables its constituent parts to leak out, it is thought that the hydroxyl (OH) groups are to blame for this. The fact that these compounds' structural makeup significantly influences how well they operate as antimicrobials attests to this reality (Quinto et al., 2019). The type of essential oil present, the type of food being utilized, and the type of microbe all affect the antibacterial action of herbs and spices. The most potent essential oils that include phenolic groups are from clove, oregano, rosemary, thyme, sage, and vanillin. Rather than Gram-negative bacteria, they have inhibitory effect against Gram-positive bacteria. Essential oils can penetrate pathogenic microorganisms through gas or liquid phases due to their high vapour pressure. *Pseudomonas* growth in sausages is prevented by using thymol, an antibacterial compound derived from thyme and oregano. Extracts of rosemary, marjoram, mustard, cinnamon, lemon grass, and other herbs are antagonistic against *E. coli* O157:H7, *S. typhi*, and *Listeria* (Arshad and Batool, 2017).

6.3.1.2 Fruits and vegetables

Due to their high phenolic and organic acid content, many fruits and vegetables are now well established to have an antibacterial impact against various pathogenic and spoilage microbes. Antimicrobial chemicals are also present in fruit peels, which are typically discarded.

Capsicum annum extracts in minced beef had an inhibiting impact on *S. typhimurium* and *Pseudomonas. E. coli* development is inhibited by pomegranate extract. Pomegranate peel includes a variety of phenols and flavonoids with strong antibacterial properties against Gram-positive bacteria. Peel extracts, at a dosage of 0.01%, show inhibitory effects on *S. aureus* and *B. cereus.*

Inhibitors of food pathogens may include garlic. When included in processed foods, garlic can lengthen their shelf lives and reduce the risk of deterioration and food poisoning. Due to the presence of the organic sulphur component allicin, which inhibits the growth of both Gram-positive and Gram-negative bacteria such *E. coli*, *Salmonella*, *Streptococcus*, and *Staphylococcus*, garlic extract possesses antimicrobial properties (Arshad and Batool, 2017).

6.3.2 Animals

Antimicrobial qualities are found in some animal defence mechanisms, which can break down cell membranes and kill Gram-negative and Gram-positive bacteria. Animal-derived peptides, polysaccharides, and lipids contain both antibacterial and antiviral properties.

6.3.2.1 Peptides

Winter flounder (*Pleuronectes americanus*) skin secretions contain a peptide called pleurocidin that has antimicrobial properties. It is potent against Gram-positive and Gram-negative bacteria, including *Vibrio parahemolyticus*, *Saccharomyces cerevisiae*, *L. monocytogenes*, *E. coli* O157:H7, and *Penicillium expansum*. Protamine, magainin, and casocidin are antimicrobial peptides in addition to defensins. The phagocytes of mammals and epithelial cells in vertebrates generate defensins, which have antibacterial capabilities against bacteria, fungi, and viruses. Protamine is found in the sperm of vertebrate animals; the antimicrobial effect of protamine, a cationic peptide, is likely a result of its electrostatic attraction to bacteria which are negatively charged. Magainin for its antibacterial activity founding that, at low concentrations, inhibited the growth of numerous species of bacteria (*E. coli*, *Staphylococcus epidermidis*, *Pseudomonas aeruginosa*, etc.) and fungi (*Saccharomyces cerevisiae*, *Cryptococcus neoformans*, and *Candida albicans*) and induce osmotic lysis of protozoa (*Amoeba proteus*, *Euglena gracilis*, and *Paramecium caudatum*). Another peptide with antibacterial efficacy against *E. coli* and *Staphylococcus carnosus* is casocidin, which is derived from bovine milk. Thirty-nine bovine s2-casein amino acids make up the main structure of casocidin (Quinto et al., 2019).

The glycoprotein enzyme lactoperoxidase is found in raw milk, colostrum, saliva, and other secretions. The lactoperoxidase system (LPS), which has antibacterial properties, is created when lactoperoxidase combines with thiocyanate and hydrogen peroxide. Since milk doesn't naturally contain hydrogen peroxide, it must be added or obtained through lactic acid bacteria activity or enzymatic action. Ovotransferrin, also known as conalbumin, is a different glycoprotein found in egg albumin that has an inhibitory effect against both Gram-positive and Gram-negative bacteria as well as some yeasts. Despite the fact that its ability to bind the iron needed for growth by microbes is assumed to be the cause of ovotransferrin's antimicrobial function, ovotransferrin's protective functions are comparable to those of the homologous lactoferrin found in mammals, indicating a "direct link between egg consumption and human health" (Quinto et al., 2019).

An enzyme called lysozyme can be found in milk, egg whites, and even blood. Lysozyme's enzymatic activity is primarily responsible for its antibacterial properties against bacteria; these properties include cell lysis and peptidoglycan breakdown (Quinto et al., 2019).

6.3.2.2 Polysaccharides

Chitosan, which has the capacity to stop the growth of bacteria, yeast, and mould in food, is made from chitin, a by-product obtained from the exoskeletons of crustaceans and arthropods. *E. coli*, *Cronobacter sakazakii*, *Salmonella*, *Staphylococcus*, *L. monocytogenes*, and *B. cereus* could not grow on the antimicrobial chitosan films that were employed for packing purposes or in conjunction with ethanolic extract and polypropylene to contact the food's surface. Chitosan's molecular weight, function as an enzyme inhibitor and water-binding agent, enhancement of membrane permeability, ability to compete with bacterial nutrition as a bio-absorbent, and ability to bind to DNA and prevent the creation of mRNA and proteins have all been described (Quinto et al., 2019).

6.3.2.3 Lipids

Food lipids, such as those found in milk, can prevent the growth of some microbes, such as fungi and Gram-positive and Gram-negative bacteria. Their initial response is the destruction of the electron transport chain and oxidative phosphorylation of the cell membrane. The mechanism of action also suggests peroxidation, nutrient absorption damage, and the cell's suppression of enzyme function. An increase in antibacterial activity was observed when fatty acids and monoglycerides were combined, according to research. They are particularly effective agents against microbes for a variety of applications in food safety, boosting food quality through its preservation, thanks to their broad spectrum of activity (Quinto et al., 2019).

6.3.3 Microorganisms

6.3.3.1 Bacteriocins

These proteins, known as bacteriocins, are made by some lactic acid bacteria and have the ability to suppress both pathogenic and spoilage bacteria without altering the physicochemical properties of the food. Despite not being harmful to the host, bacteriocins have antimicrobial properties. Most bacteriocins inhibit Gram-positive bacteria. Following the administration of chelating chemicals or hydrostatic pressure, Gram-negative bacteria may become more susceptible to bacteriocins. Bacteriocins can be utilized in food in a variety of ways:

a) Nisin is the only commercially available bacteriocin, and it can be applied directly to foods to prevent the growth of harmful and spoilage bacteria. Foods like milk, cheese, canned food, mayonnaise, and others all contain nisin. It stops *Listeria* spp. from growing on surfaces once it has been absorbed there and prevents the development of biofilms. Pediocins, which are used in salads, salad dressings, cream, cheese, and meats, also prevent the growth of *L. monocytogenes.*

b) To increase food safety by adding bacteriocin-producing bacteria to non-fermented foods or using them as starting cultures.
c) Using starter cultures of bacteria that produce bacteriocin to control the fermentation is a third method (Quinto et al., 2019).

6.3.3.2 Controlled acidification

The production of acid from lactic acid bacteria (LAB) under controlled acidification conditions is a preservation method that is crucial in the food industry; factors to be considered include the pH of the food, the traits of the targeted microorganism, the fermented carbohydrate that will be used by the bio preservatives, the kinetics of LAB growth, the temperature, and the targeted pathogens. *Clostridium botulinum* cannot thrive at pH levels below 4.6, thus LAB is added to the food to produce acid after the addition of a fermentable carbohydrate, which lowers the pH. This method of biopreservation prevents the development of botulinum toxin (Quinto et al., 2019).

6.3.3.3 Bacteriophages

Low quantities of non-growing pathogens and high populations of native bacteria are characteristics of the use of bacteriophages to manage foodborne bacteria. Additionally, the continued usage of bacteriophages in food may result in resistance (Quinto et al., 2019).

6.4 IMPROVING APPLICABILITY OF FOOD ANTIMICROBIALS THROUGH ENCAPSULATION

Encapsulation is the technique of encasing an active ingredient (material) into a carrier fluid to create minute particles that release their contents under precise conditions and at controlled rates over time (Nedovic et al., 2011). Preventing deterioration, solubilization, or unfavourable interactions with packaging materials, boosting the antimicrobial's availability, improving compatibility between the packaging polymer and the antimicrobial, supplying a controlled release or/and stimulus response, and encapsulating antimicrobial compounds improve the effectiveness of packaging materials for antimicrobial food packaging (Chausali et al., 2022). To overcome several issues that restrict their use in packaging applications, encapsulating particular antimicrobial chemical types has become necessary. Encapsulation is necessary for EOs, for instance, in order to reduce losses from volatilization or degradation during packaging manufacturing or storage, to increase their solubility and improve compatibility with biopolymers, and/or to reduce the organoleptic impact they have on food products due to their potent odour (Zanetti et al., 2021; Prakash et al., 2018). To overcome several issues that restrict their use in packaging applications, encapsulating particular antimicrobial chemical types has become necessary. The food and pharmaceutical industries have created a wide variety of delivery systems or carriers to encapsulate bioactive substances, including cyclodextrins, liposomes, emulsions, nanoparticles, and microcapsules. The food and pharmaceutical industries have created a wide variety of delivery systems or carriers to encapsulate bioactive substances, including cyclodextrins, liposomes, emulsions, nanoparticles, and microcapsules (Pisoschi et al., 2015).

6.5 FOOD ANTIMICROBIAL NANOCARRIERS

Throughout food preparation and storage, encapsulation techniques improve the solubility, reduce movement, and retain the stability of the bioactive metabolites, shielding them from chemical changes and undesirable interactions with other food components. Antimicrobial agents can be incorporated into food packaging materials with the use of nanoencapsulation methods. By shielding antimicrobials from harmful elements, enhancing their stability, and guiding them to the site of action, nanocarriers can control the release of antimicrobials, lowering the concentration needed to see an antimicrobial impact.

There are restrictions on all nanoencapsulation methods for application in the food sector linked to their high manufacturing costs and lack of permitted ingredients (Adriana et al., 2014).

6.5.1 Nanoliposomes

The hydrophobic hydrocarbon tails of phospholipids are linked together to form a bilayer in nanoliposomes, which have a spherical core and polar head groups that are directed towards the watery phases of the inner and outer media. Amphiphilic, lipid-soluble, and water-soluble molecules can all be delivered to and released from liposome formations. Nanoliposomes can be created for use in the food business by employing natural sources that include phospholipids with biological activity, such as milk, soy, or egg. The use of micro fluidization and heating are techniques for producing nanoliposomes without the use of hazardous solvents. Adsorption on the cell surface, fusing with the cell membrane, release of active by micropinocytosis, and specific or non-specific endocytosis are possible ways that liposome structures interact with their target cells. A high-pressure homogenization technique called the microfluidizer produced nanoliposomes of nisin in soybean-lecithin. Nisin was enclosed in the lamellar and core phases of nanoliposomes because of its amphiphilic nature (Adriana et al., 2014).

6.5.2 Nanofibres

The process of electrospinning, which creates continuous polymer fibres by applying an external electric field to a polymeric solution or melt, is the main way that nanofibers are made. Nanofibers are extremely thin structures with diameters under 100 nm. Proteins, carbohydrates, and lipids are examples of materials that can be used. The use of cationic cellulose derivatives in the textile, food, cosmetic, and pharmaceutical industries was made possible by their high hydrophilicity, biodegradability, and antibacterial characteristics. PVA, Polyquaternium-4 cellulose (PQ-4), and a copolymer of hydroxyethyl cellulose diallyl dimethyl ammonium chloride were used to create the electrospun nanofibers. *E. coli* and *S. aureus* were effectively combatted by the nanofibers' antibacterial properties. Bacterial cellulose (BC), a chemically pure type of cellulose produced by *Acetobacter xylinum*, is utilized in the processing of food due to its high fibre content, low cost, high water holding capacity, and tensile strength (Adriana et al., 2014).

6.6 FOOD APPLICATIONS OF NATURAL ANTIMICROBIAL COMPOUNDS

To prevent the growth of dangerous bacteria and spoilage, the food industry has used a range of preservation techniques, involving thermal treatment, seasoning, acidification, and drying. To stop the development of unwanted bacteria in food, the antimicrobials can be directly included in the ingredients, coated on the surface of the product, or included in the packaging material. Bacterial populations are immediately but transiently reduced when active ingredients are added directly to food, but their effectiveness is sustained for a considerable amount of time by antimicrobial coatings (Del Nobile Matteo Alessandro et al., 2012; Lucera et al., 2012).

6.6.1 Dairy Products

Ready-to-eat fresh dairy products are susceptible to contamination by unfavourable microbes. Some of these are pathogens that impair product safety, while others are spoilage bacteria that can give cheese an undesirable appearance and reduce its market value. Sinigaglia et al. (2008) investigated the impact of lysozyme and EDTA on the microbiological shelf life of mozzarella cheese. Brine containing lysozyme and varying concentrations of EDTA was used to package the mozzarella, which was then kept at 4 ±1°C for eight days. Without harming the normal lactic acid bacteria, the packaging approach dramatically reduced the growth of coliforms and Pseudomonadaceae.

In 2007 Conte et al. tested the effects of adding various quantities of lemon extract to a gel solution consisting of sodium alginate and the brine of mozzarella. Tests on shelf life were conducted at 15°C to mimic thermal abuse. All active packaged mozzarella cheese's shelf life was extended, indicating that the drug under investigation may have an inhibitory effect on the microorganisms that cause spoiling without having an adverse effect on the regular dairy microbial population (Del Nobile Matteo Alessandro et al., 2012; Lucera et al., 2012).

6.6.2 Meat-based Products

Meat products' spoilage causes their texture to degrade and their flavour and colour to change. To extend the shelf life of meat, natural antibacterial ingredients such bacteriocins, essential oils, salts, and extracts of herbs and spices are utilized.

- Mastromatteo et al. (2010) recommended combining MAP with natural antimicrobials like lemon and thymol to extend the shelf life of sausages with lower levels of pork backfat.
- The development of *Pseudomonas* spp., the cause of the unacceptability of sausages, was specifically restricted by the use of thymol and thymol-MAP.
- In cooked Swedish-style meatballs, the antioxidant and antibacterial benefits of rosemary, orange, and lemon extracts were also studied.

Turkish-style meatballs treated with lactoferrin both separately and in conjunction with nisin dramatically decreased spoiling bacterial counts and increased shelf life to 10 days (Del Nobile Matteo Alessandro et al., 2012; Lucera et al., 2012).

6.6.3 Fish-based Products

The primary factor in deterioration is the activity of bacteria that induce spoiling in seafood, which results in the loss of important fatty acids, fat-soluble vitamins, and protein functioning as well as the development of biogenic amines and off-flavours.

- Another improvement was the quality of a fish known as hot-smoked rainbow that was vacuum-sealed and given a thymol and garlic oil (1%) treatment.
- Fresh Mediterranean swordfish fillets' shelf life was significantly extended when MAP and thyme essential oil (0.1%) were combined.
- The sea bream fillet's shelf life was increased by more than 17 days by oregano essential oil (0.8%).
- A further study showed that MAP further improved the effects of the naturally occurring active chemicals by combining them with thymol, grapefruit seed extract, and lemon extract.

Coatings containing nisin and EDTA, both separately and together, were employed to regulate the quality of northern snakehead fish fillets stored at refrigerated temperature (Del Nobile Matteo Alessandro et al., 2012; Lucera et al., 2012).

6.6.4 Cereal-based Products

Bread often has a short shelf life due to stalling phenomena and fungus deterioration, specifically moulds.

- By limiting microbial development and delaying oxidation and stalling, chitosan coating increased bread quality.
- Spraying citrus peel essential oil produced the most inhibitory effect against germs and mould.
- For bread made from wheat and rye that had been experimentally infested with moulds, the combination of MAP and mustard oil was suggested.

Anise, black cumin, rosemary, and sage were among the natural substances whose antioxidant and antibacterial properties were examined in an effort to extend the shelf life of particular bakery products (Del Nobile Matteo Alessandro et al., 2012; Lucera et al., 2012).

6.7 COMMERCIAL ASPECTS OF ANTIMICROBIALS IN FOOD PRESERVATIONS

6.7.1 In Food Industry

Essential oils of basil, thyme, and oregano show antimicrobial activity against *Bacillus cereus* when added to rice (Quinto et al., 2019). Mayonnaise with 1 percent fresh garlic is effective against *Escherichia coli* O157 and *Salmonella enterica* serovar Enteritidis (Leuschner and Zamparini, 2002). It was discovered that ground cinnamon in pasteurized apple juice and essential oils of cinnamon, bark, and clove in semi-skimmed milk had an inhibitory effect on *L. monocytogenes* (Yuste and Fung, 2002; Cava et al., 2007). *Listeria monocytogenes* in cheese were successfully eradicated by the use of clove, cinnamon, thyme, and bay essential oils (Quinto et al., 2019). The total mesophilic population of cabbage and radish sprouts was tested for inhibition using essential oils of thyme, oregano, and lemongrass along with changed environment packing. The results showed that almost all of the microorganisms were inhibited (Hyun et al., 2015). Natural antimicrobial components found in extracts or essential oils have been used to extend the storage life of meat and meat products (Quinto et al., 2019).

6.7.2 For Packaging

Natural antimicrobials in edible films and coatings have gained the attention of the food packaging industry as a means to increase product safety and quality. Consumers' worries about plastic packaging are alleviated in the process (Park et al., 2004). *Streptococcus faecalis* (reduction of 3.8 log cycles) and *Escherichia coli* (reduction of 1.6 log cycles) were both inhibited by chitosan films containing 60% lysozyme (reductions of 2.7 log cycles) (Leuschner, and Zamparini, 2002). It was discovered that combining lysozyme and methoxy pectin results in a film that is both edible and antibacterial. Although the formation of these complexes greatly reduces lysozyme's antimicrobial activity, the release of lysozyme was reportedly enhanced after its use in the production of an edible antimicrobial film, allowing for the use of the film to protect foods from microorganisms sensitive to lysozyme activity (Bayarri et al., 2014). Chitosan and sodium caseinate were mixed together to make films, and the authors tested how effective these films were against the microbiota of cheese, salami, and carrots (Moreira et al., 2011). In tests conducted with film-forming solutions applied to meals, there was evidence of a considerable antibacterial effect against populations of mesophilic and psychrotrophic microorganisms, as well as yeasts and moulds, which resulted in decreases. It has been found that films made from chitosan are particularly successful at extending the shelf life of a variety of different products, including fruits and vegetables, as well as various meats and products (Quinto et al., 2019) and reported that an increase in activity against several bacteria in minimally processed pear after adding nisin and peptide P34 to chitosan films. These bacteria include *E. coli*, *B. cereus*, *Clostridium perfringens*, *Lactobacillus acidophilus*, *L. monocytogenes*, *S. aureus*, and *S. enteritidis*, as well as *Aspergillus phoenicis* and *Penicillium stolonif.* The inhibition shown by films containing natamycin was comparable to that shown by films containing chitosan alone.

6.7.3 In Nanotechnology

The application of nanotechnology is becoming more widespread in the food business, and numerous studies on the topic have been carried out in recent years. Because of the intricacy of the required technology and the food matrix, applications of nanotechnology to deliver naturally occurring antimicrobial chemicals in foods are quite restricted at this time (Quinto et al., 2019).

It was discovered that the lysozyme in hen egg white accelerated the synthesis of silver nanoparticles, which were able to keep the hydrolase function of the enzyme intact. These nanoparticles were efficient against *E. coli*, *Bacillus anthracis*, *S. aureus*, and *Candida albicans*. These nanoparticles had significant action against antibiotic- and silver-resistant strains of *E. coli* and *Proteus mirabilis*, as well as against a strain of *E. coli* that was resistant to both antibiotics and silver. Liposome nanoparticles loaded with nisin were tested for their sustained antibacterial stability against the bacteria *Listeria monocytogenes* and *Staphylococcus aureus*. The minimum inhibitory concentration (MIC) of the nanoparticles against both of the bacteria was 320 UI/mL, and after 48 and 72 hours of incubation, respectively, their populations were reduced by more than 6 log CFU/mL (Zou et al., 2012) soybean phosphatidylcholine nanovesicles with the pediocin peptide. Half of pediocin's antibacterial activity against *Listeria monocytogenes*, *L. innocua*, and *L. ivanovii* was preserved in nanovesicles for 13 days at 4 C. Mice that were immunosuppressed were treated with gold nanoparticles laced with the bacteriocin produced by *Lactobacillus acidophilus* CH1. This treatment was successful in eliminating intestinal microsporidiosis. The antimicrosporidia activity of the bacteriocin was increased, and the results showed a sustained reduction in both the number of spores shed in faeces and the number of spores in the intestinal tract (Mossallam et al., 2014).

6.7.4 In Agriculture

Different agrochemical treatments were used to keep the food's quality intact and prevent spoilage after harvest. These agrochemical preservation processes guarantee a product's security, quality, and extended shelf life. However, these have also caused severe health and environmental issues due to their chemical composition. There have been reports of health problems associated with ingesting agrochemical residues, including cardiac arrest, liver and kidney failure, and neurological diseases. Environmental and ecological concerns are further exacerbated when these leftovers are released into the environment, most often to water, soil, and air. Reports have indicated that biomaterials offer a viable alternative to agrochemicals for postharvest food quality preservation. Plant extracts such as essential oils, phytohormones, antioxidants, organic acids, and salts are also included in this category of biomaterials. Other popular examples include chitosan, carboxyl methylcellulose, polysaccharides, alginate, and agar. Post-harvest food preservation through the research and application of such biomaterials contributes to the long-term viability of agricultural output, the safety of food supplies around the world, and the stability of our planet's ecosystems (Chunpeng et al., 2022).

6.7.5 Hurdle Technology

The term "hurdle technology" refers to the practice of combining many preservation methods in order to preserve an organism without compromising its biological integrity (Leistner, 2000). Natural antimicrobials have been combined with high pressure, pulsed electric fields, and other antimicrobial therapies, both thermal and non-thermal. This might be the best way to guarantee the amount of security and conservation that you need (Gaare et al., 2013).

Synergistic or additive effects could result from the combination of natural antimicrobials, which would enable greater inactivation. Nisin had a minimum inhibitory concentration (MIC) of 12.5 international units per millilitre, and garlic essential oil had a MIC of 100 micrograms per millilitre against *L. monocytogenes*. When combined with 75 g/mL of essential oil, nisin at a concentration of 6.25 IU/mL was found to have the same inhibitory effect as when taken alone. This demonstrated the need for only a modest concentration when using natural antimicrobials in conjunction with one another (Razavi et al., 2011). Synergistic activity was demonstrated against naturally occurring microbes in milk when a pulsed electrical field, mild heat, natural antimicrobials, nisin, and lysozyme were combined. This resulted in substantially higher microbial inactivation than the total of the individual decreases (Smith et al., 2002).

When combined with bacteriocin, high hydrostatic pressure (HHP) treatment was more effective than either treatment alone. Treatment with 250 MPa of HHP resulted in a 2.2-log reduction of *S. aureus* in reconstituted skim milk, while testing with 10,000 AU/ml of lacticin alone resulted in a 1-log reduction. Even while each treatment was effective on its own, the combination resulted in a reduction of the pathogen of more than 6 log cycles, which was greater than the reductions produced by any treatment alone.

The addition of potassium sorbate, sodium benzoate, or eugenol to the recovery medium resulted in a decrease of 67% and 33%, respectively, in the amount of ultrasonic treatment intensity and time that was necessary to inhibit *Zygosaccharomyces rouxii* (Arce-Garcia et al., 2002). Similarly, the presence of vanillin or citral significantly boosted the bactericidal impact of thermos sonication on *L. monocytogenes* and altered the distribution of inactivation duration in orange juice (Ferrante et al., 2007).

6.8 CONCLUSION AND FUTURE DIRECTIONS

Essential oils, enzymes, bacteriocins, chitosans, and organic acids have all shown promise as preservatives in food and may provide a viable alternative to synthetic food additives, as seen by their widespread trial use with a wide variety of fresh perishable foods. There are a number of crucial variables that must be addressed if natural active chemicals are to be more widely used in industrial settings. Knowing how antibacterial work is a prerequisite for developing effective treatments. These details are currently unavailable for a large number of naturally occurring substances (Park et al., 2004).

Nanoencapsulation has been used mostly in the pharmaceutical business, but it has found rapid adoption in the food market in recent years. The creation of antimicrobial active packaging is one of the most important applications of nanoencapsulation in the food industry. Global Trends & Forecasting projects that the market for nanoencapsulated food additives will increase from 2013 to 2018 at a compound annual growth rate (CAGR) of 6.2%, reaching $26,208.3 million by 2018. Understanding how nanosystems interact with food matrices, how food components are released from nanocapsules, and the results of toxicological research are all needed. Applying nanocarriers in the food business requires further study to build scalable procedures and identify low-cost components. There is some evidence that antimicrobial food nanocarriers could be useful for preventing the growth of harmful germs and preventing food from spoiling (Cava et al., 2007).

REFERENCES

Adriana, P.B., Karen, M., Soto, M., Hernández, I. and Sandra, M. 2014. Food antimicrobials nanocarriers. *The Sci. World J.* Article ID 837215, 11 pages.

Arce-Garcia, M.R., Jimenez-Munguia, M.T., Palou, E. and Lopez-Malo, A. 2002. Ultrasound treatments and antimicrobial agent effect on *Zygosaccharomyces rouxii. Int. J. Food Microbiol.* 166(1): 155–162.

Arshad, M.S. and Batool, S.A. 2017. Natural antimicrobials, their sources and food safety. *Food Add.* DOI: 10.5772/intechopen.70197.

Bayarri, M., Oulahal, N., Degraeve, P. and Gharsallaoui, A. 2014. Properties of lysozyme/low methoxyl (LM) pectin complexes for antimicrobial edible food packaging. *J. Food Eng.* 11: 18–25.

Cava, R., Nowak, E., Taboada, A. and Marin-Iniesta, F. 2007. Antimicrobial activity of clove and cinnamon essential oils against Listeria monocytogenes in pasteurized milk. *J. Food Prot.* 70(12): 2757–2763.

Chausali, N., Saxena, J. and Prasad, R. 2022. Recent trends in nanotechnology applications of bio-based packaging. *Journal of Agriculture and Food Research*, https://doi.org/10.1016/j.jafr.2021.100257.

Chunpeng, W., Rengasamy, K.R.R. and Chuying, C. 2022. Biomaterials for food preservation. *J. Food Qual.* Article ID 9805934, 3 pages.

Del Nobile, M.A., Lucera, A., Costa, C. and Conte, A. 2012. Food applications of natural antimicrobial compounds. *Front. Microbiol.* 3: 1–13.

Ferrante, S., Guerrero, S. and Alzamora, S.M. 2007. Combined use of ultrasound and natural antimicrobials to inactivate Listeria monocytogenes in orange juice. *J. Food Prot.* 70(8): 1850–1856.

Gaare, M., Hussain, A., Mishra, M. and Chand, R. 2013. Natural antimicrobials for preservation of food. In *Dairy and food processing industry recent trends-part* (1st ed.). Chapter 13. New Delhi: Biotech Books.

Hyun, J.E., Bae, Y.M., Yoon, J.H. and Lee, S.Y. 2015. Preservative effectiveness of essential oils in vapor phase combined with modified atmosphere packaging against spoilage bacteria on fresh cabbage. *Food Control* 51: 307–331.

Ibrahim, S.A., Ayivi, R.D., Zimmerman, T., Siddiqui, S.A., Altemimi, A.B., Fidan, H., Esatbeyoglu, T. and Bakhshayesh, R.V. 2021. Lactic acid bacteria as antimicrobial agents: Food safety and microbial food spoilage prevention. *Foods (Basel, Switzerland)* 10(12): 311.

Leistner, L. 2000. Basic aspects of food preservation by hurdle technology. *Int. J. Food Microbiol.* 55(1–3): 181–186.

Leuschner, R.G.K. and Zamparini, J. 2002. Effects of spices on growth and survival of Escherichia coli 0157 and Salmonella enterica serovar Enteritidis in broth model systems and mayonnaise. *Food Control* 1(6–7): 399–404.

Lucera, A., Costa, C., Conte, A. and Del Nobile, M.A. 2012. Food applications of natural antimicrobial compounds. *Front. Microbiol.* 3: 287–295.

Mastromatteo, M., Danza, A., Conte, A., Muratore, G. and Del Nobile, M.A. 2010. Shelf life of ready to use peeled shrimps as affected by thymol essential oil and modified atmosphere packaging. *Int J Food Microbiol.* 144(2): 250–256.

Moreira, M.D.R., Pereda, M., Marcovich, N.E. and Roura, S.I. 2011. Antimicrobial effectiveness of bioactive packaging materials from edible chitosan and casein polymers: Assessment on carrot, cheese, and salami. *J. Food Sci.* 76: 54–63.

Mossallam, S.F., Amer, E.I. and Diab, R.G. 2014. Potentiated anti-microsporidial activity of *Lactobacillus acidophilus* CH bacteriocin using gold nanoparticles. *Exp. Parasitol.* 144: 14–21.

Nedovic, V., Kalusevic, A., Manojlovic, V., Levic, S. and Bugarski, B. 2011. An overview of encapsulation technologies for food applications. *Procedia Food Sci.* 1: 1806–1815.

Park, S.I., Daeschel, M.A. and Zhao, Y. 2004. Functional properties of antimicrobial lysozyme-chitosan composite films. *J. Food Sci.* 69(8): 215–221.

Pisoschi, A.M., Pop, A., Cimpeanu, C., Turcuş, V., Predoi, G. and Iordache, F. 2015. Nanoencapsulation techniques for compounds and products with antioxidant and antimicrobial activity. *Eur. J. Med. Chem.* 5(157): 1326–1345.

Poonia, A. 2017. Antimicrobial edible films and coatings for fruits and vegetables. *Food Sci. Nut.* DOI: 10.4018/978-1-5225-0591-4.ch014.

Prakash, B., Kujur, A., Yadav, A., Kumar, A., Singh, P.P. and Dubey, N.K. 2018. Nanoencapsulation: An efficient technology to boost the antimicrobial potential of plant essential oils in food system. *Food Control* 1(89): 1–11.

Quint, O.E.J., Caro, I., Villalobos-Delgado, L.H., Mateo, J., De-Mateo-Silleras, B. and Redondo-Del-Río, M.P. 2019. Food safety through natural antimicrobials. *Antibiot.* 8(4): 208.

Ramos, M.S., Attene, S.-O., Irais, G.-A., Blanca, E., Santos-López, E.M., Amaro-Reyes, A., Barboza-Corona, J.E. and Regalado, C. 2014. Antimicrobial edible films and coatings for meat and meat products preservation. *Sci. World J.* 2014: 248935.

Razavi Rohani, S.M., Moradi, M., Mehdizadeh, T., Saei-Dehkordi, S.S. and Griffiths, M.W. 2011. The effect of nisin and garlic (Allium sativum L.) essential oil separately and in combination on the growth of Listeria monocytogenes. *Food Sci. Technol.* 44(10): 2260–2265.

Sinigaglia, M., Bevilacqua, A., Corbo, M.R., Pati, S. and Del Nobile, M.A. (2008). Use of active compounds for prolonging the shelf life of mozzarella cheese. *Int. Dairy J.* 18: 624–630.

Smith, K., Mittal, G.S. and Griffiths, M.W. 2002. Pasteurization of milk using pulsed electrical field and antimicrobials. *J. Food Sci.* 67(6): 2304–2308.

Yuste, J. and Fung, D.Y.C. 2002. Inactivation of Listeria monocytogenes Scott A 49594 in apple juice supplemented with cinnamon. *J. Food Prot.* 65(10): 1663–1666.

Zanetti, M., Carniel, T.K., Dalcanton, F., dos Anjos, R.S., Gracher Riella, H., de Araújo, P.H.H., de Oliveira, D. and Antônio Fiori, M. 2021. Use of encapsulated natural compounds as antimicrobial additives in food packaging. *Trends Food Sci. Technol.* DOI: 10.1016/j.tifs.2018.09.003.

Zou, Y., Lee, H.Y., Seo, Y.C. and Ahn, J. 2012. Enhanced antimicrobial activity of nisin-loaded liposomal nanoparticles against foodborne pathogens. *J. Food Sci.* 77(3): 165–170.

7 Plants as antimicrobial agents in food technology

Sana Sheikh, Jyothi Miranda, Akshitha R. Amin, and Bhagyalakshmi

7.1 INTRODUCTION

For several thousands of years, plant oils and plant extracts have been utilized, serving many purposes, such as food preservatives and medical therapeutic agents (El-Ghorab et al. 2002). Due to the increasing awareness about the risks associated with the use of synthetically manufactured additives and preservatives in the food industry, the use of naturally occurring antimicrobials from plant sources has been a concern for consumers. Natural compounds present in herb- and spice-derived extracts, essential oils, and other secondary metabolites from plants are used for the same purpose. Their use as replacements for synthetic additives can open new frontiers in safety and quality preservation in food, as they are relatively safer and do not pose health risks to consumers (Batiha et al. 2021). Plant-derived compounds from herbs and spices have been used since ancient times for flavouring food, as traditional medicine, and as preservatives. In addition to being used in food to impart flavour, pungency, and colour, herbs and spices also have antioxidant, antimicrobial, nutritional, and pharmaceutical properties. Several researchers are working to isolate and identify plant compounds with antibacterial activity against foodborne bacteria which can be effectively used in food products. Moreover, the use of plant antimicrobials in foods is challenging due to their interaction with the food constituents which can decrease their activity. To tackle this problem, various new approaches and technologies have come up wherein plant antimicrobials could be used in foods. The potential antibacterials from plants are used against foodborne pathogens, and the effect of various food processing techniques on the activity of plant antimicrobials are worked out. Further, several methods are adopted to incorporate plant biopreservatives into the food system (Permeant et al. 2022). Although a number of studies have demonstrated the potential of using plant-derived antimicrobials as preservatives in foods, more investigations are required to optimize the quantities to be used in food products to be effectively used as antimicrobials without influencing the organoleptic properties of foods.

7.2 PLANT PRODUCTS USED AS ANTIMICROBIALS

Finding natural antimicrobial agents that inhibit bacterial and fungal growth for better quality and shelf-life has been of considerable interest in recent years. Herbs and spices have been recognized to contain a broad spectrum of active constituents that are responsible for antibacterial, antifungal, antiparasitic, and/or antiviral activities. Essential oils have been used for centuries as part of natural traditional medicine that are aromatic oily liquids of plant materials. The principal components responsible for essential oils' effective antimicrobial activity include saponins, tannins, flavonoids, alkaloids, alkenyl phenols, glycoalkaloids, carvacrol, thymol, citral, eugenol, linalool, phenolic acids, terpenes, and their precursors (Burt 2004).

 DOI: 10.1201/9781003268949-7

7.2.1 Herbs and Spices

Spices and herbs contain aromatic and volatile compounds that are effective against microbial spoilage (Hyldgaard et al. 2012). Most of these compounds are found in different parts of plants, namely roots, barks, and flowers, and are primarily produced as a defence against attacks from plant predators (Mendonca et al. 2018). In addition to the aforesaid, natural compounds in plants not only act as antimicrobials and antioxidants but also enhance colour and flavour. For instance, natural extracts from cinnamon stick, oregano, clove, and grape seed had an immediate effect on the colour parameters when applied to raw pork and flavour (Corbo et al. 2009).

The efficacy of spices and herbs and their extracts as natural antimicrobial compounds have been stated in past studies (Martínez-Graci´a et al. 2015). Aside from the antimicrobial nature of these plant extracts, they also exhibit great antioxidant capacity, which helps to inhibit oxidative spoilage (Pisoschi et al. 2018). Most antimicrobials derived from plants are regarded as safe (Burt 2004). Shan et al. (2009) studied the inhibitory impact of five spices and herbs like clove, peel of pomegranate plant, and extracts of grape seed, cinnamon stick, and oregano on common spoilage and pathogenic bacteria affecting uncooked pork meat stored at 20 °C over a 9 day period. Results from this study indicated the potency of herbs-derived extracts significantly reduced the microbial load and oxidative spoilage in uncooked meat stored at room temperature which further highlights their importance. Clove was reportedly the most efficient in curbing spoilage caused by microorganisms and oxidative rancidity (Shan et al. 2009). Interestingly, Batiha et al. (2020) stated that clove is used as a spice in food cooking and used commercially for many medicinal purposes and in the perfume industry and that it is considered one of the spices that can potentially be used for preservative purposes in many foodstuffs (Batiha et al. 2020), like meat, to replace the chemical preservatives because of their excellent antioxidant and antimicrobial properties, although consumer acceptability remains a cause of concern for their use. Similarly, Aziz and Karboune (2018) stated the importance of some extracts of spices and herbs in enhancing quality and preventing spoilage. Some of these natural biopreservatives can serve the crucial function of overcoming difficulties associated with resistance of microorganisms and also keep food products safe without side effects that could be detrimental to consumers (Pisoschi et al. 2018). These attributes make spices- and herbs-derived extracts a possible replacement for currently existing synthetic additives. Hence, their use is further encouraged in food preservation.

7.2.1.1 Antimicrobial peptides

Plants also produce a variety of antimicrobial peptides, many of which can be grouped in different classes: thionins, defensins, lipid transfer proteins, cyclotides, and snakins (Padovan et al. 2010). Some of them could possibly be exploited for food biopreservation.

7.2.1.2 Oils

In olive oil, distinctive antimicrobial compounds include oleuropein, oleuropein aglycon, elenoic acid and oleocanthal (in addition to hydroxytyrosol and tyrosol) (Cicerale et al. 2012)

7.2.2 Essential Oils

Essential oils (also known as volatile oils) are concentrated oily liquids containing volatile aroma compounds derived from plant materials such as flowers, buds, leaves, and bark (Burt 2004). Typically, essential oils are natural antimicrobial complex mixtures of volatile substances, including terpenes, terpenoids (oxygenated terpenes), and phenols, that are produced by many plants to protect themselves from infection and have great potential for utilization in the food industry to combat foodborne pathogens and spoilage organisms (Rivera Calo et al. 2015). Indeed, some essential oils, such as clove oil, thyme oil, tea tree oil, and cinnamon oil, exhibit broad-spectrum antimicrobial properties. Essential oils are usually complex mixtures of many different kinds of chemical compounds, typically containing approximately 20–80 different molecular species. Terpenes, terpenoids, and phenols are the major constituents that display antimicrobial activity (Burt 2004). The antimicrobial activity of a specific essential oil is mainly determined by its concentration and chemical composition.

Essential oils (EOs) are secondary oily aromatic metabolites extracted from different plant parts, including leaves, bark, flowers, and roots by various methods, which include solvent extraction (solvent, supercritical CO_2, subcritical water), distillation (steam, hydrodistillation, hydrodiffusion), solvent-free (microwave), and combination method (solvent + steam) (Mendonca et al. 2018). EOs have been reported to possess antioxidant and antimicrobial effects, mostly reducing the population of pathogenic bacteria and lipid oxidation, by inhibiting the growth of foodborne pathogens and acting as antioxidants, thereby preventing food deterioration and degradation (Mendonca et al. 2018; Shan et al. 2009). The presence of biologically active compounds, such as phenolic acids, terpenes, aldehydes, terpenoids, and ketones, able to inhibit the growth of microbial pathogens, makes this achievement possible (Tiwari et al. 2009).

Essential oils are complex natural blends consisting of 20–60 components present in different concentrations. Generally, the predominant components (or major compounds) determine their biological properties using more than one specific mechanism and the effects occur in several cellular sites. However, the components present in lower concentrations or traits can act synergistically with the majority increasing the antimicrobial activity of essential oils (Burt 2004). The antimicrobial action of EOs occurs in several ways, due to the wide variety of components present in their constitution. However, the main target in microbial cells is the cytoplasmic membrane (Nazzaro et al. 2013), whose permeability in bacterial cells is influenced by its composition and the hydrophobicity of the compounds that pass through it. Thus, due to the hydrophobic nature of EOs, the action of these compounds triggers structural and functional damage to the cell membrane of bacterial cells (Burt 2004) changing its structure and fluidity (Pasqua et al. 2007), and as a consequence, there is a dissipation of proton force, with regard to the reduction of the ATP pool, internal pH disorder, electrical potential, and loss of metabolites and ions (Lambert et al. 2001).

For example, EOs from underutilized herbs and spices, such as goraka (*Garcinia quaesita*), galangal (*Alpinia galanga*), lemon iron bark (*Eucalyptus staigerana*), and mountain pepper (*Tasmannia lanceolata*), have great potentials as antimicrobials against foodborne pathogenic and spoilage bacteria (Weerakkody et al. 2010). Apart from their strong antioxidant and antimicrobial properties, they also influenced the sensory and organoleptic attributes of foods (Shan et al. 2009), so that they can also be combined with new packaging and preservation technologies. Moreover, terpenoids and terpene compounds, like thymol, carvacrol, geraniol, and menthol, have also been reported to have good antimicrobial potency against *Aspergillus* spp., *E. coli*, *L. monocytogenes*, *S. sonnei*, *Enterobacter aerogenes*, *S. flexneri*, and *E. coli*.

Linalool, the monoterpene compound isolated from lavender EO, was reported to possess potent antimicrobial activity against resistant *K. pneumoniae* (Yang et al. 2019). Torres-Romero et al. (2011) reported the inhibitory effect of 1α-acetoxy-6β, 9β-dibenzoyloxy-dihydro-β-agarofuran and dihydro-β agarofuran sesquiterpenes against the growth of *Bacillus* spp. (Torres-Romero et al. 2011). Farnesol, the isoprenoid natural acyclic sesquiterpene alcohol, has also revealed good antibacterial activities against *S. mutans*, *S. sobrinus*, *S. epidermidis*, and *S. aureus* biofilm formation (Mahizan et al. 2019). Furthermore, the combination treatment of farnesol and xylitol showed promising effects against atopic dermatitis caused by *S. aureus*, without affecting the microbial flora, and effectively suppressing the *S. aureus* biofilm production (Masako et al. 2005). Oleanic acid and bonianic acid A and B are two triterpenoids isolated from *Radermachera boniana* and found to be effective against *Mycobacterium tuberculosis*. Moreover, ergosterol peroxide and ursolic acid combination treatment resulted in synergistic effect against *M. tuberculosis* (Jim´enez-Arellanes et al. 2013). Also, Cunha et al. reported that oleanic and ursolic acids extracted from *Miconia ligustroides* resulted in remarkable antibacterial effect against *Vibrio cholerae*, *Bacillus cereus*, *S. pneumoniae*, *Klebsiella pneumoniae*, and *S. choleraesuis* (Cunha et al. 2010).

Garlic EOs have a strong antimicrobial effect against *E. coli*, *S. aureus*, and *S. typhimurium* resulting in the destruction of bacterial cell membrane (Gyawali & Ibrahim 2014). Several other EOs from citrus peel, basil, rosemary plant, and lemon grass are also strong against microorganisms that cause spoilage in foods (Savoia 2012).

EOs also have inhibitory activity against microorganisms in dairy and fruit juice products. For example, cinnamon EOs have a good inhibitory power against *L. monocytogenes* in heat-treated apple juice (Cava et al. 2007). EOs from cinnamon and clove have also been used as preservatives against semi-skimmed milk (Hyun et al. 2015). In the same wise, EOs from thyme, cinnamon, bay, and clove are effective against *L. monocytogenes* in cheese (Khorshidian et al. 2018).

Numerous publications have described the correlation between the chemical constituents and antimicrobial activities of essential oils (Aligianniset al. 2001, Baydar et al. 2004, Diao et al. 2014). In general, essential oils and their chemical constituents have been shown to have broad-spectrum antimicrobial activities against many foodborne pathogens, including Gram-positive, Gram-negative, and spoilage microorganisms.

The major components in an essential oil typically constitute up to 85% of the system, whereas other components are present only in trace amounts. A number of reports have indicated that the antimicrobial activity of an essential oil is closely related to its major constituents as well as their interactions with some of the minor constituents (Angioni et al. 2004, Hussain et al. 2008, Kordaliet al. 2005). Most of the constituents that exhibit high antimicrobial efficacy are phenols, followed by oxygenated terpenoids. Terpenes and other constituents in essential oils, including ketones and esters, such as β-myrcene, α-thujone, and geranyl acetate, show much weaker antimicrobial activity than do phenols and terpenoids (Bassole & Juliani 2012).

7.2.3 Terpenes

Terpenes are the primary constituents of essential oils. Examples of monoterpenes include limonene, myrcene, α-pinene, and β-pinene, which can be found mainly in citrus, rosemary, sage, and lemongrass oils. A number of studies have shown that terpenes isolated from essential oils show weaker antimicrobial activity than the intact essential oils against both Gram-positive (e.g., *Staphylococcus aureus*) and Gram-negative (e.g. *Escherichia coli*, *Pseudomonas aeruginosa*, and *Shewanella* sp.) strains (Chalchat et al. 2000, Mastelic et al.2005, O'Bryan et al. 2008). For example, the minimum inhibitory concentrations (MICs) of rosemary oil and its isolated α-pinene were measured against three Gram-positive bacteria and Gram-negative bacteria. The rosemary oil had much higher antibacterial activity (MIC = 0.03% to 1%) than α-pinene (MIC=0.3% to 4%) across all bacteria tested (Jiang et al. 2011).

7.3.1 Terpenoids

Terpenoids are a large and diverse class of naturally occurring compounds in essential oils with antimicrobial constituents that are active against a broad spectrum of microorganisms (Dorman & Deans 2000). Terpenoids have been shown to exhibit higher antimicrobial activities than terpenes. For example, the antimicrobial activity of *Helichrysum italicum* essential oil and its terpenoid fractions were higher than that of the terpene fractions against *S. aureus* and *Candida albicans* growth (Mastelic et al. 2005).

7.3.2 Phenols

Examples of phenols include thymol, eugenol, and carvacrol, which can be found in thyme, clove, cinnamon, and oregano oils. Numerous studies have reported that essential oils (pure or mixed) containing high levels of phenols exhibit a broad spectrum of antimicrobial efficacy (Bassole & Juliani 2012, Cosentino et al. 1999, Dorman & Deans 2000, Rota et al. 2008).

7.3.2.1 Phenolic-rich plant extracts

Plants synthesize polyphenols in response to stress because of their self-defence from diseases mainly induced by microorganisms. A genuine reason for considering them a promising source of antimicrobials with healthy features is like antimicrobials are fundamentally different from all other medicines; they have both an individual effect (curing the infection) and a population effect (selection for resistance). Many natural phenolics that are widespread in nature, especially those that are extracted from edible plants and that have proven antimicrobial activities, could be used as potential food preservatives.

Polyphenols have a wide variation in their chemical structure, which reflect in their mode of action, which in turn leads to multimechanisms of antimicrobial activity. Studies carried out involve the use of extracts, which are a mixture of polyphenols belonging to different groups suggesting multidimensional activity.

a) Inhibition of cell envelope synthesis

One of the important components of bacterial cell walls is peptidoglycan. Conventionally, most of the antimicrobial compounds' site of action is the peptidoglycan layer. By inhibiting the synthesis of peptidoglycan, antimicrobial compounds cause the death of bacterial cells. Various studies have been conducted and it has been found that polyphenols interact with the bacterial cell wall, inhibiting the formation of the cell wall and thereby killing the bacterial cells (Qian & Zhou 2021). The hydroxyl group of the phenolic compounds interacts with the cell membrane by hydrogen bonding. This interaction results in disruption of the membrane structure leading to leaking of cellular contents. It also causes delocalization of electrons resulting in depolarization of bacteria which in turn affects the proton motive force, reducing the pH gradient across cell membrane. These changes then lead to the death of the cell.

b) Inhibition of nucleic acid synthesis

Polyphenols are known to disrupt the functioning of the enzymes. Gyrase, a very important enzyme which is associated with DNA synthesis, is inhibited by polyphenols. The enzyme is responsible for introducing negative supercoils into DNA. The enzyme gyrase is composed of two subunits: Gyrase A and Gyrase B. While Gyrase A is involved in inducing DNA breakage, Gyrase B catalyses ATP hydrolysis. Polyphenols are known to inhibit Gyrase B by binding to it, preventing hydrolysis of ATP. The hydroxyl group present in the polyphenols play a key role in the inactivation of microbial enzymes such as ATPase, histidine decarboxylase, amylase, and protease (Qian & Zhou 2021).

c) Interaction with proteins

The phenolic compounds when oxidized result in the formation of free radicals. Thus by inducing the formation of ROS, the bacterial cells undergo oxidative stress. Polyphenols are also involved in the biosynthesis of proteins, the inhibition of which results in a metabolic disorder; these metabolic disorders can result in cell death. Phloretin is an apple flavonoid and it is stated that they produce biofilm in *E. coli* O157:H7 through a mechanism that results in inhibition of curli genes, i.e., csgA and csgB. The curli genes are involved in the production of fimbriae. Bromelain, another protein-degrading polyphenol, is found in pineapple. Bromelain is known to function by weakening the outer membrane by disrupting the surface membrane proteins in Gram-negative bacteria which in turn leads to swelling, leaking, and damaging of cells (Kumar et. al.,2021). Polyphenols are also known to inhibit enzymes like dihydroorotate dehydrogenase, Dihydrofolate reductase, fumarate reductase, and flavoprotein NADH-dependent enoyl-ACP reductase.

7.4 ANTIMICROBIAL PACKAGING WITH ESSENTIAL OILS AND PLANT EXTRACTS

Antimicrobial agents derived from essential oils are interesting candidates for the development of activated films or packagings. To alleviate the deficiencies in using plant phenolic compounds in foods, polyphenols can be added to the immediate zone of foods in direct contact through slow release from edible coatings or packaging materials. Such systems could maintain an efficient concentration of antimicrobial plant phenolics in the superficial zone of foods over time. This approach could be advantageous for foods such as raw muscle foods (fish fillets, meat pieces) or some fruit and vegetables for which most microbial contamination occurs in their superficial zone. One advantage of edible coatings or food-contact packaging materials with antimicrobial plant phenolics over their direct spraying is their controlled release over time. Various studies on the use of active films and edible coatings to deliver antimicrobial agents to the surface of a wide range of foods in contact (fruit, vegetables, and meat products) have been conducted (Ye et al. 2008).

Generally, the active components of plant essential oils inhibit microorganisms through disturbance of the cytoplasmic membrane, disrupting the proton motive force, electron flow, active transport, and inhibition of protein synthesis. Examples of plant extracts and essential oils that are most widely incorporated into food packaging are linalool, thymol, carvacrol, clove oil, cinnamaldehyde, and basil essential oils.

There are plenty of reports found related to the AM effectiveness of plant extract components in food packaging (Ha et al. 2001, Sivarooban et al. 2008). Other essential oils-coated films that have been studied extensively include allyl isothiocyanate (AIT) (Nadarajah et al. 2005), rosemary (Han et al. 2007), garlic oil (Pranoto et al. 2005; Gamage et al. 2009), and cinnamaldehyde (Han et al. 2007).

7.4.1 Applications of Antimicrobials in Food Packaging

Antimicrobial agents derived from essential oils are interesting candidates for development of activated films or packagings. Plants also produce a variety of antimicrobial peptides, many of which can be grouped into different classes: thionins, defensins, lipid transfer proteins, cyclotides, and snakins (Padovan et al. 2010). Antimicrobial phenolics can be added directly to the formulation of perishable food products or incorporated into food-contact materials to release them in the immediate zone of perishable foods. Edible coatings or active food packaging materials can thus be used as carriers of plant bioactive compounds. Antimicrobial plant extracts or molecules, such as an extract of moso bamboo (TakeguardTM) launched by Takex Labo (Osaka, Japan) or a mixture of different natural antimicrobial extracts (BioviaTM YM10) including green tea extract launched by Danisco DuPont, have been proposed as alternatives to chemical ones (Chibanea et al. 2019).

7.4.1.1 Encapsulation

In order to maintain their biological activity, reduce their volatility, increase their effective utilization rate, and minimize the adverse effects of their door, EOs need to be encapsulated in a delivery system that is suitable for food applications (Buranasuksombat et al. 2011). This not only reduces the instability of EOs but also increases their contact area with water-soluble foods. Furthermore, encapsulated EOs can be more easily dispersed on the surface of foods where microorganisms grow and proliferate (Donsì & Ferrari 2016).

Microencapsulated antimicrobial agents incorporated in food packaging have been demonstrated to successfully destroy a range of microorganisms, offering a controlled release preservation technique. This application is a type of active packaging (AP), in which the conditions of the packaged food are changed to better preserve the sensory attributes, safety, and shelf life of the product. Since microbial contact primarily occurs on the surface of a packaged food, antimicrobial activity should be focused on solid or semisolid surfaces by either indirect contact using antimicrobial volatiles or via direct contact between the antimicrobial package and the food surface. Research has successfully incorporated antimicrobial peptides, such as bacteriocins, as well as phytochemicals, such as essential oils, into packaging materials (Guarda et al. 2011)

7.4.1.2 Nanoencapsulation approaches to increase the preservative potential of plant essential oils

Nanoencapsulation of EOs offers numerous advantages such as ease of handling, stability, protection against oxidation, enhanced distribution, solubility, and controlled release, with less or no adverse effect on organoleptic properties of applicable food items with enhanced bioavailability. Hence, the use of nanotechnology to develop EO-based preservatives could enhance their effectiveness in the food system (Prakash et al. 2018).

The mechanism of action of EOs depends on their inherent biochemical profile and their ratio in crude EOs and it has a cascade of reactions. In general, EOs negatively interact with biochemical and structural attributes of microbes leading to disturbance of the functioning of cell membrane, cytoplasm content, enzymes, and proteins associated with the normal functioning of cell, thereby disturbing the microbial metabolism and leading to cell death (Kujur et al. 2017). Nanotechnology is an emerging area of food science and technology that holds great promise in the food industry. The nanoparticle size with large surface area gives distinctive features to the nanomaterials and makes them more effective and suitable in the food system. Nanoencapsulated antimicrobials (phytochemicals, alkaloids, EOs, etc.) and antioxidants could reduce microbial contamination in the food system with enhanced organoleptic properties (flavour, colour, taste, and texture). In spite of tremendous opportunity and potential, the application of nano-based products in the food system is not fully approved by the regulatory authorities. Lack of knowledge about the potential unforeseen risks, consumer safety, and ethical issues are some of the major obstacles to nanomaterials. Although nanoencapsulation is one of the promising tools for the effective delivery of EOs into food, the toxicological aspects of most of the nanocarriers and their molecular target site are yet poorly explored (Prakash et al. 2018).

7.5 CONCLUSION

To preserve food safely and to prevent human beings from various degenerative diseases we can use essential oils as natural antioxidants and antimutagenic agents preferably over synthetic ones. Even though the use of antimicrobial sources of natural origin has received escalating interest, some adverse effects on organoleptic properties may also result from its use. Despite the latest advances, more studies are still needed on the optimization of the quantities to be used to effectively inhibit spoilage and pathogenic microorganisms without affecting the organoleptic properties of foods. Contemporary research needs to focus on naturally occurring organic antimicrobials over synthetic ones for a sustainable future.

REFERENCES

Angioni, A., Barra, A., Cereti, E., et al. (2004). Chemical composition, plant genetic differences, antimicrobial and antifungal activity investigation of the essential oil of Rosmarinus officinalis L. *J. Agric. Food Chem.*, 52(11), 3530–3535.

Aligiannis, N., Kalpoutzakis, E., Mitaku, S., Chinou, I. B. (2001). Composition and antimicrobial activity of the essentialoils of two *Origanum* species. *J. Agric. Food Chem.*, 49(9), 4168–4170.

Aziz, M., Karboune, S. (2018). Natural antimicrobial/antioxidant agents in meat and poultry products as well as fruits and vegetables: A review. *Crit. Rev. Food Sci. Nutr.*, 58(3), 486–511.

Batiha, G., Hussein, D. E., Algammal, A. M., et al. (2021). Application of natural antimicrobials in food preservation: Recent views. *Food Control*, 126, 108066.

Batiha, G. E.-S., Alkazmi, L. M., Wasef, L. G., Beshbishy, A. M., Nadwa, E. H., Rashwan, E. K. (2020). Syzygium aromaticum L. (Myrtaceae): Traditional uses, bioactive chemical constituents, pharmacological and toxicological activities. *Biomolecules*, 10(2), 202.

Bassole, I. H. N., Juliani, H. R. (2012). Essential oils in combination and their antimicrobial properties. *Molecules*, 17(4), 3989–4006.

Baydar, H., Sagdic, O., Ozkan, G., Karadoğan, T. (2004). Antibacterial activity and composition of essential oils from *Origanum*, *Thymbra* and *Satureja* species with commercial importance in Turkey. *Food Control*, 15(3), 169–172.

Buranasuksombat, U., Kwon, Y. J., Turner, M., Bhandari, B. (2011). Influence of emulsion droplet size on antimicrobial properties. *Food Sci. Biotechnol.*, 20(3), 793–800.

Burt, S. (2004). Essential oils: Their antibacterial properties and potential applications in foods—A review. *Int. J. Food Microbiol.*, 94(3), 223–253.

Cava, R., Nowak, E., Taboada, A., Marin-Iniesta, F. (2007). Antimicrobial activity of clove and cinnamon essential oils against *Listeria monocytogenes* in pasteurized milk. *J. Food Prot.*, 70(12), 2757–2763.

Cicerale, S., Lucas, L. J., Keast, R. S. J. (2012). Antimicrobial, antioxidant and anti-inflammatory phenolic activities in extra virgin olive oil. *Curr. Opin. Biotechnol.*, 23(2), 129–135

Chalchat, J. C., Chiron, F., Garry, R. P., Lacoste, J., Sautou, V. (2000). Photochemical hydroperoxidation of terpenes. Antimicrobial activity of α-pinene,β-pinene and limonene hydroperoxides. *J. Essent. Oil Res.*, 12(1), 125–134.

Chibanea, L. B., Degraevea, P., Ferhoutb, H., Bouajilac, J., Oulahala, N. (2019). Plant antimicrobial polyphenols as potential natural food preservatives. *J. Sci. Food Agric.* 15, 99(4), 1457–1474.

Cosentino, S., Tuberoso, C. I. G., Pisano, B., et al. (1999). In-vitro antimicrobial activity and chemical composition of Sardinian *Thymus* essential oils. *Lett. Appl. Microbiol.*, 29(2), 130–135.

Corbo, M. R., Bevilacqua, A., Campaniello, D., D'Amato, D., Speranza, B., Sinigaglia, M. (2009). Prolonging microbial shelf life of foods through the use of natural compounds and non-thermal approaches–a review. *Int. J. Food Sci. Technol.*, 44(2), 223–241.

Cunha, W. R., de Matos, G. X., Souza, M. G. M., et al. (2010). Evaluation of the antibacterial activity of the methylene chloride extract of Miconial igustroides, isolated triterpene acids, and ursolic acid derivatives. *Pharm. Biol.*, 48(2), 166–169.

Diao, W.-R., Hua, Q.-P., Zhang, H., Xu, J.-G. (2014). Chemical composition, antibacterial activity and mechanismof action of essential oil from seeds of fennel (*Foeniculum vulgare* Mill.). *Food Control*, 35(1), 109–116.

Donsì, F., Ferrari, G. (2016). Essential oil nanoemulsions as antimicrobial agents in food. *J. Biotechnol.*, 233, 106–120.

Dorman, H. J. D., Deans, S. G. (2000). Antimicrobial agents from plants: Antibacterial activity of plant volatile oils. *J. Appl. Microbiol.*, 88(2), 308–316.

El-Ghorab, A. H., Fadel, H. M., El-Massry, K. F. (2002). The Egyptian Eucalyptus camaldulensis var. brevirostris: Chemical composition of the fruit volatile oil and antioxidant activity. *Flavour Fragr. J.*, 17(4), 306–312.

Gamage, G. R., Park., H.-J., Kim, K. M. (2009). Effectiveness of antimicrobial coated oriented polypropylene/polyethylene films in sprout packaging. *Food Res. Int.*, 42(7), 832–839.

Guarda, A. , Rubilar, J. F., Miltz, J., Galotto, M. J. (2011). The antimicrobialactivity of microencapsulated thymol and carvacrol. *Int. J. Food Microbiol.*, 146(2), 144–150.

Gyawali, R., Ibrahim, S. A. (2014). Natural products as antimicrobial agents. *Food Control*, 46, 412–429.

Ha, J. U., Kim, Y. M., Lee, D. S. (2001). Multilayered antimicrobial polyethylene filmsapplied to the packaging of ground beef. *Packag. Technol. Sci.*, 14(2), 55–62. doi:10.1002/pts.537

Han, J. H., Castell-Perez, M. E., Moreira, R. G. (2007). The influence of electron beamirradiation of antimicrobial coated LDPE/polyamide films on antimicrobial activity andfilm properties. *LWT Food Sci. Technol.*, 40(9), 1545–1554.

Hyun, J.-E., Bae, Y.-M., Yoon, J.-H., Lee, S. (2015). Preservative effectiveness of essential oils in vapor phase combined with modified atmosphere packaging against spoilage bacteria on fresh cabbage. *Food Control*, 51, 307–313.

Hussain, A. I., Anwar, F., Hussain Sherazi, S. T., Przybylski, R. (2008). Chemical composition, antioxidant and antimicrobialactivities of basil (Ocimumbasilicum) essential oils depends on seasonal variations. *Food Chem.*, 108(3), 986–995.

Hyldgaard, M., Mygind, T., Meyer, R. L. (2012). Essential oils in food preservation: Mode of action, synergies, and interactions with food matrix components. *Front. Microbiol.*, 3, 12.

Jiang, Y.,Wu, N., Fu, Y.-J., et al. (2011). Chemical composition and antimicrobial activity of theessential oil of rosemary. *Environ. Toxicol. Pharmacol.*, 32(1), 63–68.

Jim´enez-Arellanes, A., Luna-Herrera, J., Cornejo-Garrido, J., et al. (2013). Ursolic and oleanolic acids as antimicrobial and immunomodulatory compounds for tuberculosis treatment. *BMC Complement. Altern. Med.*, 13, 258.

Kujur, A., Kiran, S., Dubey, N. K., Prakash, B. (2017). Microencapsulation of Gaultheria procumbens essential oil using chitosan-cinnamic acid microgel: Improvement of antimicrobial activity, stability and mode of action. *Food Sci. Technol.*, 86, 132–138.

Kumar, H., Bhardwaj, K., Cruz-Martins, N., et al. (2021). Applications of fruit polyphenols and their functionalized nanoparticles against foodborne bacteria: A mini review. *Molecules*, 26(11), 3447. doi:10.3390/molecules26113447

Khorshidian, N., Yousefi, M., Khanniri, E., Mortazavian, A. M. (2018). Potential application of essential oils as antimicrobial preservatives in cheese. *IFSET*, 45, 62–72.

Kordali, S., Kotan, R., Mavi, A., Cakir, A., Ala, A., Yildirim, A. (2005). Determination of the chemical compositionand antioxidant activity of the essential oil of *Artemisia dracunculus* and of the antifungal and antibacterial activities of Turkish *Artemisia absinthium*, *A. dracunculus*, *Artemisia santonicum*, and *Artemisia spicig*. *J. Agric. Food Chem.*, 53(24), 9452–9458.

Lambert, R. J. W., Skandamis, P. N., Coote, P. J., Nychas, G. J. (2001). A study of the minimum inhibitory concentration and mode of action of oregano essential oil, thymol and carvacrol. *J. Appl. Microbiol.*, 91(3), 453–462.

Mahizan, N. A., Yang, S.-K., Moo, C.-L., et al. (2019). Terpene derivatives as a potential agent against antimicrobial resistance (AMR) pathogens. *Molecules*, 24(14), 2631.

Masako, K., Yusuke, K., Hideyuki, I., et al. (2005). A novel method to control the balance of skin microflora: Part 2. A study to assess the effect of a cream containing farnesol and xylitol on atopic dry skin. *J. Dermatol. Sci.*, 38(3), 207–213.

Mastelic, J., Politeo, O., Jerkovic, I., Radosevic, N. (2005). Composition and antimicrobial activity of *Helichrysumitalicum* essential oils and its terpene and the terpenoid fractions. *Chem. Nat. Compd.*, 41(1), 29–32.

Martínez-Graciá, C., González-Bermúdez, C. A., Cabellero-Valcárcel, A. M., Santaella-Pascual, M., Frontela-Saseta, C. (2015). Use of herbs and spices for food preservation: Advantages and limitations. *Curr. Opin. Food Sci.*, 6, 38–43.

Mendonca, A., Jackson-Davis, A., Moutiq, R., et al. (2018). Use of natural antimicrobials of plant origin to improve the microbiological safety of foods. In *Food and Feed Safety Systems and Analysis*, eds. Steven C. Ricke, Griffiths G. Atungulu, Chase E. Rainwater, Si Hong Park, Academic Press, 249–272.

Nazzaro, F., Fratianni, F., De Martino, L., Coppola, R., De Feo, V. (2013). Effect of Essential oils on pathogenic bacteria. *Pharmaceuticals (Basel)*, 6(12), 1451–1474.

Nadarajah, D., Han, J. H., Holley, R. A. (2005). Inactivation of *Escherichia coli* O157:H7 in packaged ground beef by allyl isothiocyanate. *Int. J. Food Microbiol.* 99, 269–279.

O'Bryan, C. A., Crandall, P. G., Chalova, V. I., Ricke, S. C. (2008). Orange essential oilsantimicrobial activities against *Salmonella* spp. *J. Food Sci.*, 73(6), M264–267.

Padovan, L., Scocchi, M., Tossi, A. (2010). Structural aspects of plant antimicrobial peptides. *Curr. Protein Pept. Sci.*, 11(3), 210–219.

Pasqua, R. D., Betts, G., Hoskins, N., Edwards, M., Ercolini, D., Mauriello, G. (2007). Membrane toxicity of antimicrobial compounds from essential oils. *J. Agric. Food Chem.*, 55(12), 4863–4870.

Pisoschi, A. M., Pop, A., Georgescu, C., Turcuş, V., Olah, N. K., Mathe, E. (2018). An overview of natural antimicrobials role in food. *Eur. J. Med. Chem.*, 143, 922–935.

Prakash, B., Kujur, A., Yadav, A., Kumar, A., Singh, P. P., Dubey, N. K. (2018). Nanoencapsulation: An efficient technology to boost the antimicrobial potential of plant essential oils in food system. *Food Control*, 89, 1–11.

Pranoto, Y., Rakshit, S. K., Salokhe, V. M. (2005). Enhancing antimicrobial activityof chitosan films by incorporating garlic oil, potassium sorbate and nisin. *Swiss Society of. Food Sci. Technol.*, 38(8), 859–865. doi:10.1016/j.lwt.2004.09.014

Permeant, R., James, J., Karunasagar, I., Vankova, E., Scholtz, V. (2022). Tropical plant products asbiopreservatives and their application in food safety. *Food Control*, 141, 109–185.

Qian, W., Zhou, J. (2021). The application of polyphenols in food preservation. *Adv. Food Nutr. Res.* ISSN 1043-4526. doi:10.1016/bs.afnr.2021.02.005

Rota, M. C., Herrera, A., Martinez, R. M., Sotomayor, J. A., Jordan, M. J. (2008). Antimicrobial activity and chemicalcomposition of *Thymus vulgaris*, *Thymus zygis* and *Thymus hyemalis* essential oils. *Food Control*, 19(7), 681–687.

Rivera Calo, J., Crandalla, P. G., O'Bryan, C. A., Ricke, S. C. (2015). Essential oils as antimicrobials in food systems: A review. *Food Control*, 54, 111–119.

Sivarooban, T., Hettiarachchy, N. S., Johnson, M. G. (2008). Physical and antimicrobial properties of grape seed extract, nisin, and EDTA incorporated soy protein edible films. *Food Res. Int.*, 41(8), 781–785.

Savoia, D. (2012). Plant-derived antimicrobial compounds: Alternatives to antibiotics. *Future Microbiol.*, 7(8), 979–990.

Shan, B., Cai, Y. Z., Brooks, J. D., Corke, H. (2009). Antibacterial and antioxidant effects of five spice and herb extracts as natural preservatives of raw pork. *JSFA*, 89(11), 1879–1885.

Tiwari, B. K., Valdramidis, V. P., O'Donnell, C. P., Muthukumarappan, K., Bourke, P., Cullen, P. J. (2009). Application of natural antimicrobials for food preservation. *J. Agric. Food Chem.*, 57(14), 5987–6000.

Torres-Romero, D., Jiménez, I. A., Rojas, R., Gilman, R. H., López, M., Bazzocchi, I. L. (2011). Dihydro-β-agarofuran sesquiterpenes isolated from Celastrusvulcanicola as potential anti-Mycobacterium tuberculosis multidrug-resistant agents. *Bioorg. Med. Chem.*, 19(7), 2182–2189.

Weerakkody, N. S., Caffin, N., Turner, M. S., Dykes, G. A. (2010). In vitro antimicrobial activity of less-utilized spice and herb extracts against selected food-borne bacteria. *Food Control*, 21(10), 1408–1414.

Ye, M., Neetoo, H., Chen, H. (2008). Control of *Listeria monocytogenes* on ham steaks by antimicrobials incorporated into chitosan-coated plastic films. *Food Microbiol.*, 25(2), 260–268.

Yang, S.-K., Yusoff, K., Ajat, M., et al. (2019). Disruption of KPC-producing *Klebsiella pneumoniae* membrane via induction of oxidative stress by cinnamon bark (*Cinnamomum verum* J. Presl) essential oil. *PLOS ONE*, 14(4), Article e0214326.

8 Antimicrobial resistance and its consequential role in food security

A scientometric and bibliometric analysis

Smita Mishra, Apeksha Rathi, and Deepak Vyas

8 INTRODUCTION

Since the discovery of the antibiotic and the inception of the "antibiotic era", the use of antibiotics in medicinal and agricultural practices has significantly advanced public health and food production (Franklin et al. 2016). The use of antibiotics in agriculture has put pressure on plant pathogens' survival; in turn, microbes have become more aggressive than usual. The subtherapeutic or random use of antibiotics induces resistance in these microbes; as a result, the antibiotic, once effective, becomes ineffective; hence this leads to prolonged disease and illness in the food-producing components, threatening food security and livelihood (FAO 2017). Antimicrobial resistance (AMR) already poses a danger to global health and could harm the planet's ability to sustain its environment (Aslam et al. 2021). AMR's continued growth and dissemination could have disastrous effects (Lee Ventola 2015; Gradmann 2011). Antimicrobials are substances designed to prevent bacterial growth or kill them. They consist of parasiticides, fungicides, antiviral medicines, and antibiotics. Other medications, natural compounds, disinfectants, antiseptics, and antimicrobials may also have antimicrobial qualities.

AMR occurs when microbes such as bacteria, viruses, parasites, and fungi are, or become, resistant to antimicrobial treatments to which they were previously susceptible. When bacteria, fungi, viruses, and parasites develop resistance to antimicrobial drugs like antibiotics and antifungals, this phenomenon is known as antimicrobial resistance (AMR) (Naimi et al. 2001). When microorganisms interact with antimicrobials, this naturally occurs through selection pressure or environmental adaptation. The process speeds up when antimicrobials are used improperly or excessively (McEwen and Reid-Smith 2004). Effective antimicrobials are crucial for addressing diseases that affect humans and animals, contributing to the safety and security of food, and as such, they must be preserved. A major global threat to human and animal health, AMR exacerbates the improper and excessive use of antimicrobials (Lee Ventola 2015). The COVID-19 pandemic is a wake-up call to understand better and improve all areas of preparedness for and prevention of infectious diseases, including their environmental dimensions. Antimicrobials have contributed to reducing infectious diseases in humans, animals, and plants, saving lives and increasing productivity for decades. Their effectiveness is now in jeopardy. Resistant microbes can subsequently spread and be transmitted to humans, food animals, plants, and wildlife because of complex interconnections across nature. There is strong evidence that antimicrobials are increasingly failing to cure infections, the pipeline of novel antimicrobials to take their place has faltered, and AMR poses a significant threat to

DOI: 10.1201/9781003268949-8

human, animal, and plant health and food security (Singh et al. 2022; Gradmann 2011). The application of antibiotics and their substantial role in food security are considered in the present chapter to highlight the importance of antimicrobial resistance to the world's food security.

8.1 Different Mechanisms of Antibiotic Resistance by Microbes

Different microorganisms' use of antibiotic resistance mechanisms can be broadly divided into two categories: (i) mechanisms with a mechanistic basis for antimicrobial resistance and (ii) mechanisms with a genetic basis for microbial resistance. The development of genetic resistance results from mutations that alter antimicrobial targets, reduce medication absorption, increase chemical efflux, and alter metabolic pathways, as well as through horizontal gene transfer via transformation, transduction, and conjugation (Langendonk et al. 2021). On the other hand, mechanical resistance results from changes to antimicrobial molecules, the blocking of chemicals from reaching antimicrobial targets, the evasion of target sites, and resistance brought on by universal cell adaptation mechanisms. In addition to these critical processes, protein promiscuity has also recently been implicated as a factor in antibiotic resistance. According to traditional biochemistry, proteins use their pre-formed binding sites to bind to their ligands/substrates in a specified way. The term "promiscuous behaviours" or "protein promiscuity" refers to a number of deviations from this norm that have been seen recently (Chakraborty and Rao 2012). According to research on the medicine albicidin, even structural changes to a drug may not be sufficient to overcome drug resistance. Albicidin prevents DNA gyrase, making it effective against Gram-positive and Gram-negative bacteria. Another comparable study focused on the development of resistance to the broad-spectrum antibiotic fosfomycin. The drug's reactions with glutathione cysteine and water are catalysed by the proteins FosA, FosB, and FosX. According to the study, a protein called FosXmt may function as a progenitor for FosA and FosX. This progenitor protein was discovered to have very low glutathione transferase and epoxide hydrolase activities and catalytically promiscuous activity. (Figures 8.1–8.3)

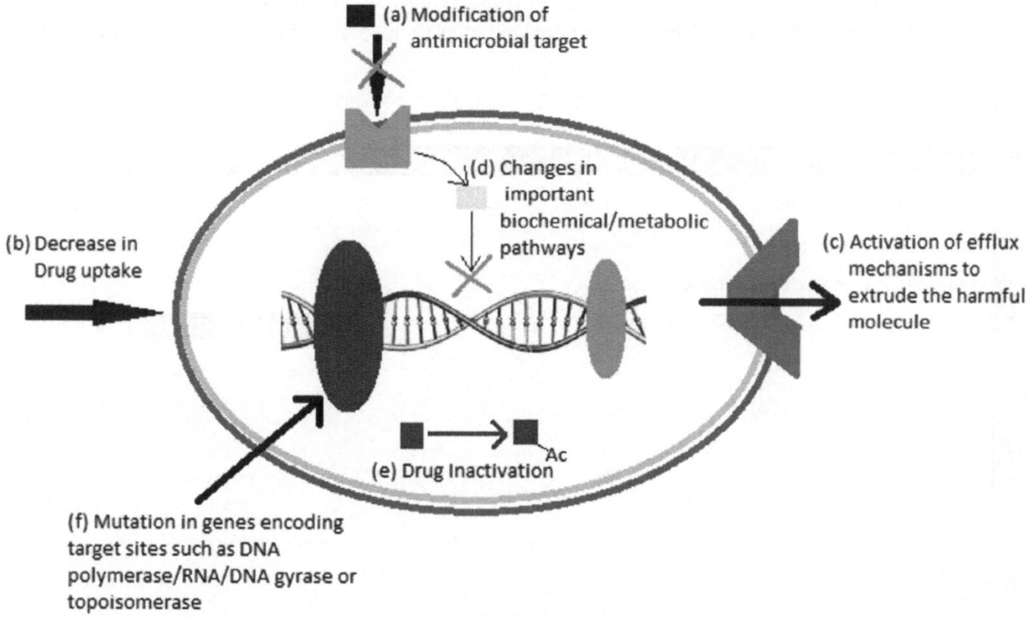

FIGURE 8.1 Mechanisms of antibiotic resistance adopted by bacteria. Source: as explained by Mann et al. 2021.

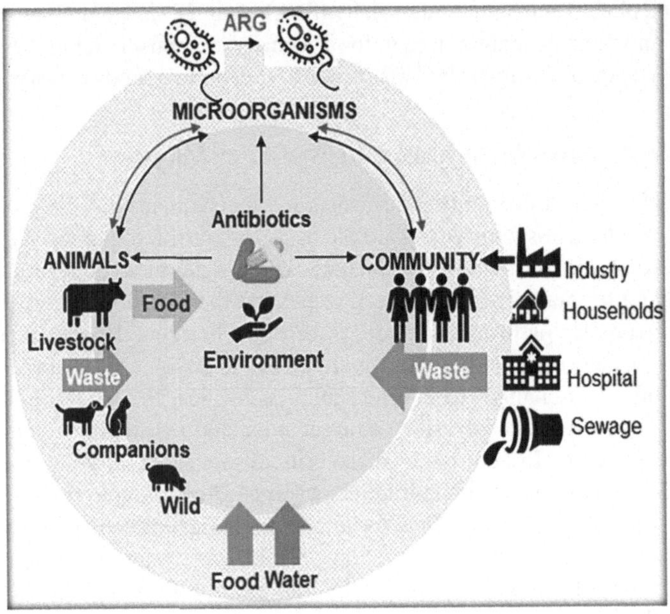

FIGURE 8.2 Antibiotic cycling in various components of the ecosystem

(a) Modification of antimicrobial targets – some bacterial membranes lack the target molecule for a particular antibiotic and thus become impermeable to the passage of such antibiotics across their cell membranes. **(b) Decrease in drug uptake** – a few bacteria modify their cell membranes in order to reduce or diminish drug uptake. **(c) Activation of efflux mechanisms to extrude the harmful molecules** – these mechanisms increase the efflux of antibiotics at a faster pace such that they are not able to produce much effect on the cell. **(d) Changes in important metabolic pathways** – the effect of antibiotics can be produced through many biochemical pathways, and hence by incorporating changes in these pathways, bacteria rescue themselves from the effect of antibiotics;

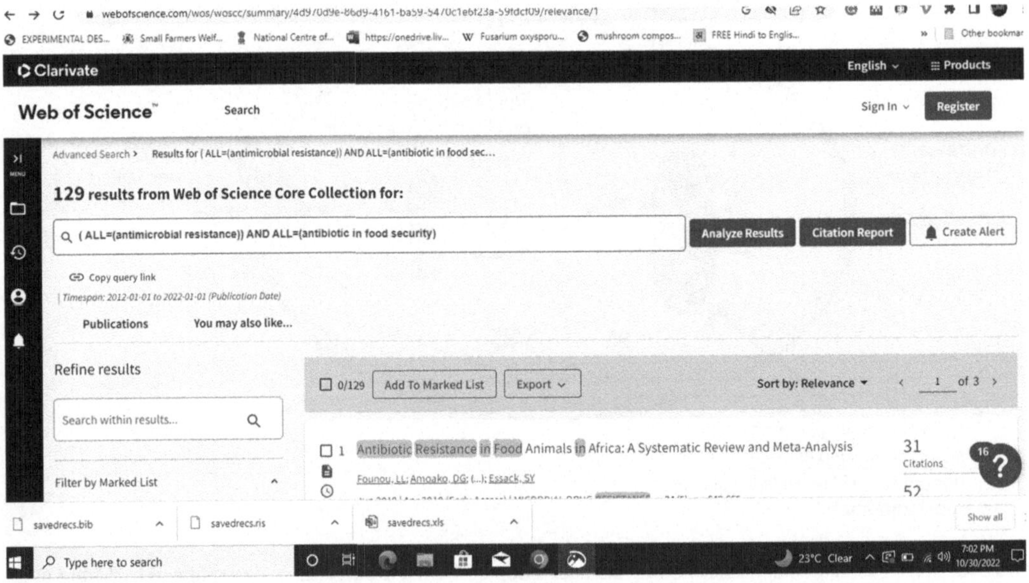

FIGURE 8.3 A screenshot of the Window tab of the Web of Science core collection

TABLE 8.1
Antibiotics grouping by the mechanism.

Cell wall synthesis	Penicillins
	Cephalosporins
	Vancomycin
	Beta-lactamase inhibitors
	Carbapenems
	Aztreonam
	Polymycin
	Bacitracin
Protein synthesis inhibitors	Aminoglycosides (gentamicin) – Inhibit 30s Subunit
	Tetracyclines (Inhibit 50s Subunit)
	Macrolides
	Chloramphenicol
	Clindamycin
	Linezolid
	Streptogramins
DNA synthesis inhibitors	Fluoroquinolones
	Metronidazole
RNA synthesis inhibitors	Rifampin
Mycolic acid synthesis inhibitors	Isoniazid
Folic acid synthesis inhibitors	Sulphonamides
	Trimethoprim

TABLE 8.2
Estimated global average consumption of the antimicrobial for the production of 1 Kg of meat.

45 mg of antimicrobials	are used to produce 1 kg of beef
148 mg of antimicrobials	are used to produce 1 kg of chicken
172 mg of antimicrobials	are used to produce 1 kg of pork

TABLE 8.3
Type of antibiotics.

Bacteriocidal antibiotics	Antibiotics that kill bacteria include aminoglycosides, beta lactams, fluoroquinolones, glycopeptides, lipopeptides, nitroimidazoles, and nitrofurans
Bacteriostatic antibiotics	Antibiotics that inhibit bacterial growth include glycylcyclines, lincosamides, macrolides, oxazolidinones, streptogramins, and sulphonamides.

(e) **drug inactivation** – drugs are inactivated by processes such as acetylation or methylation once they enter in the cell, thus making them ineffective; **(f) mutations in genes encoding target sites** – drugs have targets such as DNA polymerase/ RNA/ DNA gyrase and topoisomerase, so the genes encoding such target molecules are mutated/suspended by some bacteria to protect them from the effect of antibiotics (Tables 8.1–).

8.2 Antibiotics in Different Ecosystems

Given the close relationship between soil, plants, and animals, no one ecosystem is exempt from the spread of drugs and antibiotic resistance in our environment. All of these routes are used by antibiotics, which establish a network via which antibiotic resistance genes (ARGs) can also move (Polianciuc et al. 2020). Antibiotics are used in crop protection to prevent and treat various illnesses, as well as in animal feed, aquaculture, and agriculture to boost growth. However, in the case of animals, these antibiotics are not completely absorbed in the stomach and are instead expelled out of the body in the form of faeces and urine, which finally results in the production of manure (Thanner et al. 2016). This manure, a rich source of nutrients and organic matter, not only aids in increasing soil fertility but also plays a role in transmitting antibiotic traces from the animal world to the soil ecosystem and, ultimately, to plants. Antibiotic resistance among soil microorganisms increases as a result of the application of manure to agricultural land. According to reports, the long-term application of fresh and composted swine manure in agricultural soils results in the development of tetracycline resistance. The use of manure near animal pens has also reported a high level of chlortetracycline resistance, and the bacterial isolates obtained from such soils have been reported to encode genes for tetracycline resistance. A few studies also support that the application of manure incorporated with antibiotics to crops can also lead to the development of antibiotic resistance among different crops.

8.2.1 Strategies to encounter antibiotic resistance

- **Alternate scientific strategies to encounter antibiotic resistance**

  Molecular methods targeting DNA, RNA, or proteins and other scientific methods targeting cell walls, cell membranes, intracellular targets, biosynthetic pathways, ribosomes, etc. present several advantages in the molecular field of medical biotechnology, including having a high potential to alleviate or reduce the phenomenon of antibiotic resistance. These methods are continuously spreading both in terms of technological advances and popularity.
- **Use of peptide antibiotics/antimicrobial peptides (AMPs) as an alternative to antibiotics**

  Peptide antibiotics/antimicrobial peptides, also known as host defence peptides, are short and generally positively charged peptides found in a wide variety of life forms like mammals, bacteria, and fungi. The epithelial surfaces in mammals can remarkably provide physiologic functions in the face of frequent microbial invasions. Studies have shown that epithelial cells of the skin, respiratory, genitourinary, and alimentary tract synthesize antimicrobial peptides. These molecules contribute to intrinsic mucosal immunity and save from simple microbial infections.

8.3 Historical Summary of Antibiotics

For thousands of years, humanity has been helplessly subjected to a variety of illnesses, many of which have reached epidemic proportions and claimed millions of lives. Since the dawn of time, infectious diseases and the subject of their causes have been on people's minds. However, the quest for methods to combat, treat, and stop the spread of contagious diseases remained fruitless for a very long period due to a lack of understanding (Mohr 2016). The word antibiotic came from the word antibiosis, a term coined in 1889 by Louis Pasteur's pupil Paul Vuillemin which means a process by which life could be used to destroy life.

8.3.1 Penicillin antibiotic – its accidental discovery

Sir Alexander Fleming, in 1928, while investigating a bacterium (*Staphylococcus*) that causes boil. But before he left for vacation, he forgot to place the culture plate of *Staphylococcus* into the incubator. To his surprise, when he was back to his lab, he noted that a mould called *Penicillium notatum* had contaminated the culture plate, which inhibited the normal growth of the bacteria, and after evaluating the effects, he named the antimicrobial substance as "penicillin" (Lee Ventola 2015; Sivaraman 2018). With the discovery of penicillin by Sir Alexander Fleming, the modern era of antibiotics began. Since that time, antibiotics have revolutionized contemporary medicine and helped save countless lives. In the 1940s, doctors began using antibiotics to treat severe infections. Infections caused by bacteria were successfully managed with penicillin in World War II soldiers. But soon after, penicillin resistance developed and became a significant clinical issue, posing a danger to many of the discoveries made in the preceding decade by the 1950s. In response, fresh beta-lactam antibiotics were identified, created, and used, regaining public trust. However, during that same decade, the first instance of methicillin-resistant *Staphylococcus aureus* (MRSA) was discovered in the UK in 1962 and the US in 1968 (Lee Ventola 2015).

8.4 Uses of Antibiotics in Agricultural Produces

Globally, the demand for animal protein for human consumption is increasing at an unheard-of rate. Antimicrobials are frequently used in modern animal production methods, which could potentially put more pressure on bacteria to evolve resistance. At the global level, China and Brazil are the larger consumers of antimicrobials (Van Boeckel et al. 2015). In order to preserve productivity and health, antimicrobials are utilized in cattle production; these procedures pose a serious risk to the public's health by facilitating the spread of infections that are resistant to treatment in both humans and livestock (Van Boeckel et al. 2015). According to projections, by 2030, the usage of antibiotics will increase by 67% globally and nearly double in South Africa, Brazil, Russia, India, and China. This increase is probably the result of both a shift to large-scale farms where the use of antibiotics is commonplace and an increase in consumer demand for livestock products in middle-income nations (Schar et al. 2020; Van Boeckel et al. 2015). India's livestock sector is one of the largest in the world, accounting for 11.6% of the global cattle population. India is the fifth-largest producer of meat in the world, producing an estimated seven million tonnes in 2015–2016, of which 54% was red meat, and the remainder was poultry meat. In addition, India is the third-largest producer of eggs in the world after China and the United States and the fourth-largest producer of chicken after China, Brazil, and the United States. Because it is less expensive than other meats and has fewer religious or cultural taboos, chicken is the most popular meat in India (Sivaraman 2018). Due to its affordability, buffalo meat, often known as carabeef, ranks second among animal proteins consumed in India after chicken meat. Carabeef is consumed per person annually in amounts of about 2 kg compared to 3.6 kg for chicken. In comparison to the global average of about 18 kg, India's yearly per capita consumption of fish and shrimp is projected to be only 6 kg. The amount of antimicrobials used in food animals will rise globally, according to the Organisation for Economic Co-operation and Development (OECD), from 63,151 tonnes in 2010 to 105,596 tonnes in 2030 – an increase of 67% (Sivaraman 2018; Durso and Cook 2019).

8.4.1 Animal husbandry (livestock farming)

Animal husbandry is a large-scale business. The animals are bred, cared for, reared, and sheltered in a farm or region, which is specially built for them. Animal husbandry involves poultry, milk farms, apiculture, aquaculture, etc. With increased rates of antibiotic resistance (ABR) in common infections and a rising number of treatment failures, India has recently become a global hotspot for ABR. In India, the widespread use of antibiotics in the food-animal production industry is another area of concern in addition to the human health sector. In India, there are few laws limiting the use of antibiotics on livestock, poultry, and pigs produced for human consumption, and even when there are laws, they are rarely followed strictly. Numerous studies demonstrate the widespread usage of antimicrobials as growth promoters. In the production of chicken, non-therapeutic antibiotic use has been particularly prevalent.

The reason for using antibiotics in food animals includes:

- Used for the treatment of animals that show clinical signs of an infectious disease
- Used as metaphylaxis to treat a group of clinically healthy animals to minimize the unexpected outbreak of disease or as prophylaxis to prevent those at risk from being infected
- Used as a growth promoter to boost the weight of the animals

8.4.2 Pisciculture

Globally, aquaculture provides 8% of the animal protein used by humans, and per capita consumption is rising more quickly than that of meat and dairy. According to reports, antibiotic usage in the fast-growing aquaculture sector has been observed. This use could be a factor in the development of antimicrobial resistance, which could have negative effects on the health of humans, animals, and ecosystems (Cabello et al. 2017).

Particularly in low- and middle-income nations, aquaculture is a significant and quickly expanding source of animal protein (LMICs). Seventeen per cent of all animal protein consumed worldwide comes from aquatic creatures, and fish account for roughly 20% of animal protein consumed per person for more than 40% of the world's population. Dietary changes indicate that focussing on foods and dietary patterns is integral to the specific dietary change required for health outcomes, but social and behavioural factors will influence the achievement of these changes. Global food consumption of fish has increased by 3.2%, outpacing the combined rise of meat consumption from all terrestrial animal production sectors (2.8%), with the exception of poultry. Since 2001, the demand for animal protein in rapidly expanding nations has expanded, propelling a 5.8% yearly growth in aquaculture worldwide (Schar et al. 2020). Aquaculture now accounts for nearly half of the global supply of fisheries products for human consumption (Schar et al. 2020). Nearly 90% of the world's aquaculture production comes from Asia, and China alone accounted for 61% of global production in 2016 (Schar et al. 2020). On fish farms, antibiotics are used for illness prevention and treatment, and the same antibiotics that are used in human medicine have been utilized in fish farming, and they can be given orally or directly injected into pond water (Helliwell, Morris, and Raman 2020; Manyi-Loh et al. 2018; Schar et al. 2020).

8.4.3 Poultry

In poultry, antibiotics are often given to the entire flock and utilized for growth promotion, disease treatment (therapy), and disease prevention (metaphylaxis). All major nations that produce considerable quantities of poultry allow the use of antibiotics to prevent disease. Intestinal infections such as colibacillosis, necrotic enteritis, and other illnesses typically brought on by *Salmonella*, *E. coli*, or *Clostridium* spp. are treated with antibiotics, and these infections are a major concern among poultry leading to enormous economic losses (Mann et al. 2021).

8.4.4 Antibiotics in agriculture

Antibiotics are widely used in agriculture, livestock, poultry, fisheries, and animal husbandry. In agriculture, antibiotics are most commonly used to prevent and cure various diseases in crops, whereas in livestock and animal husbandry, these are most commonly used as growth-promoting agents and in preventing/curing infections. There are at least 30 different antibiotics that are commonly used in agriculture and livestock, among which macrolides, penicillins, and tetracyclines are the major ones.

8.5 Antimicrobial Resistance Impact on Food Security

Antimicrobials' use in animal production dates back to 1910 when workers carried out protests and riots across America due to a shortage of meat products. The use of antibiotics and other antimicrobial agents resulted from scientists' efforts at the time to find ways to produce more meat for relatively less money. Antibiotics used outside of their intended therapeutic application in animal husbandry have been outlawed in certain nations due to the growing threat of antibiotic resistance and treatment failures. Between 1986 (to promote growth) and 1988 (for prophylaxis), Sweden is credited as being the first nation to outlaw the use of antibiotics for non-therapeutic purposes (Agyare et al. 2018; Van Boeckel et al. 2015; Williams-Nguyen et al. 2016).

8.6 Sientometric and Bibliometric Analysis on the Antimicrobial Resistance

It is imperative to adopt scientometric and bibliometric analysis to understand the amount of work carried out on a pertinent subject or a specific term; this gives an overview of what has been done and what is needed to be done on a particular problem. Bibliometric and scientometric analysis is a well-liked and rigorous technique for investigating and analysing vast amounts of scientific data. It allows us to explore the nuances of a particular field's evolutionary history while illuminating its frontiers. Scholars use bibliometric and scientometric analysis for various reasons, such as to uncover emerging trends in article and journal performance, collaboration patterns, and research constituents and to explore the intellectual structure of a specific domain in the extant literature.

It enables and empowers scholars to:

a. Gain a one-stop overview
b. Identify knowledge gaps
c. Derive novel ideas for investigation
d. Position their intended contributions to the field

Web of Science database was searched using the term "Antimicrobial resistance" and "Antibiotic in food security" ranging from 2012-01-01 to 2022-01-01 (YY-MM-DD), and all types of documents were considered, including closed access and closed accesses articles were considered; a total of 129 documents were found and exported with all information in the different text file to run it in VOSviewer software.

The following map was created:

a. Map based on the number of terms of occurrence in 129 document
b. Map based on density visualization of category in the Web of Science published in the last ten years

VOSviewer is a software which creates visualization and explores the science's bibliometric maps (van Eck and Waltman 2010). Using VOSviewer, term maps, density visualization map, and other co-author maps are created. Term maps are a representation in a two-dimensional reach field in which terms are strongly related to each other. The weaker the relationship between terms, the bigger the distance is between them. Thus, term maps provide overviews for identifying the structure of a topic (Costa et al. 2019). This analysis reflects the need for the current research and shows the research gap and specific terms to be worked upon in the respective field; only the article indexed in Web of Science has been incorporated into this chapter. Moreover, this analysis was also carried out to reflect the most extensive terms used in the AMR field using the key term "Antimicrobial resistance" and "Antibiotic in food security" (Figure 8.3); this analysis also helps to recognize the work of potential authors, and also the top countries involve in such type of research in a specific domain.

Based on the data analysed using the Web of Science tools, articles published on the term "Antimicrobial resistance" and "Antibiotic in food security", we retrieved 129 articles, out of which 24 articles, i.e., 18.605% of 129, were published on the subject domain of Microbiology, the most contributing subject domain on the term "Antimicrobial resistance" and "Antibiotic in food security" search for the last ten year (2012–2022), followed by the subject domain Food Science Technology contributing 16.279% of 129 and then Infectious Diseases ≥ Pharmacology Pharmacy > Multidisciplinary Sciences > Veterinary Sciences > Biotechnology & Applied Microbiology > Agriculture, Dairy & Animal Science ≥ Environmental Sciences > Public, Environmental & Occupational Health > Zoology > Engineering Environmental > Tropical Medicine (Table 8.4). After analysing Table 8.4, it can be established that the threat due to antimicrobial resistance is only considered greatly in some of the subject domains, which would not be sufficient to curb the growing risk of AMR at the global level under the banner of One Health. Interdisciplinary and integrated efforts from every domain of the subject must put effort to reduce the global threat of AMR. A number of publications in journals indexed in Web of Science have been consistent for the last ten years. After the outbreak of COVID-19, research in antibiotics has geared up, leading to a high publication of articles in the years 2020–2022 (Figure 8.4). The increasing number of articles itself indicates how the scientific world is worried about the increasing rate of AMR, however, limited to some countries.

It can be seen from Figure 8.5 that among the 25 top productive countries contributing the quality article on the term "Antimicrobial resistance" and "Antibiotic in food security" the USA contributes the highest number of publications and the least by Switzerland. During this course of study and the publication, most terms used by authors for research are visualized in Figure 8.6 using VOSviewer software; the most used words in articles in 129 documents are mapped, and terms (words) of occurrence network build, where the lines connecting terms from one cluster to other cluster are indicated by a different colour and also the year of their use are indicated in the rectangular box in Figure 8.6.

The density visualization of the term category used in the Web of Science (Figure 8.7) published in the last ten years has been constructed to understand the domain and field or branches that deal with AMR-related research. The density of kernel colour indicates the number of publications, and the closeness of kernel to other kernel indicates the similarities among them; the highly dense kernel has a relatively higher number of publications to other kernels; thus on evaluation, Figure 8.7 indicates that "Microbiology" domain has the highest publication of articles related to AMR.

After evaluating and analysing the bibliometric and scientometric analysis, it can be emphasized that more collaborated research, open access publications, and research funding should be appreciated; moreover, all countries should participate in such research and must look into the AMR in food security.

8.7 CONCLUSION

The use of antibiotics, either naturally produced or semi-synthetically, is significant in agriculture, veterinary, and clinical settings. In the lives of humans and animals, these compounds are endowed with bacteriostatic or bactericidal activity required for therapeutic and preventative purposes to fight and prevent diseases. Despite all the beneficial roles of antibiotic, it has horrible effects too. The application of these substances at the subtherapeutic level as a growth promoter in livestock over long periods leads to antimicrobial pollution, which remains in the form of residue in animal-derived products, including meat, chicken, fishes, eggs, and other edible tissue; when consumed by the people cause health risk of toxicity and the induce development and emergence of antibiotic-resistant microbes and also leads to the therapeutical failure of contagious disease caused by microbes. The developing resistance in disease-causing microbes is of great concern to public health worldwide. Therefore, looking into developing multidrug resistance in the microbes and their role in food security, this chapter reflects the consequential effects of antibiotic resistance and explains the research and the pattern of research from every domain where antibiotics used are analysed using bibliometric and scientometric techniques. The situation of antibiotic resistance is compounded gradually. Every stakeholder should look

TABLE 8.4
Subject domain in Web of Science database where article published related to AMR

Sl.no	Web of Science categories	Record count	% of 129
1	Microbiology	24	18.605
2	Food Science Technology	21	16.279
3	Infectious Diseases	19	14.729
4	Pharmacology Pharmacy	19	14.729
5	Multidisciplinary Sciences	15	11.628
6	Veterinary Sciences	14	10.853
7	Biotechnology & Applied Microbiology	10	7.752
8	Agriculture, Dairy, & Animal Science	9	6.977
9	Environmental Sciences	9	6.977
10	Public, Environmental & Occupational Health	8	6.202
11	Zoology	6	4.651
12	Engineering Environmental	4	3.101
13	Medicine Research & Experimental	3	2.326
14	Water Resources	3	2.326
15	Biochemical Research Methods	2	1.550
16	Biochemistry & Molecular Biology	2	1.550
17	Medicinal Chemistry	2	1.550
18	Multidisciplinary Chemistry	2	1.550
19	Organic Chemistry	2	1.550
20	Fisheries	2	1.550
21	Immunology	2	1.550
22	Plant Sciences	2	1.550
23	Polymer Science	2	1.550
24	Agricultural Economics & Policy	1	0.775
25	Agronomy	1	0.775
26	Biology	1	0.775
27	Physical Chemistry	1	0.775
28	Ecology	1	0.775
29	Economics	1	0.775
30	Chemical Engineering	1	0.775
31	Entomology	1	0.775
32	Environmental Studies	1	0.775
33	Evolutionary Biology	1	0.775
34	Geography	1	0.775
35	Green Sustainable Science & Technology	1	0.775
36	Marine & Freshwater Biology	1	0.775
37	Materials Science Composites	1	0.775
38	Materials Science Multidisciplinary	1	0.775
39	Mathematical Computational Biology	1	0.775
40	Medicine General Internal	1	0.775
41	Metallurgy Metallurgical Engineering	1	0.775
42	Operations Research Management Science	1	0.775
43	Parasitology	1	0.775
44	Physics Applied	1	0.775
45	Physics Condensed Matter	1	0.775
46	Regional Urban Planning	1	0.775
47	Surgery	1	0.775
48	Tropical Medicine	1	0.775

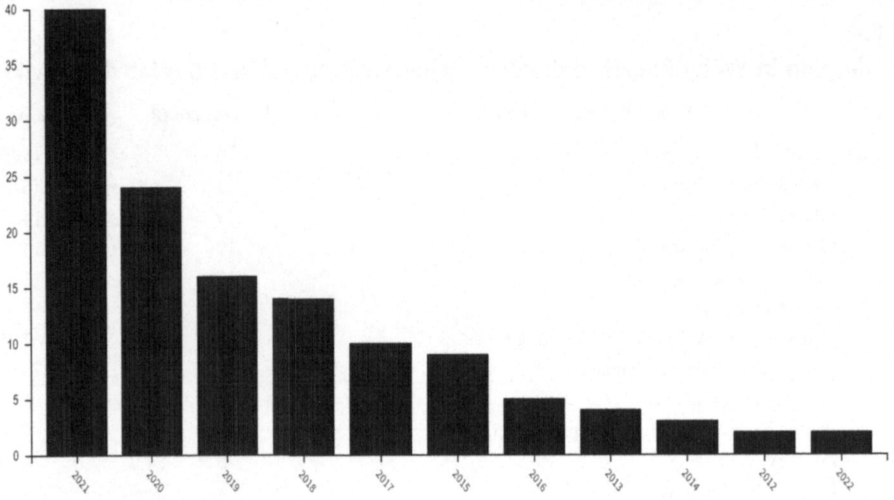

FIGURE 8.4 Number of publications in Web of Science indexed journal from 2012 to 2022 (January 2012 to January 2022).

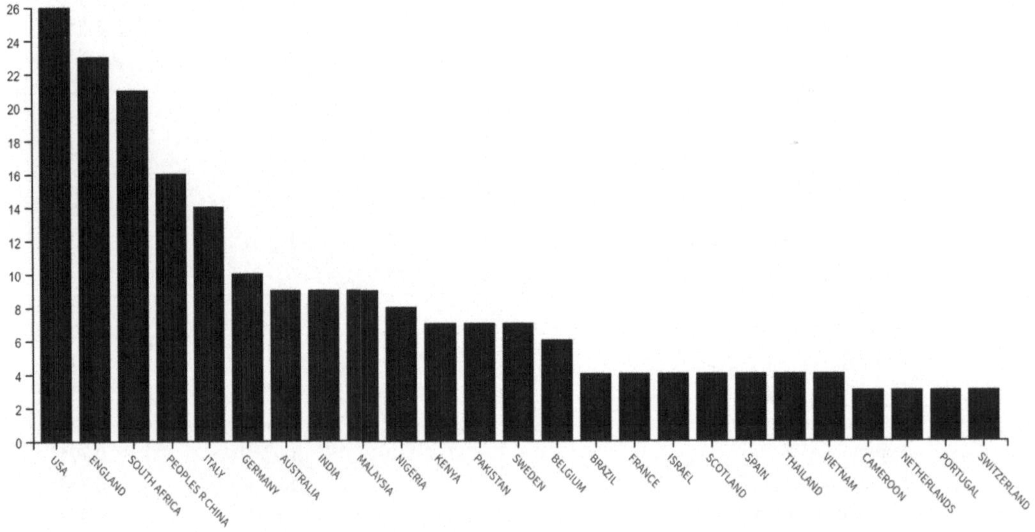

FIGURE 8.5 Top 25 country published articles on AMR in Web of Science indexed journal from 2012 to 2022 (January 2012 to January 2022)

at this matter seriously and focus on the policy and management of antibiotic application in agriculture to secure food health security.

CONFLICT OF INTEREST

There is no conflict of interest to declare.

ACKNOWLEDGEMENT

The authors are thankful to the Librarian of the Dr. Harisingh Gour University for providing authenticated access to the Web of Science database, and the authors also express their gratitude towards the Head of the Botany Department for providing us the moral strength to accomplish this task.

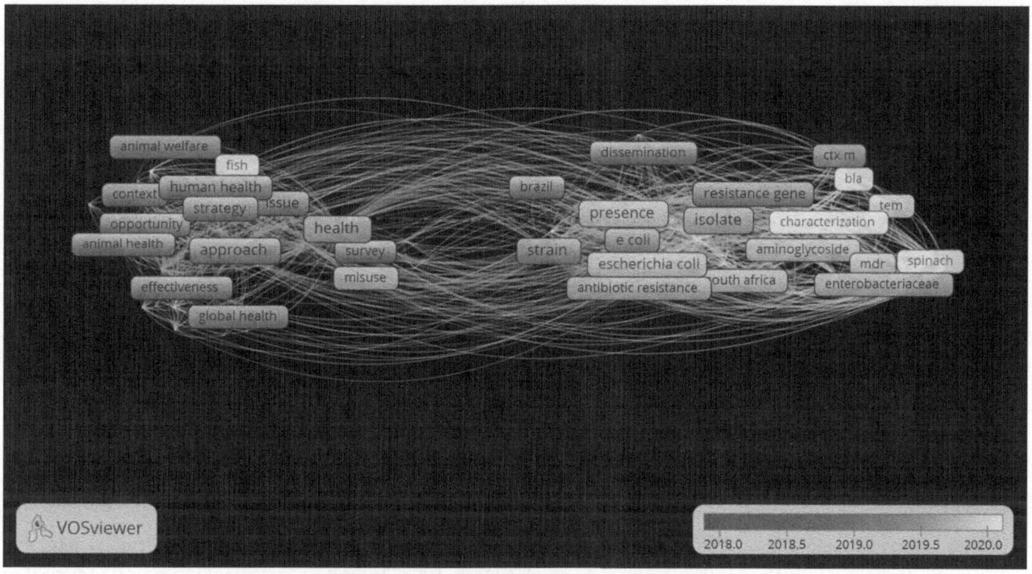

FIGURE 8.6 Overlay visualization Map of Terms occurrence related to antimicrobial resistance

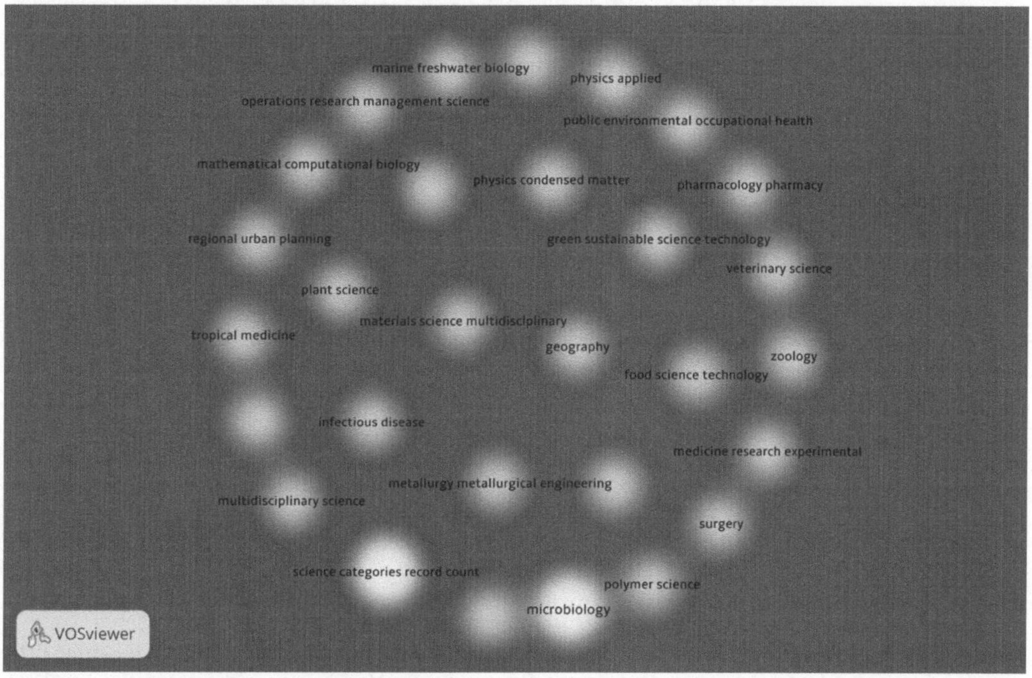

FIGURE 8.7 Density visualization of term category in the Web of Science published in the last ten years (2012–2022)

REFERENCES

Agyare, Christian, Vivian Etsiapa Boamah, Crystal Ngofi Zumbi, and Frank Osei Boateng. 2018. Antibiotic Use in Poultry Production and Its Effects on Bacterial Resistance. In *Antimicrobial Resistance*, edited by Yashwant Kumar, 33–51. IntechOpen. https://doi.org/10.5772/intechopen.79371.

Aslam, B., M. Khurshid, M.I. Arshad, S. Muzammil, M. Rasool, N. Yasmeen, T. Shah, et al. 2021. Antibiotic Resistance: One Health One World Outlook. *Frontiers in Cellular and Infection Microbiology* 11: 771510. https://doi.org/10.3389/fcimb.2021.771510.

Boeckel, Thomas P. Van, Charles Brower, Marius Gilbert, Bryan T. Grenfell, Simon A. Levin, Timothy P. Robinson, Aude Teillant, and Ramanan Laxminarayan. 2015. Global Trends in Antimicrobial Use in Food Animals. *Proceedings of the National Academy of Sciences of the United States of America* 112(18): 5649–54. https://doi.org/10.1073/pnas.1503141112.

Cabello, Felipe C., Alexandra Tomova, Larisa Ivanova, and Henry P. Godfrey. 2017. Aquaculture and MCR Colistin Resistance Determinants. *mBio* 8(5): e01229-17. https://doi.org/10.1128/mbio.01229-17.

Chakraborty, S., and B.J. Rao. 2012. A Measure of the Promiscuity of Proteins and Characteristics of Residues in the Vicinity of the Catalytic Site That Regulate Promiscuity. *PLOS ONE* 7(2): 32011. https://doi.org/10.1371/journal.pone.0032011.

Costa, Corrado, Ulrich Schurr, Francesco Loreto, Paolo Menesatti, and Sebastien Carpentier. 2019. Plant Phenotyping Research Trends, a Science Mapping Approach. *Frontiers in Plant Science* 9(January): 1–11. https://doi.org/10.3389/fpls.2018.01933.

Durso, Lisa M., and Kimberly L. Cook. 2019. One Health and Antibiotic Resistance in Agroecosystems. *EcoHealth* 16(3): 414–19. https://doi.org/10.1007/s10393-018-1324-7.

Eck, Nees Jan van, and Ludo Waltman. 2010. Software Survey: VOSviewer, a Computer Program for Bibliometric Mapping. *Scientometrics* 84(2): 523–38. https://doi.org/10.1007/s11192-009-0146-3.

FAO. 2017. Antimicrobial Resistance in Food and Agriculture AMR in the Food and Agriculture Sectors Food Chain Crisis I Emergency Prevention System I Information Sheets. http://www.fao.org/antimicrobial-resistance.

Franklin, Alison M., Diana S. Aga, Eddie Cytryn, Lisa M. Durso, Jean E. McLain, Amy Pruden, Marilyn C. Roberts, et al. 2016. Antibiotics in Agroecosystems: Introduction to the Special Section. *Journal of Environmental Quality* 45(2): 377–93. https://doi.org/10.2134/jeq2016.01.0023.

Gradmann, Christoph. 2011. Magic Bullets and Moving Targets: Antibiotic Resistance and Experimental Chemotherapy, 1900–1940. *Dynamis* 31(2): 305–21. https://doi.org/10.4321/S0211-95362011000200003.

Helliwell, Richard, Carol Morris, and Sujatha Raman. 2020. Antibiotic Stewardship and Its Implications for Agricultural Animal-Human Relationships: Insights from an Intensive Dairy Farm in England. *Journal of Rural Studies* 78(August): 447–56. https://doi.org/10.1016/j.jrurstud.2020.07.008.

Langendonk, R.F., Daniel R. Neill, and Joanne L. Fothergill. 2021. The Building Blocks of Antimicrobial Resistance in Pseudomonas Aeruginosa: Implications for Current Resistance-Breaking Therapies. *Frontiers in Cellular and Infection Microbiology* 11(April): 1–22. https://doi.org/10.3389/fcimb.2021.665759.

Lee Ventola, C. 2015. The Antibiotic Resistance Crisis Part 1: Causes and Threats. *P and T* 40(4): 277–83. https://doi.org/10.1016/B978-1-4831-9711-1.50022-3.

Mann, Avantika, Kiran Nehra, J.S. Rana, and Twinkle Dahiya. 2021. Antibiotic Resistance in Agriculture: Perspectives on Upcoming Strategies to Overcome Upsurge in Resistance. *Current Research in Microbial Sciences* 2. https://doi.org/10.1016/j.crmicr.2021.100030.

Manyi-Loh, Christy, Sampson Mamphweli, Edson Meyer, and Anthony Okoh. 2018. Antibiotic Use in Agriculture and Its Consequential Resistance in Environmental Sources: Potential Public Health Implications. *Molecules* 23(4): 795. https://doi.org/10.3390/MOLECULES23040795.

McEwen, Scott A., and Richard Reid-Smith. 2004. Antimicrobial Resistance in Food. *Canadian Journal of Infectious Diseases* 15(4): 201–3. https://doi.org/10.1155/2004/985638.

Mohr, Kathrin I. 2016. History of Antibiotics Research. In *How to Overcome the Antibiotic Crisis.Current Topics in Microbiology and Immunology*, edited by P. Stadler, and M. Dersch, 398:237–72. Springer, Cham. https://doi.org/10.1007/82_2016_499.

Naimi, Timothy, Pascal Ringwald, Richard Besser, Sharon Thompson, and David Bell. 2001. Antimicrobial Resistance. *Emerging Infectious Diseases* 7(3): 548. https://doi.org/10.3201/eid0707.017727.

Polianciuc, Svetlana Iuliana, Anca Elena Gurzău, Bela Kiss, Maria Georgia Şaria , and Felicia Loghin. 2020. Antibiotics in the Environment: Causes and Consequences. *Medicine and Pharmacy Reports* 93(3): 231–40. https://doi.org/10.15386/mpr-1742.

Schar, Daniel, Eili Y. Klein, Ramanan Laxminarayan, Marius Gilbert, and Thomas P. Van Boeckel. 2020. Global Trends in Antimicrobial Use in Aquaculture. *Scientific Reports 2020* 10(1): 1–9. https://doi.org/10.1038/s41598-020-78849-3.

Singh, Chandan, Abhishek Sharma, Smita Mishra, and Deepak Vyas. 2022. Removal of Pesticides and Reclamation of Soil through Biochar Application as a Soil Amendment. *Journal of Agricultural Science and Engineering Innovation* 2(2): 35–40. https://doi.org/10.5281/zenodo.6823476.

Sivaraman, Satya. 2018. *Antibiotic Use in Food Animals: India Overview.* ReAct Asia-Pacific Christian Medical College, Vellore, Tamil Nadu, India. www.reactgroup.org.

Thanner, Sophie, David Drissner, and Fiona Walsh. 2016. Antimicrobial Resistance in Agriculture. *mBio* 7(2): 1–7. https://doi.org/10.1128/mBio.02227-15.

Williams-Nguyen, Jessica, J. Brett Sallach, Shannon Bartelt-Hunt, Alistair B. Boxall, Lisa M. Durso, Jean E. McLain, Randall S. Singer, Daniel D. Snow, and Julie L. Zilles. 2016. Antibiotics and Antibiotic Resistance in Agroecosystems: State of the Science. *Journal of Environmental Quality* 45(2): 394–406. https://doi.org/10.2134/jeq2015.07.0336.

9 Potential uses of natural antimicrobial agents and their applications as bio-preservatives

P. Bhuvaneswari, P.F. Steffi, B. Thirumalaiyammal, and P.F. Mishel

9.1 INTRODUCTION

Adulteration occurring at various stages of the food chain is one of the major causes of food decay, leading to food waste, increased food safety risks, and significant financial losses. To combat microbial spoilage and extend the shelf life of products, artificial preservatives have been widely used. However, concerns have been raised about the negative environmental impacts of these synthetic preservatives (1). In response, natural antimicrobials have gained attention from researchers and food manufacturers due to their safety and harmless nature. These natural preservatives can be derived from plants, mammals, and bacteria, and their antimicrobial properties can be extracted using various advanced methods. Natural preservatives like nisin, essential oils, and natamycin have demonstrated strong potential against decay-causing and pathogenic microorganisms. However, the regulations and guidelines governing the use of these naturally occurring preservatives may be unclear in some developing countries. This review aims to provide an overview of these preservatives, their antimicrobial mechanisms, their role in food preservation, and the current understanding of the topic (2). Chemical preservatives such as benzoate, propionate, sorbate, nitrate, nitrite, and sulphites have been established as effective in controlling microbial growth in food. Various traditional food preservation methods, including freezing, chilling, reduction of water activity, modified atmosphere packaging, acidification, mineral limitation, fermentation, and the addition of synthetic antimicrobials, have been employed to control spoilage microorganisms. However, concerns have emerged regarding the long-term use of synthetic preservatives, as they have been linked to health issues. Consumers are becoming increasingly aware of the connection between food-related health problems and the consumption of artificial antimicrobials. Prolonged use of these preservatives has been associated with liver damage, asthma, allergies, and even cancer. As a result, scientists are actively seeking alternative organic antimicrobial agents for food preservation. Two main types of naturally occurring antimicrobials are recognized: a combination of compounds derived from various plants, mammals, or microorganisms, and individual antimicrobial agents with distinct properties. Essential oils, bacteriocins, protamine, endolysins, lysozyme, lactoferrin, flavor compounds, phenolic compounds, chitosan, isothiocyanates, and probiotics have been used in the development and treatment of food products. Probiotics have been particularly effective in reducing the presence of pathogens in in vivo studies. The use of specific natural antimicrobials can enhance consumer confidence in the use of food products. However, the regulation of natural antimicrobial agents for food preservation varies across different countries, and some concerns remain regarding their health effects, especially when consumed in large quantities (3).

 DOI: 10.1201/9781003268949-9

9.2 SOURCES AND KINDS OF NATURAL ANTIMICROBIAL AGENTS

9.2.1 Sources of Antimicrobials

The use of synthetic antimicrobials can lead to side effects, and foodborne pathogens are developing resistance to multiple drugs. As a result, there is a growing trend towards natural alternatives. Antimicrobial compounds extracted from plants, animals, and microorganisms are being utilized to ensure food safety, increase shelf life, and serve as suitable alternatives to antibiotics (Figure 9.1).

Natural antimicrobials exert their action through various mechanisms, including cell membrane disruption, interference with nucleic acid processes, disruption of the Proton Motive Force (PMF), and depletion of ATP (3, 4). Plant-derived antimicrobials such as polyphenols and essential oils (obtained from onion, garlic, spices, and hops), animal-derived compounds like lysozyme, lactoperoxidase, and lactoferrin, and microorganism-derived substances such as bacteriocins and other metabolites are used to inhibit the growth of foodborne pathogens (Table 9.1).

Natural preservatives can be obtained from different sources. Plants, mammals, and microorganisms are considered major sources of these valuable compounds. These natural derivatives have broad applications in the development and treatment of food products, ensuring decay prevention, prolonged shelf life, and crop safety. Plants, herbs, and spices are rich sources of aldehydes, organic terpenoids, phenolics, and sulfur-containing compounds. These natural compounds are typically found in the roots, flowers, leaves, seeds, and bulbs of plants (5). They possess protective properties and are effective in inactivating or preventing the growth of various microorganisms, including bacteria, yeasts, and molds. Essential oils (EOs) derived from different plants have extensive use as food preservatives and are considered viable alternatives to synthetic products. Spices are particularly rich sources of antimicrobial compounds. Different parts of plants, such as flowers, bark, herbs, tree leaves, buds, twigs, fruits, and roots, serve as valuable sources of essential oils. Essential oils can be obtained from plants and spices using various extraction methods. Steam distillation or hydrodistillation is commonly used to extract essential oils, although other techniques such as microwave-assisted extraction or supercritical fluid extraction may also be employed (6).

9.2.2 Plant Extracts as Natural Antimicrobial Agents

Extracts of plants, herbs, and spices are Generally Recognized as Safe (GRAS) substances that are used for specific periods or quantities in food production for their durability and flavor-enhancing properties. These plant extracts and flavor compounds have excellent antimicrobial activity. Numerous studies have demonstrated the antimicrobial potential of essential oils and components from plants such as thyme, oregano, clove, and cinnamon, including cinnamaldehyde, eugenol,

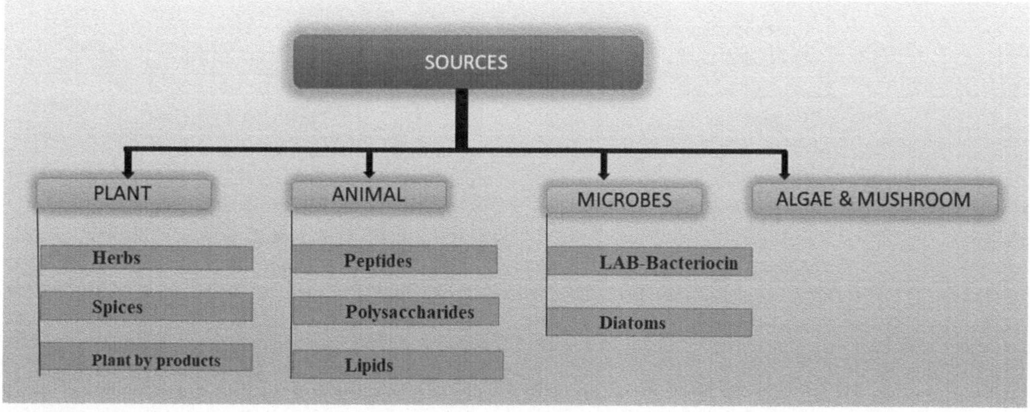

FIGURE 9.1 Sources of antimicrobials.

TABLE 9.1
Antimicrobials and their sources

Source	Examples with Active components	Active against
Plants and herbs	Oregano (Thymol and carvacrol) Onion (Allicin, quercetin, fisetin) Garlic (Allicin) Parsley (Myristicin and limonene) Sage (Thujone) Coriander (linalool, neryl acetate, gamma terpinene) Rosemary (Rosmarinic acid, carnosol and carnosic acid) Lemongrass (myrcene, limonene, citral, gerniol, citronellol) Hops (acylphloroglucinols and xanthohumol)	*Aeromonas* spp., *B. cereus*, *Brochothrix thermosphacta*, *Campylobacter jejuni*, *Escherichia coli*, *Enterobacter faecalis*, *Lactobacillus plantarum*, *Listeria monocytogenes*
Spices	Cinnamon (Cinnamic aldehyde) Clove (Eugenol)	*Aeromonas hyrophila*, *Bacillus sp.*, *Campylobacter jejuni*, *Listeria monocytogens, Salmonella, Shigella, staphylococcus aureus, Aspergillus, Candida*
Plant by products	Coffee husks, pulps, peels of pomegranate, seeds and unused flesh (tannins, flavonols, flavandiols, flavonoids, and phenol acids)	*Enterobacteriaceae* and spoilage microorganisms Inhibit both Gram negative and Gram positive bacteria (enhancing the shelf-life of chicken products)
Animal	Peptides Pleurocidin Dermaseptin Defensins Protamine Magainin Casocidin Lactoferrin Avidin Ovotransferrin or conalbumin Polysaccharides Chitosan Lipid Eicosapentaenoic acid Docosahexaenoic acid	Gram-positive and -negative bacteria
Microorganisms	Lactic acid bacteria Bacteriocin Microalgae Diatoms	*Listeria monocytogens*, *Campylobacter jejuni*
Macroalgae	*Himanthalia elongata*	*S. aureus* and *E. coli*
Mushroom	*Agaricus* *Cantharellus cibarius*, *Clavaria vermiculris*, *Lycoperdon perlatum*, *Marasmius oreades*, *Pleurotus pulmonarius*, *Ramaria formosa*	*E. coli*, *B. subtilis*, *P. aeruginosa*, and *S. aureus*

carvacrol, and phenol (7). Essential oils derived from spices contain active compounds that exhibit remarkable antimicrobial potential, such as 3-phenylprop-2-enal, 5-isopropyl-2-alkyl phenol, and so on. These compounds have shown antimicrobial activity against *fungus genus* spp., *enteric bacteria coli*, true bacteria monocytogenes, *enterobacteria sonnei*, and *enterobacteria flexneri*. *E. coli* and enterohemorrhagic *E. coli* are establishing additional alert garlic extract than what? Garlic extract, in particular, has shown strong antimicrobial potential against *Staphylococcus aureus* and *Salmonella typhimurium*. Additionally, essential oils or extracts from basil, eucalyptus, citrus, thyme, lemongrass, rosemary, mint, and tea plants have demonstrated antimicrobial effects against specific pathogens. The composition of essential oils can vary depending on the geographic location and harvesting time. Some plants contain up to 85% essential oil, while others have only trace amounts (7, 8). The minor components present in essential oils also play a significant role as antimicrobial agents through synergistic effects. The essential oils and minor components disrupt the cell membrane of microorganisms. The hydrophobic nature of essential oils impairs the growth of microorganisms, preventing them from adapting and reproducing.

9.2.3 Antimicrobials from Plants

9.2.3.1 Eugenol

Eugenol is a notable phenolic compound that can exhibit conflicting properties. It is the most abundant component (70-90%) in clove essential oil, which is derived from the buds and leaves of the clove plant. Eugenol plays a recognized role in dental and oral hygiene products. It is used as a flavoring agent, an antiseptic, and an analgesic, and it can provide a soothing effect (9). Dental products containing eugenol are widely used in clinical dentistry and have demonstrated effectiveness against pathogens such as *Shigella*, *Clostridia botulinum*, certifiable minuscule life forms monocytogenes, and *E. coli*.

9.2.3.2 Thymol

Thymol is a highly prevalent compound found in thyme. It is the main monoterpene phenol present in thyme essential oil. Thymol possesses immunomodulatory, antioxidant, anti-inflammatory, and antifungal properties. As a phenol, it exhibits activity against enteric bacteria and staphylococci. Its inhibitory effect is attributed to its ability to disrupt the integrity of the microbial cell membrane, which affects hydrogen ion concentration, osmotic balance, and the equilibrium of inorganic ions (10).

9.2.3.3 Aldehydes

Green plant sources contain compounds such as hexanal, 2-(E)-hexenal, trans-2-hexenal, and hexyl acetate derived from the lipoxygenase pathway in plants. These compounds have strong antimicrobial properties against both Gram-negative and Gram-positive bacteria. α, β-unsaturated aldehydes, including these compounds, have a broad antimicrobial spectrum and exhibit similar activity against both Gram-positive and Gram-negative bacteria (11).

9.2.3.4 Carvacrol

Carvacrol, a phenolic compound, is considered one of the key components of certain essential oils that exhibit antimicrobial activity. It is found in plants such as savory, thyme, and oregano. Carvacrol has been reported to have significant effects on the cell membrane and intracellular adenosine triphosphate (ATP) content of *E. coli* O157:H7. Numerous studies have recognized the potency of carvacrol and its antimicrobial efficacy (12).

9.2.3.5 Vanillin

Vanillin is a phenolic compound found in vanilla beans. It has numerous applications in the food, pharmaceutical, confectionery, fragrance, and nutraceutical industries. It exhibits strong antimicrobial activity against various bacteria and food spoilage microorganisms, including species from

Escherichia, *Klebsiella*, *Salmonella*, *Bacillus*, *Serratia*, *Staphylococcus*, and *Listeria*. Vanillin is also used as an additive in fruits and vegetables, applied as coating agents or in modified atmosphere packaging to extend the shelf life of products (13).

9.2.3.6 Allicin

Allicin is a sulfur-containing natural compound. It is responsible for the characteristic smell and taste of freshly cut or crushed garlic. Allicin is primarily extracted from garlic for industrial purposes. The main physiological role of garlic is its antimicrobial, antioxidant, anticancer, antifibrinolytic, and antiplatelet aggregatory activities, which have been extensively studied (14).

9.2.3.7 Cinnamaldehyde

Cinnamaldehyde is the major and characteristic constituent in cinnamon. The essential oil of cinnamon bark is pale yellow, viscous liquid occurs in the bark of cinnamon trees and other species of the genus *Cinnamomum* which is about 90% cinnamaldehyde (15).

9.2.3.8 Alkaloids

Alkaloids are a group of naturally occurring chemical compounds that primarily contain basic nitrogen atoms (15, 16).

9.2.3.9 Against microbial amide

Plant antimicrobial peptides act as natural defense compounds against various microorganisms (pAMPs) and were first discovered in 1942. Examples of plant antimicrobial peptides include potato defensin, hevein, thionins, and snakins. These peptides exhibit membrane-active antifungal, antibacterial, and antiviral properties (17).

9.2.3.10 Citral

Citral is a terpenoid that is a byproduct of terpenes, a class of compounds derived from a combination of two chemical structures. The trans-isomer is known as geranial or Citral A, while the cis-isomer is called neral or Citral B. Citral possesses antifungal properties and its antifungal effects, along with eugenol, have been studied in various research (18).

9.2.3.11 Saponins

Saponins are high molecular weight glycosides that are found in a wide variety of plants and some marine organisms. They possess antiviral, antimicrobial, antitumor, and growth-promoting properties when included in animal feeds as a stimulatory supplement (19).

9.2.3.12 Flavonoids

Flavonoids are a class of hydroxylated phenolic substances and occur as a C6-C3 unit attached to an aromatic ring. They are commonly found in fruits, vegetables, and medicinal plants. Flavonoids exhibit a wide range of biological effects, including antimicrobial properties. They have been studied as natural antimicrobial substances against various microorganisms. Flavonoids also demonstrate antimutagenic, anti-inflammatory, anticancer, and antiviral activities (19, 20).

9.2.3.13 Quinones

Quinones are aromatic compounds with two carbonyl groups. They are widely distributed in nature. Quinones can also act as substrates inaccessible to microorganisms. Tertiary butylhydroquinone (TBHQ) is commonly used as a food antioxidant to prevent rancidity in fats, oils, and protein-based foods (21).

9.2.3.14 Tannins

Tannins are polymeric phenolic substances capable of tanning animal skins or precipitating gelatin from solutions, a property known as astringency. Ellagitannin, with molecular weights ranging from 500 to 3000, is a prominent example of tannin. Ellagitannins are commonly found in various parts of plants, including bark, wood, leaves, fruits, and roots. Tannins have demonstrated antimicrobial activities against *E. coli*, *S. aureus*, Enteric *Typhimurium*, *B. subtilis*, enterobacteria sonnei, MDR *E. coli*, *C. albicans*, and *K. pneumonia*. They act by disrupting the cell membrane of microorganisms and inhibiting their growth (22).

9.2.3.15 Coumarins

Coumarins are phenolic compounds derived from the fusion of an aromatic hydrocarbon and an alpha-pyrone ring. They exhibit antimicrobial activity against fungi and also have an impact on bacteria (22).

9.2.3.16 Caffeic acid

Caffeic acid (3,4-dihydroxycinnamic acid) is a natural phenolic acid derived from hydroxycinnamic acid. It possesses notable properties, including antimicrobial, fungicidal, and antioxidant activities. It has been found to exhibit antibacterial activity against *S. epidermidis*, *S. aureus*, and *K. pneumoniae* (23).

9.2.4 Main Antimicrobials from Animal Origin

9.2.4.1 Chitosan

Chitosan is obtained from the partial deacetylation of chitin and is also known as deacetylated chitin. It is a natural polycationic linear polysaccharide primarily found in the shells of marine crustaceans. Due to its non-toxicity, biodegradability, and low allergenicity, chitosan has a wide range of applications. It exhibits antineoplastic, antifungal, antimicrobial, and antioxidant activities. It is effective against Gram-negative bacteria such as *Bacteroides fragilis*, *cholera,* enteric bacteria dysenteries, *E. coli*, and *Vibrio.* Chitosan also possesses antimicrobial resistance to swelling and antioxidant potential (24).

9.2.4.2 Defensin

Defensins are small cationic peptides primarily known for their antimicrobial activities, particularly against bacteria and fungi. They are found in all vertebrate cells and tissues, abundant in leukocytes (25).

9.2.4.3 Lactoperoxidase

Lactoperoxidase (LP) belongs to the oxidase family and its primary function is to catalyze the oxidation of certain molecules. It is a group of natural enzymes found in plants and animals, including humans. Lactoperoxidase is secreted by ductal epithelial cells of the mammary gland. The concentration of lactoperoxidase in bovine milk is approximately twenty times higher than that in human milk and varies throughout the postnatal period. The presence of salt, which is abundant in secretions, milk, and airway secretions, is required for the antimicrobial activity of lactoperoxidase. Bacteria including *Salmonella*, *Shigella*, *Pseudomonads*, and *Coliforms* can be inhibited or killed by lactoperoxidase (26).

9.2.4.4 Lysozyme

Lysozyme exhibits strong antimicrobial activity and causes the death of bacteria by cleaving glycosidic linkages in the peptidoglycan of bacterial cell walls. Lysozyme is an important defense mechanism and is considered a component of the innate immune system in most mammals. It is also a significant component of human breast milk. Large amounts of lysozyme can be found in egg white.

9.2.5 Main Antimicrobials from Microbial Origin

9.2.5.1 Natamycin

Natamycin, which has a relative molecular mass of 665.7 Da, has been widely used for food preservation to combat food spoilage organisms such as yeast and mold. It is produced by Actinomycete natalensis and has proven effectiveness against a wide range of molds and yeasts. However, it has been found to have limited or no activity against several pathogenic microorganisms. Due to its strong antifungal properties, natamycin is commonly used in various products including dairy, meats, and more. It has demonstrated effectiveness in inhibiting the growth of yeasts and molds in juices, both pasteurized and unpasteurized (27).

9.2.5.2 Reuterin

Reuterin is an antimicrobial compound produced by *Lactobacillus reuteri*. It is a water-soluble non-protein molecule with a broad antimicrobial spectrum. It is effective against both Gram-negative and Gram-positive bacteria, filamentous molds, and yeasts. Reuterin exhibits activity over a wide range of pH and is resistant to various enzymes such as proteases and lipases. It shows significant inhibitory activity against several pathogenic microorganisms, particularly *L. monocytogenes* (28).

9.2.5.3 Bacteriophages

Bacteriophages are natural bio-preservatives derived from animal and plant sources. They are considered as alternatives to chemical preservatives due to their high safety and ability to extend the shelf life of food products. Both bacteriophages and bacteria can be used as preservatives in food applications. Bacteriophages can be easily propagated and have the ability to target specific bacteria, making them favorable for use as bio-preservatives (29).

9.2.5.4 Lactic acid bacteria (LAB)

Lactic acid bacteria (LAB) are important probiotics that provide numerous health benefits, including a protective role in foods. They act as preservatives by inhibiting the growth of many pathogenic bacteria. LAB inhibit bacterial growth by producing antimicrobial agents such as organic acids and bacteriocins (antimicrobial peptides). Various strains of LAB have shown effectiveness against several pathogens (30).

9.2.5.5 Bacteriocins

Bacteriocins are major antimicrobial compounds produced by many Gram-positive bacteria. These are metabolites of LAB that are produced during their growth. Bacteriocins are polypeptides that give the producing organism a competitive advantage over other microorganisms. They are classified based on their chemical nature and are primarily produced by Gram-positive bacteria (31).

9.2.6 Methods for the Extraction of Natural Antimicrobial Agents

The extraction of plant-based antimicrobials using solvents (such as hydrochloric acid, salt, ethanol, methanol, and acetone) can be time-consuming and cumbersome. These methods require large amounts of solvents and may not be cost-effective from an economic perspective. Additionally, heat treatments can alter the activity of bioactive compounds and affect their natural characteristics, functionality, total content, and activity. Proposed methods such as direct, aqueous, and juice extraction have been widely used to evaluate the antimicrobial activity of plant extracts (32).

9.2.7 Mechanisms of Action of Natural Antimicrobial Agents

The action of natural antimicrobial agents is not fully understood. Different natural antimicrobials act in different ways. The following list presents potential actions of natural antimicrobial agents, which target the pathogenic organism in one or more of the following ways: membrane disruption, direct pH reduction of the substrate, inhibition of NADH oxidation by organic acids, interaction with the microbial membrane, and production of structural and functional damage to the microbial plasma membrane by essential oils (33).

9.2.8 Methods for Application

The application of natural antimicrobial agents to food is influenced by various factors, including food preparation, the chemical properties of both the food and the agents, food composition, different processing operations, storage conditions, and the specific strains of spoilage and foodborne microorganisms targeted. The choice of antimicrobial agents can also impact the sensory qualities, quality, and safety of the food. In some cases, natural antimicrobial agents from different sources may impart odors and flavors to the food. It has been observed that certain food constituents such as proteins, lipids, complex carbohydrates, and sugars can reduce the antimicrobial activity of these agents. Various strategies are available for the application of these agents, including the use of edible films, encapsulation techniques, and direct application methods such as spraying, dusting, and dipping (33).

9.3 ANTIMICROBIALS AS BIO-PRESERVATIVES

Antimicrobials can be derived from various sources such as peptides, enzymes, bacteriocins, bacteriophages, plant extracts, essential oils, and fermented compounds. These natural substances offer alternatives to chemical antimicrobials. Plant extracts and essential oils contain terpenes, flavonoids, aldehydes, and phenolic chemicals that exhibit antibacterial and antioxidant properties. LAB-derived chemicals have synergistic effects that inhibit the growth of bacteria and fungi. These LAB chemicals can also reduce mycotoxins, enhance food safety and nutritional value, and even have potential anticancer properties. Bacteriocins, which are antibacterial compounds, are produced by various microbes. Additionally, antimicrobial peptides can be used as bio-preservatives either on their own or when combined with other substances (34).

9.3.1 Plant Antimicrobials

Phytochemicals, also known as plant antimicrobials, play a vital role in the defense and functioning of plants. They serve as defense mechanisms against microbes and predators, as well as regulate important processes such as fertilization, pollination, and growth. Examples of phytochemicals include alkaloids, lectins, polypeptides, phenolic compounds, terpenoids, and essential oils. Phenolic compounds, such as quinones, flavonoids, tannins, simple phenols, and phenolic acids, are particularly abundant in plants. When used in food, these phenolic compounds enhance sensory qualities and provide antioxidant and antibacterial properties that can help extend the shelf life of food products (34).

Polyphenols, a type of phenolic compound, have been associated with various health benefits, including antibacterial and other biological activities. Glucosinolates, a subgroup of chemicals found in plants, have demonstrated antifungal, antibacterial, and antioxidant properties. Other chemical groups like polyamines, glucosinolates, and glucosides have also shown potential as natural antimicrobials. Allyl isothiocyanate, a potent antibacterial compound produced through the hydrolysis of glucosinolates, is used as a preservative in the food industry. In summary, antimicrobials are substances or chemicals that can inhibit microbial growth, cause microbial death, or extend the shelf life of food products when incorporated into a food matrix. They can be classified as "naturals" and encompass both conventional and new chemicals. Natural antimicrobials can be derived from plant-based materials, fruits, vegetables, herbs/spices, or microorganisms, and they serve as effective bio-preservatives (35).

9.3.2 Lactic Acid Bacteria

Lactic acid bacteria (LAB) are a diverse group of microorganisms characterized by being Gram-positive, non-spore forming, non-motile, aerotolerant, and having rod- or coccus-shaped cells. They play a crucial role in fermenting carbohydrates and produce lactic acid as a key byproduct. Examples of LAB include Leuconostoc, Pediococcus, Streptococcus, and Lactobacillus. LAB are widely used in fermentation processes and have probiotic properties that can provide health benefits. LAB, due to their competitive metabolism, are often used as starter cultures in fermented products to preserve their nutritional quality and inhibit the growth of spoilage and pathogenic microorganisms. LAB are known to produce various metabolites, including bacteriocins, organic acids, diacetyl, and lactic acid. These metabolites lower the pH of the environment, creating an antibacterial effect that extends the shelf life of the product and enhances its safety (36).

Moreover, LAB-produced metabolites have been found to reduce antibiotic resistance. These substances have broader implications for preserving the ecological balance as they can act against both Gram-positive and Gram-negative bacteria. Some LAB strains also possess probiotic properties and contribute to promoting health. Genera like Lactobacillus and Enterococcus are commonly utilized for their probiotic potential.

9.3.4 Chemical Antimicrobial Activity of Essential Oil

Terpenes, terpenoids, phenylpropenes, and other sulfur- and nitrogen-containing compounds found in essential oils have antibacterial properties. However, the addition of essential oils (EO) to foods can result in unpleasant sensory effects due to their high volatility. Additionally, the interaction between EO and dietary constituents like meat fat requires higher quantities of EO in food to achieve the same inhibitory effect compared to experimental media. To mitigate these sensory effects, using essential oils derived from plants commonly used as food seasonings is a strategy to minimize unpleasant sensory effects. For example, mustard oil can lower the pH inside bacterial cells such as *E. coli* O157:H7 and *Salmonella*, while garlic can enhance the shelf life of food products due to its antibacterial and antioxidant effects. In the case of fresh sausages, combining essential oils with other natural preservatives like peptides (e.g., nisin) can lower their concentration while preserving their antimicrobial effectiveness. This approach allows for effective control of bacterial growth and spoilage while minimizing the sensory impact of essential oils. By combining essential oils with other natural preservatives, the overall concentration of antimicrobial agents can be reduced while still maintaining their effectiveness.

Targeting lactic acid bacteria that contribute to food spoilage. By reducing the growth of these bacteria, nisin can help prevent food decay without altering the sensory characteristics of the dish. Consequently, this natural approach may effectively inhibit the growth of microorganisms, enhance product safety, and prolong the shelf life of food items utilizing EO as bio-preservatives. Experimental studies have already demonstrated the synergistic antibacterial activity of EO and nisin (NI), further confirming the efficacy of EO as an antimicrobial agent (37).

9.4 BIOACTIVE COMPOUNDS

A bioactive compound is a substance that possesses biological activity and directly affects physiological or cellular processes in a living organism. This term stems from the combination of the Greek word "bios," meaning life, and the Latin word "activus," meaning dynamic or full of energy. The biological effects of a bioactive compound can have positive or negative outcomes, depending on the nature of the substance, its dosage, and its bioavailability. In the context of nutrition, bioactive food components refer to the constituents found in foods or dietary supplements that go beyond meeting basic nutritional requirements and have the potential to modify the health status of humans or animals upon consumption. While the term "bioactive compound" is often associated with compounds derived from foods, such as those found in plants or food crops, it can encompass a broader range of sources within the food chain.

In the context of plants, the term "plant bioactive compound" typically excludes nutrients. Instead, bioactive compounds in plants refer to secondary metabolites that are not essential for the plant's basic functions, such as growth, but serve important roles in competition, defense, attraction, and signaling. These compounds can be defined as secondary plant metabolites that elicit pharmacological or physiological effects in humans and animals (38).

9.4.1 Bioactive Compounds and Useful Foods

The economic, cultural, and scientific advancements in our society have significantly influenced lifestyle and dietary patterns. In many developing countries, there is a common consumption of highly caloric and imbalanced diets. Coupled with a decrease in physical activity, this has led to a rise in the prevalence of cardiovascular diseases, diabetes, obesity, and other health issues. Considering the increasing life expectancy, it becomes evident that alternative solutions need to be explored to mitigate the anticipated healthcare costs in the near future.

Is it possible to transform a regular food into a functional food? Once again, there is no single answer, as multiple approaches can be employed to enhance the beneficial effects of a specific food. These approaches range from more or less advanced biotechnological processes to various methods aimed at removing or increasing the content of a particular compound. In many cases, a functional food is achieved by adding an element or a series of ingredients that are either absent in the corresponding regular food or present at lower concentrations. These ingredients are known as functional ingredients and are primarily micronutrients, such as omega-3 fatty acids, linoleic acids, phytosterols, soluble fiber (such as inulin and fructooligosaccharides, known as prebiotics), probiotics (microorganisms capable of improving gut health and the immune system), carotenoids, polyphenols, vitamins, etc., which can exert specific health-promoting effects on the body (39).

Algae can be found in almost any aquatic and terrestrial environment, exhibiting a wide range of diversity and various morphologies, ranging from microscopic species to large brown algae. Algae are photosynthetic organisms that possess simple reproductive structures. The exact number of algal species remains unknown but has been estimated to be between one and ten million. As mentioned, algae can exist as tiny microscopic organisms (microalgae) or as larger cellular structures (macroalgae). For example, microalgae utilize light energy and carbon dioxide with higher photosynthetic efficiency compared to plants, making them a potential source for biofuel production, wastewater purification, extraction of high-value food and pharmaceutical products, and a food source for aquaculture (40).

Moreover, another necessary side to be thought of is that the development of acceptable, fast, Furthermore, another important aspect to be considered is the development of suitable, rapid, efficient, and environmentally friendly extraction methods capable of isolating the desired compounds from these natural sources. In this chapter, green extraction techniques such as supercritical fluid extraction (SFE) and pressurized liquid extraction (PLE), along with ultrasound-assisted extraction (UAE) and microwave-assisted extraction (MAE), are presented, with a focus on their applications in extracting bioactives from algae. A review of the various types of bioactives described in algae is provided, including compounds such as lipids, carotenoids, proteins, phenolics, vitamins, polysaccharides, etc. Additionally, a brief description of methods for rapid screening of bioactivity, primarily antioxidant activity, is included, considering both chemical and biological approaches. Finally, future research trends and research needs for the extraction of bioactives from algae are critically discussed.

9.4.2 Bioactive Compounds from Algae and Microalgae

Algae are important sources of various bioactive compounds with different physiological effects, which can range from toxic to curative, on human health. Many of these compounds exhibit antioxidant, antimicrobial, and antiviral activities, which play a crucial role in protecting algal cells against stress conditions. The development of new analytical methods and techniques is essential for studying the metabolites present in algae and similar organisms, particularly in relation to their applications in pharmacology and the food industry (41).

9.4.3 Green Extraction Techniques for Bioactive Compounds

Today, there is a wide range of classical or traditional extraction techniques that have been conventionally used for extracting interesting compounds from natural matrices, including algae. These techniques include Soxhlet extraction, liquid-liquid extraction (LLE), solid-liquid extraction (SLE), and other methods that involve the use of organic solvents. However, these techniques have several well-known limitations. They are time-consuming, labor-intensive, lack automation, and are prone to low reliability. Moreover, they may have low sustainability and provide low extraction yields. These drawbacks can be partially or completely addressed by utilizing newly developed advanced extraction techniques. These new extraction techniques offer faster extraction times, improved selectivity towards the target compounds, and importantly, they are more environmentally friendly. In fact, the use of hazardous solvents is greatly reduced when employing advanced extraction techniques. In the following sections, the most important advanced extraction techniques that have been employed for extracting bioactive compounds from algae will be discussed (41, 42).

9.4.4 Fast Screening for Bioactivity

In general terms, the bioactivity of algae and microalgal extracts can be evaluated using two main groups of techniques: chemical and biological methods. Since there is no universal method to assess bioactivity, marine extracts are commonly evaluated using multiple approaches. Bioactive compounds found in algae and microalgae are known to possess antioxidant activity, therefore many of the chemical methods described in this section focus on measuring various parameters related to antioxidant activity. On the other hand, marine compounds are associated with a wide range of bioactivities, primarily pharmacological activities, which can be tested using biological or biochemical methods (43).

9.5 CHEMICAL WAYS

9.5.1 Inhibitor Activity

Interest in natural antioxidants for both health and improved food stabilization has increased significantly since the late 20th century. The potential health benefits of natural antioxidants stem from their ability to counteract free radicals and oxidative stress, which play a role in various physiological functions and pathological conditions. Natural antioxidants offer food, pharmaceutical, nutraceutical, and cosmetic industries a "green" label, minimal regulatory interference, and the potential for multiple actions to enhance and extend the stability of food and pharmaceutical products. Determining antioxidant capacity has become a highly active research topic, and numerous antioxidant assay methods are currently in use. However, there are no standardized methods due to the large volume of claims and the frequent contradictory results regarding the "antioxidant activities" of many products.

9.6 BIOLOGICAL WAYS

9.6.1 Antihelmintic, Antifungal, and Medicament Activity

Several compounds found in algae extracts have been described for their anthelmintic and antifungal activity. These compounds include phenols, indoles, peptides, hormone glycosides, terpenes, and fatty acids. The basic method involves exposing an organism to an extract or compound. For example, Mendiola et al. used a broth microdilution method to determine the minimum inhibitory concentration (MIC) of Spirulina extracts and to assess the growth of various microorganisms and fungi. Assays were performed in microtiter plates by adding the test substance to culture broth and different dilutions of the extract to each well. After incubation, the MIC of each extract was determined by visually examining the bottom of the well, where bacterial growth was indicated by the presence of a white "pellet." The lowest concentration of extract that suppressed bacterial growth was recorded as the MIC. Minimum bactericidal and fungicidal concentrations were determined by subculturing clear wells that showed no growth. Among anthelmintic compounds derived from algae, sesquiterpenes such as b-bisabolene are the most active. The most common method for assessing their activity is to culture helminths (worms) such as *Nocardia brasiliensis* in the presence of algae extracts (44).

9.7 CONCLUSION

The regulation and new methods of application of natural antimicrobial agents are important factors that should be addressed in this book chapter. Improving application methods and regulation can enhance consumer confidence. Greater efficiency is required for the application of natural antimicrobial agents to various food products. Researchers are still grappling with how to use natural antimicrobials on fruits and vegetables without negatively impacting their sensory properties. The concentrations of natural antimicrobial agents needed to inhibit spoilage or eliminate pathogenic microorganisms are often very high. These high concentrations can have a negative impact on human health as well as sensory qualities. It is important to conduct research on the synergistic combinations of natural antimicrobial agents. A combination of different treatments ensures that food products are safe and of high quality. We have presented some of the bioactive compounds derived from algae (macro- and microalgae) that have the potential to be used as functional food ingredients. Given the wide variety of algae and the strong influence of growing conditions on bioactive formation, the list of compounds and combinations could be endless. On the other hand, we aim to provide an overview of the significant potential of algae as natural reactors capable of synthesizing a diverse range of compounds with variable polarities and physiological effects on human health. Many of these compounds, such as proteins, lipids, and carbohydrates, as well as other minor components (metabolites), are produced to protect algae cells from stress conditions. Most of them are beneficial for the food industry as macronutrients (fiber, proteins, etc.), while others have a promising future as functional ingredients to prevent or even improve human health. In this chapter, we also presented new technologies for extracting valuable compounds from algae, all of which share a "green" label, the ability to enhance efficiency through process improvement, the elimination of toxic solvents, improved cost efficiency, and the enhancement of sustainability and isolation steps. The text includes several examples to demonstrate the utility and advantages of such processes over traditional extraction methods. However, this step should not be viewed in isolation, but rather as part of a larger picture of what should be a sustainable process utilizing algae as a material. In this sense, algae (primarily microalgae) can be considered as (1) a sustainable source of mass and energy, as their metabolism meets energy efficiency requirements (conversion, biomass growth); (2) a future source of clean energy if oil production is obtained, which could be used for large-scale biodiesel production; (3) an efficient carbon dioxide sequestrant for greenhouse gas emissions control (Kyoto Protocol); and (4) a valuable source of biodiversity.

REFERENCES

1. Carvalho F, Coimbra AT, Silva L, Duarte AP, Ferreira S. Melissa officinalis Essential Oil as an Antimicrobial Agent Against Listeria monocytogenes in Watermelon Juice. *Food Microbiol* 2023 February;109:104105.
2. Farjami A, Jalilzadeh S, Siahi-Shadbad M, Lotfipour F. The Anti-biofilm Activity of Hydrogen Peroxide against Escherichia coli strain FL-Tbz Isolated from a Pharmaceutical Water System. *J Water Health* 2022 October;20(10):1497.
3. Musse SL, Nielsen GB, Stege H, Weber NR, Houe H. Prevalences and Excretion Levels of Lawsonia intracellularis, Brachyspira pilosicoli and Escherichia coli F4 and F18 in Fecal Sock Samples from Danish Weaner and Finisher Pig Batches and the Association with Diarrhea. *Porcine Health Manag* 2022 October 28;8(1):44.

4. Adelusi OA, Gbashi S, Adebiyi JA, Makhuvele R, Adebo OA, Aasa AO, Targuma S, Kah G, Njobeh PB. Variability in Metabolites Produced by Talaromyces Pinophilus SPJ22 Cultured on Different Substrates. *Fungal Biol Biotechnol* 2022 October 28;9(1):15.
5. da Silva Bruni AR, de Souza Alves FJ, de Jesus GAM, da Silva AE, da Costa JCM, Souza PR, de Oliveira Santos JO, Bonafe EG. Characterization and Application of Active Films Based on Commercial Polysaccharides Incorporating ZnONPs. *Int J Biol Macromol* 2023 224: 1322–1336.
6. Prabhakar P, Sen RK, Patel M, Shruti, DN, Singh S, Kumar P, Chouhan M, Yadav AK, Mondal DP, Solanki PR, Srivastava AK, Dhand C. Development of Copper Impregnated Bio-inspired Hydrophobic Antibacterial Nanocoatings for Textiles. *Colloids Surf B Biointerfaces* 2022 October 10;220:112913.
7. Lee HJ, Tokle IF, Lunestad BT, Lerfall J, Hoel S, Jakobsen AN. The Effect of Food Processing Factors on the Growth Kinetics of Aeromonas Strains Isolated from Ready-to-Eat Seafood. *Int J Food Microbiol* 2022 October 21;384:109985.
8. Koutsoumanis K, Allende A, Alvarez-Ordonez A, Bolton D, Bover-Cid S, Chemaly M, Davies R, De CA, Herman L, Hilbert F, Lindqvist R, Nauta M, Ru G, Simmons M, Skandamis P, Suffredini E, Arguello-Rodriguez H, Dohmen W, Magistrali CF, Padalino B, Tenhagen BA, Threlfall J, Garcia-Fierro R, Guerra B, Liebana E et al. Transmission of Antimicrobial Resistance (AMR) during Animal Transport. *EFSA J* 2022 October;20(10):e07586.
9. Zhang J, Xu J, Lei H, Liang H, Li X, Li B. The Development of Variation-Based Rifampicin Resistance in Staphylococcus aureus Deciphered through Genomic and Transcriptomic Study. *J Hazard Mater* 2023 January 15;442:130112.
10. Caballero GN, Manetsberger J, Benomar N, Castillo GS, Abriouel H. Antibacterial and Antibiofilm Effects of Essential Oil Components, EDTA and HLE Disinfectant Solution on Enterococcus, Pseudomonas and Staphylococcus sp. Multiresistant Strains Isolated along the Meat Production Chain. *Front Microbiol* 2022;13:1014169.
11. Ledormand P, Desmasures N, Schlusselhuber M, Sesboue A, Ledauphin J, Dalmasso M. Phages Shape Microbial Dynamics and Metabolism of a Model Community Mimicking Cider, a Fermented Beverage. *Viruses* 2022 October 17;14(10): 2283.
12. Kim H, Maeng JY, Lee DB, Kim KH, Chung MS. Antiviral Activities of Asarones and Rhizomes of Acorus gramineus on Murine Norovirus. *Viruses* 2022 October 10;14(10).
13. Ravlo E, Evensen L, Sanson G, Hildonen S, Ianevski A, Skjervold PO, Ji P, Wang W, Kaarbo M, Kaynova GD, Kainov DE, Bjoras M. Antiviral Immunoglobulins of Chicken Egg Yolk for Potential Prevention of SARS-CoV-2 Infection. *Viruses* 2022 September 26;14(10).
14. Bao Q, Zhang Z, Yu B, Sun H, Leung PH, Tao X. Synthesis of Polylactic Acid Oligomers for Broad-Spectrum Antimicrobials. *Polymers (Basel)* 2022 October 18;14(20).
15. Jamroz E, Janik M, Marangoni L, Jr., Vieira RP, Tkaczewska J, Kawecka A, Szuwarzynski M, Mazur T, Jasinska JM, Krzysciak P, Juszczak L. Double-Layered Films Based on Furcellaran, Chitosan, and Gelatin Hydrolysates Enriched with AgNPs in Yerba Mate Extract, Montmorillonite, and Curcumin with Rosemary Essential Oil. *Polymers (Basel)* 2022 October 12;14(20).
16. Maurizzi E, Bigi F, Quartieri A, De LR, Volpelli LA, Pulvirenti A. The Green Era of Food Packaging: General Considerations and New Trends. *Polymers (Basel)* 2022 October 11;14(20).
17. Efthimiou I, Vlastos D, Triantafyllidis V, Eleftherianos A, Antonopoulou M. Investigation of the Genotoxicological Profile of Aqueous Betula pendula Extracts. *Plants (Basel)* 2022 October 11;11(20).
18. Perez-Guillen I, Domenech O, Botet-Carreras A, Merlos A, Sierra JM, Albericio F, de la Torre BG, Montero MT, Vinas M, Borrell JH. Studying Lipid Membrane Interactions of a Super-Cationic Peptide in Model Membranes and Living Bacteria. *Pharmaceutics* 2022 October 14;14(10).
19. Sanad YM, Aljahdali NH, Khajanchi BK, Nayak R, Khan A, Foley SL. Microbial Genetics and Clonal Dissemination of Salmonella enterica Serotype Javiana Isolated from Human Populations in Arkansas, USA. *Pathogens* 2022 October 16;11(10).
20. Wong CWY, Wang S. Efficacy of Repeated Applications of Bacteriophages on Salmonella enterica-Infected Alfalfa Sprouts during Germination. *Pathogens* 2022 October 6;11(10).
21. Wisniewski P, Zakrzewski AJ, Zadernowska A, Chajecka-Wierzchowska W. Antimicrobial Resistance and Virulence Characterization of Listeria monocytogenes Strains Isolated from Food and Food Processing Environments. *Pathogens* 2022 September 25;11(10).
22. Abril AG, Pazos M, Villa TG, Calo-Mata P, Barros-Velazquez J, Carrera M. Proteomics Characterization of Food-Derived Bioactive Peptides with Anti-allergic and Anti-Inflammatory Properties. *Nutrients* 2022 October 20;14(20).

23. Dai Z, Li S, Meng Y, Zhao Q, Zhang Y, Suonan Z, Sun Y, Shen Q, Liao X, Xue Y. Capsaicin Ameliorates High-Fat Diet-Induced Atherosclerosis in ApoE(-/-) Mice via Remodeling Gut Microbiota. *Nutrients* 2022 October 17;14(20).
24. Donnelly J, Appathurai A, Yeoh HL, Driscoll K, Faisal W. Vitamin E in Cancer Treatment: A Review of Clinical Applications in Randomized Control Trials. *Nutrients* 2022 October 16;14(20).
25. Ureta-Velasco N, Keller K, Escuder-Vieco D, Serrano JCE, Garcia-Lara NR, Pallas-Alonso CR. Assessment of Iodine Concentration in Human Milk from Donors: Implications for Preterm Infants. *Nutrients* 2022 October 14;14(20).
26. Mannino F, D'Angelo T, Pallio G, Ieni A, Pirrotta I, Giorgi DA, Scarfone A, Mazziotti S, Booz C, Bitto A, Squadrito F, Irrera N. The Nutraceutical Genistein-Lycopene Combination Improves Bone Damage Induced by Glucocorticoids by Stimulating the Osteoblast Formation Process. *Nutrients* 2022 October 14;14(20).
27. McKay S, Teitsma-Jansen A, Floris E, Dekker T, Smids B, Khurshid R, Calame W, Kardinaal A, Lutter R, Albers R. Effects of Dietary Supplementation with Carrot-Derived Rhamnogalacturonan-I (cRG-I) on Accelerated Protective Immune Responses and Quality of Life in Healthy Volunteers Challenged with Rhinovirus in a Randomized Trial. *Nutrients* 2022 October 12;14(20).
28. Yuan Y, Liu Q, Huang Y, Qi M, Yan H, Li W, Zhuang H. Antibacterial Efficacy and Mechanisms of Curcumin-Based Photodynamic Treatment against Staphylococcus aureus and Its Application in Juices. *Molecules* 2022 October 21;27(20).
29. Albaqami JJ, Benny TP, Hamdi H, Altemimi AB, Kuttithodi AM, Job JT, Sasidharan A, Narayanankutty A. Phytochemical Composition and In Vitro Antioxidant, Anti-Inflammatory, Anticancer, and Enzyme-Inhibitory Activities of Artemisia nilagirica (C.B. Clarke) Pamp. *Molecules* 2022 October 21;27(20).
30. Ammar H, Touihri I, Kholif AE, M'Rabet Y, Jaouadi R, Chahine M, Marti MEH, Vargas-Bello-Perez E, Hosni K. Chemical Composition, Antioxidant, and Antimicrobial Activities of Leaves of Ajuga Iva. *Molecules* 2022 October 20;27(20).
31. Ibrahim HM, Thangavelu M, Khalifa A. Honey-Propolis-Engineered Collagen Peptides as Promising Wound-Healing Matrix in Mouse Model. *Molecules* 2022 October 20;27(20).
32. Volkov AG, Hairston JS, Taengwa G, Roberts J, Liburd L, Patel D. Redox Reactions of Biologically Active Molecules upon Cold Atmospheric Pressure Plasma Treatment of Aqueous Solutions. *Molecules* 2022 October 19;27(20).
33. Shady NH, Altemani AH, Altemani FH, Maher SA, Elrehany MA, Saber EA, Badawi AM, El-Mordy FMA, Mohamed NM, Abourehab MAS, Sayed AM, Abdelmohsen UR, Mohamad SA. The Potential of Corchorus Olitorius Seeds Buccal Films for Treatment of Recurrent Minor Aphthous Ulcerations in Human Volunteers. *Molecules* 2022 October 18;27(20).
34. Ekrami M, Ekrami A, Hosseini MA, Emam-Djomeh Z. Characterization and Optimization of Salep Mucilage Bionanocomposite Films Containing Allium Jesdianum Boiss. Nanoliposomes for Antibacterial Food Packaging Utilization. *Molecules* 2022 October 18;27(20).
35. de Moraes Motta Machado MC, Lepaus BM, Bernardes PC, de Sao Jose JFB. Ultrasound, Acetic Acid, and Peracetic Acid as Alternatives Sanitizers to Chlorine Compounds for Fresh-Cut Kale Decontamination. *Molecules* 2022 October 18;27(20).
36. Silva S, Costa EM, Oliveira H, Freitas V, Morais RM, Calhau C, Pintado M. Impact of a Purified Blueberry Extract on In Vitro Probiotic Mucin-Adhesion and Its Effect on Probiotic/Intestinal Pathogen Systems. *Molecules* 2022 October 18;27(20).
37. Fan Y, Yu R, Chen Y, Sun Y, Waterhouse GIN, Xu Z. A Capillary Electrophoresis Method Based on Molecularly Imprinted Solid-Phase Extraction for Selective and Sensitive Detection of Histamine in Foods. *Molecules* 2022 October 17;27(20).
38. Palani V, Chinnaraj S, Shanmugasundaram M, Malaisamy A, Maluventhen V, Arumugam VA, Rengasamy KRR, Balasubramanian B, Liu WC, Arumugam M. Derivation, Functionalization of (S)-Goniothalamin from Goniothalamus wightii and Their Derivative Targets SARS-CoV-2 M(Pro), S(Pro), and RdRp: A Pharmacological Perspective. *Molecules* 2022 October 17;27(20).
39. Valdemir da CS, do Nascimento TG, Mergulhao NLON, Freitas JD, Duarte IFB, de Bulhoes LCG, Dornelas CB, de Araujo JXJ, Dos SJ, Silva ACA, Basilio IJ, Goulart MOF. Development of a Polymeric Membrane Impregnated with Poly-Lactic Acid (PLA) Nanoparticles Loaded with Red Propolis (RP). *Molecules* 2022 October 17;27(20).
40. Begum R, Mamun-Or-Rashid ANM, Lucy TT, Pramanik MK, Sil BK, Mukerjee N, Tagde P, Yagi M, Yonei Y. Potential Therapeutic Approach of Melatonin against Omicron and Some Other Variants of SARS-CoV-2. *Molecules* 2022 October 16;27(20).

41. Yang L, Afshari MJ, Ge J, Kou D, Chen L, Zhou D, Li C, Wu S, Zhang L, Zeng J, Zhong J, Stauber RH, Gao M. Functionalized Ultrasmall Iron Oxide Nanoparticles for T1-Weighted Magnetic Resonance Imaging of Tumor Hypoxia. *Molecules* 2022 October 15;27(20).
42. Chen Y, Huang W, Chen Y, Wu M, Jia R, You L. Influence of Molecular Weight of Polysaccharides from Laminaria Japonica to LJP-Based Hydrogels: Anti-Inflammatory Activity in the Wound Healing Process. *Molecules* 2022 October 15;27(20).
43. Gong Y, Luo W, Chen H, Ren B, Hu W, Li L. Systematical Ingredient Investigations of Ficus Tikoua Bur. Fruit and Immunoregulatory and Antioxidant Effects of Different Fractions. *Molecules* 2022 October 14;27(20).
44. Hrichi S, Chaabane-Banaoues R, Alibrando F, Altemimi AB, Babba O, Majdoub YOE, Nasri H, Mondello L, Babba H, Mighri Z, Cacciola F. Chemical Composition, Antifungal and Anti-Biofilm Activities of Volatile Fractions of Convolvulus Althaeoides L. Roots from Tunisia. *Molecules* 2022 October 12;27(20).

Index

L

M

N

P

Q

R

S

T

W

Y